Methods in Microbiology
Volume 52

Recent titles in the series

Methods in Microbiology
Volume 52

Genome Engineering

Edited by

Volker Gurtler

School of Science,
College of Science Engineering and Health,
RMIT University, Bundoora, VIC, Australia

Michael Calcutt

Department of Veterinary Pathobiology,
College of Veterinary Medicine, University of Missouri,
Columbia, MO, United States

ACADEMIC PRESS

An imprint of Elsevier

ELSEVIER

Academic Press is an imprint of Elsevier
125 London Wall, London, EC2Y 5AS, United Kingdom
The Boulevard, Langford Lane, Kidlington, Oxford OX5 1GB, United Kingdom
525 B Street, Suite 1650, San Diego, CA 92101, United States
50 Hampshire Street, 5th Floor, Cambridge, MA 02139, United States

First edition 2023

Notices
Knowledge and best practice in this field are constantly changing. As new research and experience broaden
our understanding, changes in research methods, professional practices, or medical treatment may
become necessary.

Practitioners and researchers must always rely on their own experience and knowledge in evaluating and
using any information, methods, compounds, or experiments described herein. In using such information
or methods they should be mindful of their own safety and the safety of others, including parties for whom
they have a professional responsibility.

To the fullest extent of the law, neither the Publisher nor the authors, contributors, or editors, assume any
liability for any injury and/or damage to persons or property as a matter of products liability, negligence or
otherwise, or from any use or operation of any methods, products, instructions, or ideas contained in the
material herein.

ISBN: 978-0-12-823540-9
ISSN: 0580-9517 (Series)

For information on all Academic Press publications
visit our website at https://www.elsevier.com/books-and-journals

Publisher: Zoe Kruze
Acquisitions Editor: Mariana Kuhl
Developmental Editor: Federico Paulo Mendoza
Production Project Manager: Abdulla Sait
Cover Designer: Mark Rogers

Typeset by STRAIVE, India

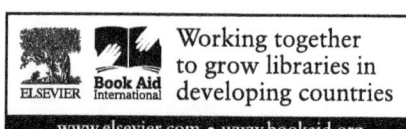

Contents

SECTION II Recombineering and engineering

SECTION III   CRISPR

CHAPTER 5 Applications of CRISPR/Cas9 in the field of microbiology...**155**

Iqra Bano and Adnan Ali

CHAPTER 8 Natural transformation as a tool in *Acinetobacter baylyi*: Streamlined engineering and mutational analysis 207

Stacy R. Bedore, Ellen L. Neidle, Isabel Pardo, Jin Luo,
Alyssa C. Baugh, Chantel V. Duscent-Maitland,
Melissa P. Tumen-Velasquez, Ville Santala, and
Suvi Santala

Contributors

Adnan Ali
Faculty of Veterinary & Animal Sciences, Pir Mahar Ali Shah Arid Agriculture University, Rawalpindi, Punjab, Pakistan

Rubén D. Arroyo-Olarte
Unidad de Biomedicina, Facultad de Estudios Superiores-Iztacala, Universidad Nacional Autónoma de México, Tlalnepantla, México

Asheemita Bagchi
Department of Biochemical Engineering and Biotechnology, Indian Institute of Technology Delhi, Delhi, India

Iqra Bano
SBBUVAS, Sakrand Faculty of Bio-sciences, Sakrand, Sindh, Pakistan

Alyssa C. Baugh
Department of Microbiology, University of Georgia, Athens, GA, United States

Stacy R. Bedore
Department of Microbiology, University of Georgia, Athens, GA, United States

Virendra Swarup Bisaria
Department of Biochemical Engineering and Biotechnology, Indian Institute of Technology Delhi, Delhi, India

Hongzhi Dong
Key Laboratory for Northern Urban Agriculture of Ministry of Agriculture and Rural Affairs; College of Bioscience and Resources Environment, Beijing University of Agriculture, Beijing, China

Chantel V. Duscent-Maitland
Department of Microbiology, University of Georgia, Athens, GA, United States

Christopher E. French
School of Biological Sciences, University of Edinburgh, Edinburgh, United Kingdom; Joint Research Centre for Engineering Biology, Zhejiang University, Haining, Zhejiang, China

Joerg Jores
Institute of Veterinary Bacteriology; Multidisciplinary Center for Infectious Diseases (MCID), University of Bern, Bern, Switzerland

Shreyoshi Karmakar
Department of Biochemical Engineering and Biotechnology, Indian Institute of Technology Delhi, Delhi, India

Fabien Labroussaa
Institute of Veterinary Bacteriology; Multidisciplinary Center for Infectious Diseases (MCID), University of Bern, Bern, Switzerland

Jin Luo
Faculty of Engineering and Natural Sciences, Tampere University, Tampere, Finland

Edgar Morales-Ríos
Departamento de Bioquímica, Centro de Investigación y Estudios Avanzados del Instituto Politécnico Nacional (CINVESTAV), Mexico City, Mexico

Ellen L. Neidle
Department of Microbiology, University of Georgia, Athens, GA, United States

Li Pan
School of Biology and Biological Engineering; Guangdong Provincial Key Laboratory of Fermentation and Enzyme Engineering, South China University of Technology, Guangzhou, China

Isabel Pardo
Centro de Investigaciones Biológicas Margarita Salas (CIB), Spanish National Research Council (CSIC), Madrid, Spain

Karla Daniela Rodríguez-Hernández
Instituto de Química, Universidad Nacional Autónoma de México, Mexico City, México

Suvi Santala
Faculty of Engineering and Natural Sciences, Tampere University, Tampere, Finland

Ville Santala
Faculty of Engineering and Natural Sciences, Tampere University, Tampere, Finland

Preeti Srivastava
Department of Biochemical Engineering and Biotechnology, Indian Institute of Technology Delhi, Delhi, India

Sergi Torres-Puig
Institute of Veterinary Bacteriology, University of Bern, Bern, Switzerland

Melissa P. Tumen-Velasquez
Department of Microbiology, University of Georgia, Athens, GA, United States

Marcos Valenzuela-Ortega
School of Biological Sciences, University of Edinburgh, Edinburgh, United Kingdom

Florentina Winkelmann
School of Biological Sciences, University of Edinburgh, Edinburgh, United Kingdom

Preface

We are fortunate to present a number of chapters in this book dedicated to outlining detailed methods in the emerging field of genome engineering. This field has been made possible by a number of advances in the past 70 years, beginning with the elucidation of the structure of DNA by Watson, Crick, Franklin and Wilkins (Franklin & Gosling, 1953; Wilkins et al., 1953; Watson & Crick, 1953), followed by recombinant DNA technology in the 1960s, 1970s, and 1980s (by many investigators who developed a large number of techniques, including plasmids, DNA restriction enzymes, DNA polymerases, bacterial cloning, DNA sequencing and PFGE, to name only a few key technologies outlined in the study by Sambrook, Fritsch, & Maniatis, 1989); PCR in 1987 (Mullis & Falloona, 1987); whole-genome sequencing and analysis in the past decade of the 20th century (Fleischmann, Adams, White, et al., 1995); and CRISPR technology (see Section III).

The eight chapters of this book have been divided into four sections: Section I, "Genome Transformation", including Chapter 1 "Genome transplantation in *Mollicutes*"; Section II, "Recombineering and Engineering", including Chapter 2 "Genome engineering in bacteria: Current and prospective applications", Chapter 3 "Towards a circular bioeconomy: Engineering biology for effective assimilation of cellulosic biomass" and Chapter 4 "Recombineering"; Section III, "CRISPR", including Chapter 5 "Applications of CRISPR/Cas9 in the field of microbiology" and Chapter 6 "Genome engineering in *Aspergillus niger*"; and Section IV, "Transformation", including Chapter 7 "Natural transformation as a tool in *Acinetobacter baylyi*: Evolution by amplification of gene copy number" and Chapter 8 "Natural transformation as a tool in *Acinetobacter baylyi*: Streamlined engineering and mutational analysis".

We hope that these methods, protocols and reviews of methods will provide valuable resources to undergraduate students, postgraduate students and researchers.

Editors

Volker Gurtler and Michael Calcutt

Abbreviations

CRISPR   clustered regularly interspaced short palindromic repeat
PCR   polymerase chain reaction
PFGE   pulsed-field gel electrophoresis

References

Fleischmann, R. D., Adams, M. D., White, O., et al. (1995). Whole-genome random sequencing and assembly of *Haemophilus influenzae* Rd. *Science, 269*, 496–512.

Franklin, R. E., & Gosling, R. G. (1953). Molecular configuration in sodium thymonucleate. *Nature, 171*, 740–741.

Mullis, K. B., & Faloona, F. A. (1987). Specific synthesis of DNA in vitro via a polymerase-catalysed chain reaction. *Methods in Enzymology, 155*, 335–350.

Sambrook, J., Fritsch, E. F., & Maniatis, T. (1989). *Molecular cloning: A laboratory manual* (2nd ed.). Cold Spring Harbor, NY: Cold Spring Harbor Laboratory.

Watson, J. D., & Crick, F. H. C. (1953). Molecular structure of nucleic acids: A structure for deoxyribose nucleic acid. *Nature, 171*, 737–738.

Wilkins, M. H. F., Stokes, A. R., & Wilson, H. R. (1953). Molecular structure of deoxypentose nucleic acids. *Nature, 171*, 738–740.

Genome transformation

Genome transplantation in *Mollicutes*

Fabien Labroussaa[a,b,†], **Sergi Torres-Puig**[a,†], **and Joerg Jores**[a,b,*]

[a]*Institute of Veterinary Bacteriology, University of Bern, Bern, Switzerland*
[b]*Multidisciplinary Center for Infectious Diseases (MCID), University of Bern, Bern, Switzerland*
Corresponding author: e-mail address: joerg.jores@unibe.ch

Abbreviations

5-FOA	5-fluoroorotic acid
BAC	bacterial-artificial chromosome
CFU	colony forming units
CRISPR	Clustered Regularly Interspaced Short Palindromic Repeats
EDTA	ethylenediaminetetraacetic acid
gDNA	genomic DNA
GT	genome transplantation
HR	homologous recombination
JCVI	J. Craig Venter Institute
Mcap RE(−)	restriction enzyme-negative *Mcap*
Mcap	*Mycoplasma capricolum* subsp. *capricolum*
Mccp	*M. capricolum* subsp. *capripneumoniae*
Mmc	*Mycoplasma mycoides* subsp. *capri*
Mmm	*M. mycoides* subsp. *mycoides*
NETs	neutrophil extracellular traps
PEG	polyethylene glycol
PFGE	pulsed field gel electrophoresis
REXER	replicon excision enhanced recombination
RM	restriction-modification
SOB	super optimal broth
SP5	*Spiroplasma* 5 medium
TAE	tris acetate EDTA buffer
TAR	transformation associated recombination
TE	Tris-EDTA
TREC	tandem repeat coupled with endonuclease cleavage

[†]These authors contributed equally.

Methods in Microbiology, Volume 52, ISSN 0580-9517, https://doi.org/10.1016/bs.mim.2023.02.001

TREC-IN	TREC-assisted gene Knock-In
X-Gal	5-bromo-4-chloro-3-indolyl-β-D-galactoside
YAC	Yeast Artificial Chromosome
YRE	Yeast Replicative Elements

1 The historical scientific context associated with genome transplantation

The technique of genome transplantation (GT) was developed by scientists from the J. Craig Venter Institute (JCVI: https://www.jcvi.org/) in the framework of developing the first synthetic cell or, in other words, a cell controlled by a synthetic chromosome (Fig. 1). In 2007, GT was reported for the first time, and it was shown that the chromosome of *M. mycoides* subsp. *capri* (*Mmc*) could be transplanted into a recipient *M. capricolum* subsp. *capricolum* (*Mcap*) cell (Lartigue et al., 2007). Soon afterwards, researchers showed that the yeast *S. cerevisiae* could be used as a host to assemble an entire *Mycoplasma* genome as a yeast artificial chromosome (YAC) (Gibson et al., 2008). Functional transplantation of an *Mmc* chromosome isolated from *S. cerevisiae* into *Mcap* was achieved a year later (Lartigue et al., 2009). The establishment of GT from yeast paved the way for the creation of the first cell harbouring a fully synthetic chromosome (Gibson et al., 2010), a scientific achievement which marked the birth of synthetic genomics as a field within synthetic biology. Later on, the generation of the smallest synthetic, autonomously replicating organism, named JCVI-syn3.0, was reported (Hutchison 3rd et al., 2016) (Fig. 1). Since then, this minimal cell has become the epicentre to study the essential pathways required for the replication and survival of a living cell (Breuer et al., 2019; Haas et al., 2022; Hossain, Deter, Peters, & Butzin, 2021; Pelletier et al., 2021).

Besides, GT has also been used to generate isogenic mutants that allowed the identification of important virulence traits of mycoplasmas (Jores, Ma, et al., 2019; Jores, Schieck, et al., 2019; Schieck et al., 2016), and the study of unique biological functions present in mycoplasmas (Lartigue et al., 2014; Nottelet et al., 2021). Moreover, in recent years the GT technique has been successfully expanded to other bacterial species within the class *Mollicutes* (Baby et al., 2018; Labroussaa et al., 2016; Talenton et al., 2022). In the meantime, several techniques have been developed to improve the capacity to modify chromosomes of *Mollicutes* maintained as YACs in yeast (Chandran et al., 2014; Noskov, Segall-Shapiro, & Chuang, 2010; Ruiz et al., 2019; Tsarmpopoulos et al., 2016) (Fig. 1), which greatly accelerated the generation of mutants.

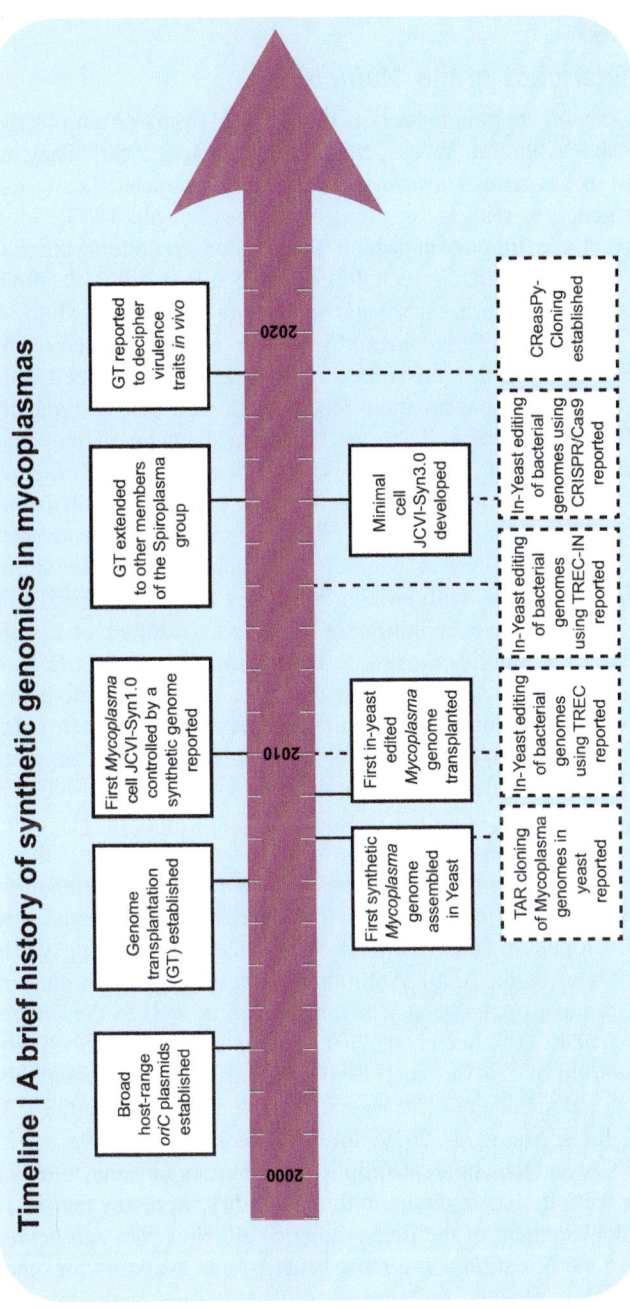

Timeline | A brief history of synthetic genomics in mycoplasmas

FIG. 1

A brief history of synthetic genomics in mycoplasmas. The timeline displays the major scientific breakthroughs resulting from the emergence of the field of synthetic genomics applied to *Mollicutes*. The establishment of the genome transplantation (GT) technique allowed for the design and creation of the first synthetic (JCVI-syn1.0) and minimal (JCVI-syn3.0) mycoplasma cells. The different in-yeast editing techniques that facilitated the design and building of engineered mycoplasma genomes are also displayed (dashed rectangles). Associated to the extension of the GT technique to other members of the '*Spiroplasma* phylogenetic group', the use of synthetic genomics paved the way to create and study several mycoplasma isogenic mutants to decipher the genetic mechanisms involved in host-mycoplasma interactions.

2 *Mollicutes*—The perfect model organisms for the establishment of GT

2.1 General characteristics of the *Mollicutes*

Bacteria of the class *Mollicutes* are minute bacteria that evolved from a Gram-positive ancestor through regressive evolution (Woese, Maniloff, & Zablen, 1980). They are phylogenetically related to the genus *Clostridium* and have a characteristically low G+C content in their genomes (Rogers et al., 1985; Woese et al., 1980). Most *Mollicutes* use an atypical genetic code in which TGA codons encode tryptophan rather than opal (stop) codons (Citti, Marechal-Drouard, Saillard, Weil, & Bove, 1992), with known exceptions such as *Acholeplasma laidlawii* (Tanaka, Muto, & Osawa, 1989) and the '*Candidatus* Phytoplasma' spp. The loss of genes encoding different metabolic pathways, like the biosynthesis of the peptidoglycan cell wall, hampers their Gram staining and causes their pleomorphic cell shapes typically observed for many *Mollicutes* species. These cells have a diameter of less than 1 μm and can pass through filters with a pore size of 0.2 μm (Razin, 1996). *Mollicutes* encompass the smallest self-replicating microorganisms that can be grown in axenic media yet discovered. The loss of multiple genes resulted in reduced chromosomes that typically encode less than 1000 genes, which is particularly exacerbated in the case of *Mycoplasmoides genitalium*, with around 500 genes and a size of 580 kbp (Fraser et al., 1995). The reductive evolution of *Mollicutes* resulted in a total dependence on their host as it supplies them with the building blocks of life such as sterols, fatty acids, amino acids, vitamins and the precursors of nucleic acids. Therefore, they are fastidious organisms that require complex media for their cultivation in vitro (Razin, 1996; Razin, Yogev, & Naot, 1998). Despite their apparent simplicity, *Mollicutes* can cause infectious diseases reported for animals, humans and plants. Important pathogenic species are especially found in the '*M. mycoides* cluster', which is part of the '*Spiroplasma* phylogenetic group'. Among these, *M. mycoides* subsp. *mycoides* (*Mmm*) and *M. capricolum* subsp. *capripneumoniae* (*Mccp*), are the causative agents of Contagious Bovine Pleuropneumonia and Contagious Caprine Pleuropneumonia, two devastating diseases affecting cattle and goats, respectively (Jores et al., 2020). Additionally, the '*M. mycoides* cluster' includes the two small ruminant pathogens *Mmc* and *Mcap*, as well as the bovine pathogen *Mycoplasma leachii* (Fischer et al., 2012; Manso-Silvan et al., 2009), which cause a wide spectrum of diseases such as mastitis, arthritis, keratoconjunctivitis, pleuropneumonia and septicemia (DaMassa, Brooks, Adler, & Watt, 1983; Djordjevic et al., 2001; Jores, Ma, et al., 2019; Jores, Schieck, et al., 2019).

The minute genome size and the apparent simplicity of these organisms, together with the ability to grow them in axenic media in the laboratory, were key considerations early on for the development of the first synthetic cell. However, *Mollicutes* appeared to be defiant to modification using most genetic tools available for other bacteria at the time, which compelled researchers to develop new methodologies to introduce targeted genomic mutations. Consequently, novel transformation methods, replicative plasmids and selection markers became available for many *Mollicutes* species, and eventually shaped the GT technology as we know it today.

2.2 Transformation methods for *Mollicutes*

The outer surface of *Mollicutes* solely consists of a plasma membrane, which is decorated with lipoproteins and can be laced with simple capsular polysaccharides (Razin et al., 1998). Therefore, in the absence of a cell wall, foreign DNA simply needs to overcome only the cell membrane to enter the bacterial cell. Nevertheless, several *Mollicutes* express DNases at the cell surface (Minion, Jarvill-Taylor, Billings, & Tigges, 1993) or encode different restriction-modification systems (Dybvig, Sitaraman, & French, 1998; King & Dybvig, 1994), which negatively impacts their transformation efficacy. Transformation of *Mollicutes* in the laboratory usually relies on either electroporation- or polyethylene glycol (PEG)-based protocols, with wide disparities in efficiencies in the different species for which these methods are reported. The use of PEG as membrane fusing agent was first described for *Spiroplasma citri*, in which the successful transfer of chromosomal DNA between bacterial cells was followed by integration of foreign DNA by homologous recombination (Barroso & Labarere, 1988). This gene transfer mechanism was DNase resistant, required cells to be in close contact and could still occur naturally without the use of any fusion agent. The DNA uptake was more efficient after a tight contact between the donor and recipient cells induced by centrifugation or the use of PEG. Therefore, it was postulated that these DNA transfer events involved a membrane fusion step (Barroso & Labarere, 1988). Ever since, transformation protocols using PEG have been widely established for many species of *Mollicutes*, while optimization often relies on different PEG concentrations and time of exposure of cells to the latter.

2.3 Transformable replicative plasmids

Most cloning vectors of bacteria use natural replicative plasmids isolated from the species of interest as a blueprint. Unfortunately, in *Mollicutes* only a very limited number of plasmids have been identified, most of them being present in species within the 'Spiroplasma phylogenetic group' (Breton et al., 2012; Hill et al., 2021; King & Dybvig, 1992; Thiaucourt et al., 2011). These plasmids are generally of low molecular weight and vary in size from 1 to 3.4 kbp (Breton et al., 2012). They generally encode only a few genetic elements essential for their replication and maintenance. However, some of these natural plasmids were found unstable when used as cloning vectors (King & Dybvig, 1992), with the notable exception of the *Mycoplasma yeatsii* plasmid pMyBK1 (Breton et al., 2012; Kent, Foecking, & Calcutt, 2012). This general limitation was overcome when plasmids containing the origin of replication (*oriC*) of mycoplasma chromosomes were constructed and tested. The *oriC* regions allowing plasmid propagation typically consist of the *dnaA* gene, involved in the early denaturation of the genome prior to replication, along with the upstream and downstream intergenic regions containing its cognate DnaA binding sites (i.e., the *dnaA* boxes). First developed in *S. citri*, *oriC* plasmids have been adapted to many other mycoplasma species (Lartigue, Blanchard, Renaudin, Thiaucourt, & Sirand-Pugnet, 2003). *oriC* plasmids gained popularity and were seen as a potential tool to generate targeted knockout mutants through gene disruption by plasmid integration. This strategy was used to perform gene-targeted mutagenesis in both *Mcap* and *Mmc* (Janis et al., 2005), but failed in *Mmm*. despite several attempts targeting various genes (Janis et al., 2008). This is likely due to generally low recombination capacity of *Mollicutes* (Allam, Reyes, Assad-Garcia, Glass, & Brown, 2010) and the high tendency for *oriC* plasmids to integrate in the host genome prevalently at the origin of replication locus (Lartigue et al., 2003). Moreover, most *oriC* plasmids have a strict host range, some being restricted to the same bacterial species the *oriC* region originates from.

However, one noticeable exception was found in *Mcap*, which seemed to support the replication of plasmids containing *oriC* regions from phylogenetically related *Mollicutes* (Lartigue et al., 2003). This feature likely played an important role in the choice of *Mcap* as the recipient cell for the successful establishment of GT. It should be noted that the capacity for a cell to replicate plasmids carrying the *oriC* region of a particular species does not necessarily correlate with its capacity to be amendable to GT, as observed for *S. citri* (Labroussaa et al., 2016). Noticeably, the first GT involved the transplantation of an *Mmc* strain GM12 chromosome marked using an integrated *oriC* plasmid containing a tetracycline resistance cassette and the β-galactosidase reporter gene (Lartigue et al., 2007).

3 Baker's yeast—An engineering platform for microbial genomes

Baker's yeast—*S. cerevisiae* can be used as an engineering workhorse (Vashee, Arfi, & Lartigue, 2020) to edit diverse viral and bacterial genomes (Benders et al., 2010; Labroussaa et al., 2021; Oldfield et al., 2017; Thi Nhu Thao et al., 2020; Vashee et al., 2017; Venter, Glass, Hutchison 3rd, & Vashee, 2022). This yeast harbours 16 chromosomes totalling a genome size of 12 Mbp. It is a powerful organism that can easily maintain foreign DNA in the form of YACs (Murray & Szostak, 1983). YACs can be stably replicated and maintained at 1–2 copies per cell even for sizes greater than 1 Mbp (Benders et al., 2010). Chromosomes of a few *Mollicutes* maintained as YACs do not significantly alter fitness of yeast and are considered 'inert,' as exemplified for the genome of *Mesoplasma florum* that was transcriptionally inactive once cloned in yeast (Baby et al., 2018). Moreover, the intrinsic ability of yeast to recombine free homologous DNA ends is an advantageous feature to rapidly assemble large DNA fragments as circular entities, an approach termed Transformation Associated Recombination (TAR) cloning (Noskov et al., 2002).

All these advantages made *S. cerevisiae* the platform of choice to clone and modify chromosomes of *Mollicutes*, most of which were considered refractory to site-directed modifications. In 2008, the first synthetic *Mycoplasma* genome, using the *Mycoplasma genitalium* chromosome as a blueprint, was assembled in yeast using TAR cloning employing 25 overlapping chemically-synthetized DNA fragments (Gibson et al., 2008). A year later, the chromosome from *Mmc* strain GM12, previously modified by a transposon insertion bearing Yeast Replicative Elements (YREs), was transformed and stably maintained in yeast as well. This chromosome (*Mmc* GM12-derived purified from yeast) was the first donor genome successfully transplanted into an *Mcap* recipient cell from yeast (Lartigue et al., 2009). In 2010, this very same chromosome (with minor modifications) was chemically synthesized in the form of 20 overlapping subgenomic fragments, assembled in yeast by TAR cloning and transplanted in a *Mcap* recipient cell, giving birth to the first bacterial cell controlled by a fully synthetic genome (Gibson et al., 2010). In parallel to the TAR cloning method, additional powerful in-yeast editing tools were developed to precisely modify *Mollicutes* genomes. First, the 'Tandem Repeats coupled with Endonuclease Cleavage' (TREC) (Noskov et al., 2010), was established to generate seamless deletions in YACs. TREC is a two-step method based on the yeast's ability to repair site-specific double-strand breaks using homologous recombination (HR) (Fig. 2). Briefly, a selection cassette containing an I-*Sce*I-based cleavage system along with the URA3 gene and a 378 bp-direct repeat of the sequence upstream of the locus to delete, is integrated by HR in the cloned *Mollicutes* chromosome at the locus of interest and selected on uracil dropout synthetic medium. Upon galactose induction, the I-*Sce*I endonuclease cleaves the *Mollicutes* chromosome and triggers a second HR event resulting in the counter-selection of the integrated cassette using 5-fluorootic acid (5-FOA) and in the complete deletion of the region of interest. An add-on feature, named TREC-IN, was later developed and allows the targeted insertion of DNA sequences by using a third HR event in combination with a split marker (Chandran et al., 2014). Briefly, a modified TREC cassette with an addition of the 5′-end of the kanamycin marker is introduced at the targeted location and its integration is selected on uracil. The complementary 3′ end of the kanamycin cassette, restoring a fully functional kanamycin gene, is integrated by HR together with the knock-in DNA sequence and is selected by using geneticin. Finally, cleavage of the TREC cassette by I-*Sce*I is triggered as previously described, which results in the knock-in DNA sequence to be integrated into the chromosome (Fig. 2). Years later, the popular CRISPR/Cas9 editing technology was adapted to efficiently modify *Mollicutes* chromosomes cloned in yeast in a single HR event (Tsarmpopoulos et al., 2016). Deletion of the gene of interest can be achieved by a double-strand break induced by Cas9 followed by the chromosomal integration of a recombination template resulting in a seamless deletion (Fig. 2). More recently, the CReasPy-Cloning method, also based on the CRISPR/Cas9 technology, was developed to generate seamless deletions of a locus of interest. However, the sequence to be deleted must be replaced by a cassette containing the YREs as well as any selection marker, allowing the cloning and engineering of chromosomes of *Mollicutes* in yeast in a single step (Ruiz et al., 2019) (Fig. 2).

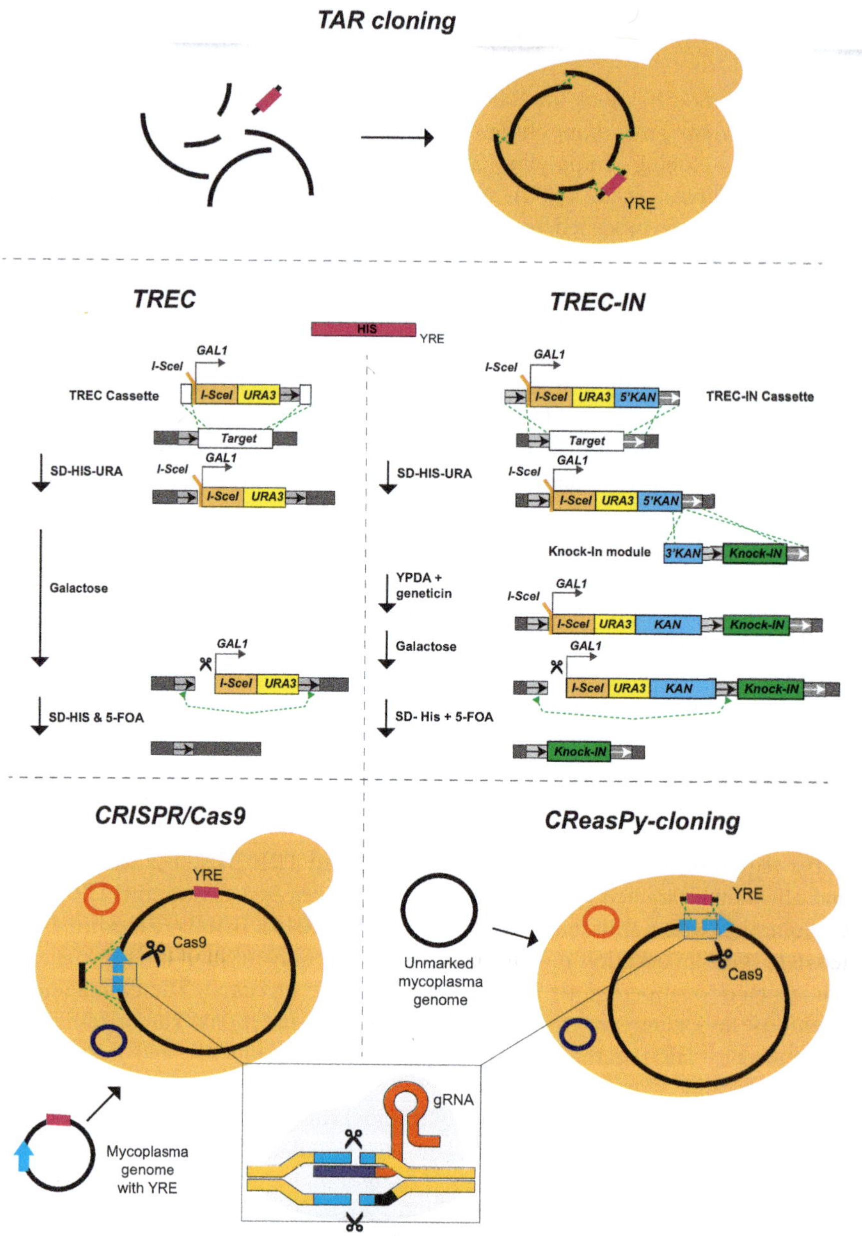

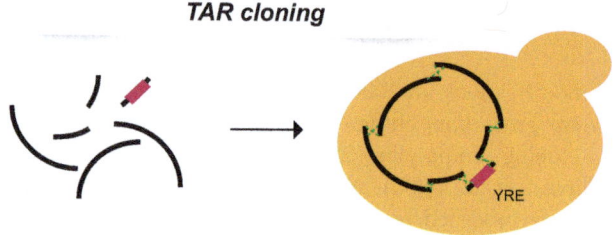

FIG. 2

In-yeast editing techniques allowing the modification of mycoplasma genomes.
Figure displaying the five different techniques developed for the engineering of the
mycoplasma genomes in Baker's yeast, *Saccharomyces cerevisiae*. The TREC, TREC-IN and
CRISPR/Cas9 techniques all require the mycoplasma genomes to be already cloned and
maintained in yeast through the integration of yeast replicative elements before being
applicable. The TAR- and CReasPy-cloning methods allow for the cloning and engineering of
mycoplasma genomes in yeast in a single step. Due to its versatility, the TAR cloning is so far
the only technique able to assemble fully synthetic mycoplasma genomes in yeast.

4 The GT protocol

The boot up of edited or fully synthetic chromosomes mainly propagated in yeast will be discussed in this chapter. A step-by-step protocol is included in this book chapter as an Appendix. It describes the preparation of intact donor chromosomes derived from *Mmc* GM12 or yeast harbouring the *Mmc* GM12 chromosome as a YAC followed by transplantation into the recipient strain *Mcap* RE(−).

4.1 Isolation of intact *Mmc* donor genomic DNA

A successful GT requires high-quality intact genomic DNA (gDNA) to be isolated either directly from a bacterial culture (Fig. 3A and Appendix, Section A.1.1) or from a yeast culture, which harbours the cloned donor chromosome (Fig. 3B and Appendix, Section A.1.2). In the case of from-bacteria GT, donor cells are first cultured in SP5 medium until late logarithmic phase. Streptomycin and chloramphenicol are generally added at sub-inhibitory concentrations to obtain compact and fully replicated genomes. *Mycoplasma* cells are then carefully washed using an isotonic buffer and finally embedded in low-melting agarose plugs using the CHEF Mammalian Genomic DNA Plug kit (BioRad). The latter step was implemented to avoid exposure of gDNA to shearing forces as the result of repetitive pipetting during conventional gDNA isolation protocols. Still embedded in agarose plugs, *Mycoplasma* cells are lysed with mild detergents while the proteins of the cells are digested using a prolonged proteinase K treatment followed by a thorough washing phase consisting of four consecutive washes in Tris-EDTA (TE) buffer. As a result, predominantly circular chromosomes, free of proteins and other cellular factors, are kept entrapped in plugs at 4 °C in 1× TE buffer until further use. These chromosomes are then released from the plugs using a β-agarase I treatment, which effectively digests the agarose at 42 °C (Appendix, Section A.1.2). The quality of the chromosome preparations can be assessed using Pulsed-Field Gel Electrophoresis (PFGE). As the quantity of donor chromosomal DNA was demonstrated to be an important factor affecting GT, a fast quantitative PCR protocol was developed to accurately measure the gDNA concentrations embedded in agarose plugs prepared from *Mycoplasma* cultures (Labroussaa et al., 2016).

The protocol to isolate high-quality *Mycoplasma* chromosomes from a yeast culture is basically identical, except that YACs need to be released from yeast cells instead of *Mycoplasma* cells (Appendix, Section A.1.2). Thus, the yeast-containing plugs are first treated with Zymolyase® to digest the cell wall of yeast (Kitamura, Kaneko, & Yamamoto, 1971) and then the cells are lysed and the proteins are digested as previously described.

In both cases, the recipient cells receiving incoming donor chromosomes must be cultivated in selective medium after transplantation. Therefore, the donor genomes need to be 'tagged' by selective marker genes conferring resistance to either

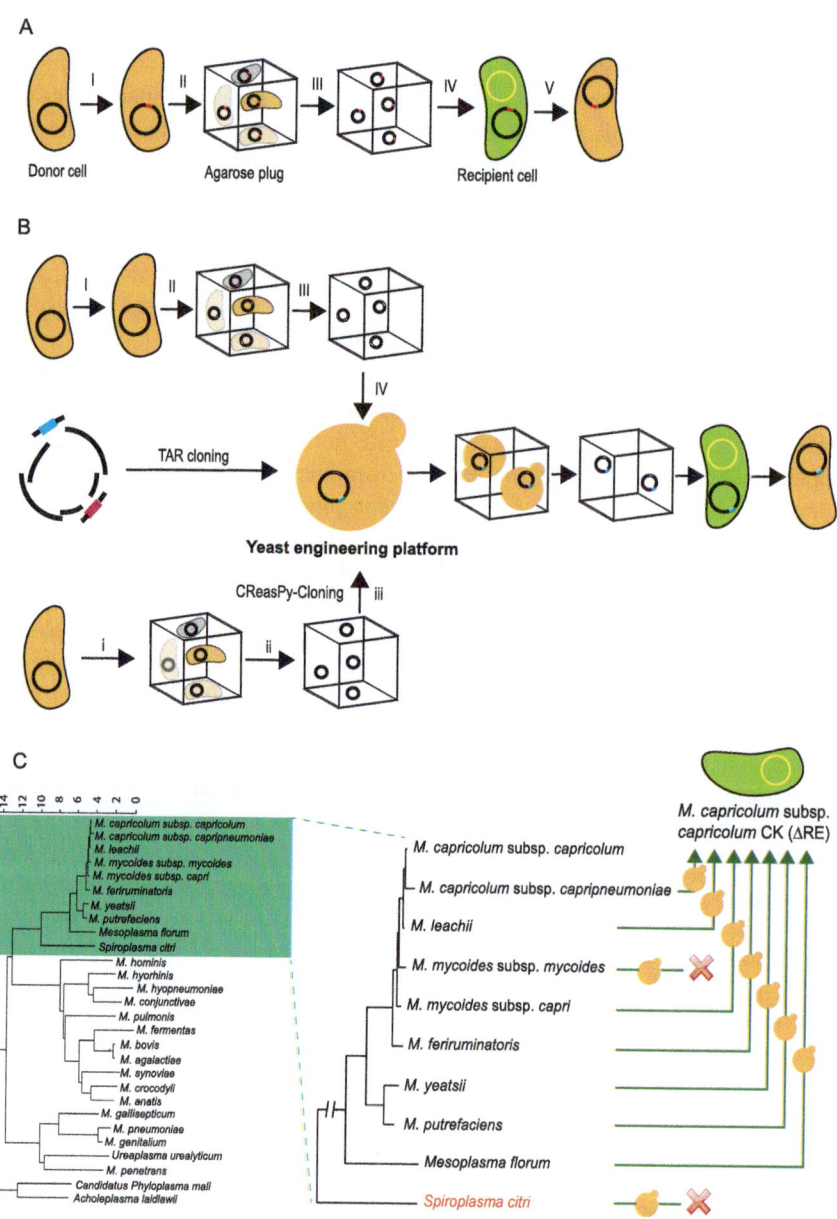

FIG. 3

General overview of the genome transplantation technology. (A) Depiction of the main steps required for GT between bacterial cells. For selection of the clones after GT, donor genomes are marked with an antibiotic cassette (red rectangle) using random mutagenesis or plasmid integration (I). *Mollicutes* donor cells (in orange) are embedded in low-melting agarose plugs (black cubes) (II) prior isolation of bacterial genomes by cell lysis and protein tetracycline via a *tetM* gene or to puromycin via the *pac* gene. In addition, a β-galactosidase gene (*lacZ*) integrated in the donor chromosome will facilitate the identification of true transplants using the classical blue-white selection on agar plates.

4.2 Preparation of recipient *Mcap* cells

All reported successful GT experiments were done using the *Mcap* type strain 'California Kid (CK)' or, more particularly, its derivate mutant strain *Mcap* RE(−), which does not possess restriction-modification systems (Lartigue et al., 2009).

The preparation of recipient *Mcap* cells for GT was optimized from previous transformation protocols (Lartigue et al., 2003) and is described in Appendix (Section A.2.1). To be an efficient recipient cell, *Mcap* requires cultivation in a sub-optimal medium called SOB(+) at 30 °C until the culture reaches late exponential/early stationary phase which correlates with a medium pH of 6–6.5 (Lartigue et al., 2009). The SOB(+) is composed of Bacto Super Optimal Broth (SOB) medium supplemented with fetal bovine serum and glucose as a carbon source. The use of complex, rather optimal *Mycoplasma* growth media such as SP5, as well as normal growth temperatures of 37 °C, were shown to decrease the efficiency of GT for unknown reasons. The use of SOB(+) medium at a suboptimal temperature is likely to favour the destabilization of the cellular membrane and therefore improves transformation rates as originally observed for *Escherichia coli* (Hanahan, 1983).

removal. Plugs are digested with β-agarase I (grey cubes) (III) and genomes are subsequently transplanted into recipient cells (in green, with their natural chromosome indicated in yellow) using PEG (IV). Transplanted cells are selected with the respective antibiotic on solid medium. The resulting strain will become genetically identical to the donor cells with marked genomes (V). (B) Illustration of the main steps required to engineer *Mollicutes* genomes in *Saccharomyces cerevisiae* and transplant them into recipient bacterial cells. Donor genomes can be marked with YREs (depicted in magenta), isolated from bacteria as in (A) and maintained in *S. cerevisiae* as YACs. Alternatively, unmarked genomes can be isolated from agarose plugs (i and ii), and introduced and marked in yeast by using the CReasPy-Cloning technique (iii); or subgenomic fragments with desired modifications (depicted in blue) can be fully assembled following a TAR cloning approach. Bacterial chromosomes in yeast can be subsequently modified by TREC, TREC-IN or CRISPR-Cas9 prior to isolation from agarose plugs before transplantation as in (A). (C) GT is only available for a limited number of mycoplasma species of the 'Spiroplasma phylogenetic group' (boxed in green). Transplantation has not yet been achieved in other species of *Mollicutes*. Genomes of most species of the 'Spiroplasma phylogenetic group' can be successfully transplanted in the *Mcap* RE(−) recipient strain, with the notable exceptions of *Mmm* and *S. citri*. The transplantation of the *Mmm* genome is limited to GT from bacteria, while the genome of *S. citri* cannot be transplanted when isolated from bacteria or yeast.

4.3 Transformation of *Mcap* with chromosomal DNA

The first *Mmc* transplants were obtained using a protocol combining $CaCl_2$ and PEG (Lartigue et al., 2007). In contrast to other mycoplasma species requiring PEG concentrations ranging between 30% and 70%, the optimal transformation of *Mcap* involves the use of PEG6000 at only 5% (King & Dybvig, 1994; Lartigue et al., 2007). The protocol also requires the use of 'serum-free SP5' medium when *Mcap* cells are mixed gently by slow pipetting using wide-bore pipette tips with the gDNA and the PEG. Sodium chloride is generally added to this medium to restore its isotonic nature. Finally, a last incubation step of 90 min at 30 °C is necessary before plating. At this suboptimal temperature, the duplication time of the *Mycoplasma* cells is significantly extended, which is likely to allocate more time to the incoming donor genome to gain control of the recipient cell. All these steps are detailed in Appendix, Section A.2.3. The GT protocol finishes with the screening of putative transplants obtained on selective plates (Appendix, Section A.2.4).

4.4 Transplanted *Mollicutes* chromosomes

As previously mentioned, *Mmc* (using a derivative of strain GM12) was the first species amenable to GT (Lartigue et al., 2007), paving the way for the adoption of the technique to other species. However, only a limited number of *Mollicutes* genomes have been successfully transplanted since GT was first established. These species all belong to the 'Spiroplasma phylogenetic group' (Fig. 3C) and include *Mcap*, *Mmm*, *M. leachii*, *Mycoplasma putrefaciens* (Labroussaa et al., 2016), *M. florum* (Baby et al., 2018) and *Mycoplasma feriruminatoris* (Talenton et al., 2022). Successful GT events were reported for all of them using chromosomes isolated from both bacteria and yeast cultures except for *Mmm*, which is still refractory to GT from *S. cerevisiae* (Labroussaa et al., 2016). Noteworthy, all successful GT events reported to date involved *Mcap* or *Mcap*-derivative recipient cells, as described below.

4.5 General comments about the GT protocol

The precise mechanism involved in the entry of the donor chromosome into the recipient cell remains unclear. However, the fusion of cell membranes of *Mollicutes* has been observed when PEG and $CaCl_2$ acted together (Barroso & Labarere, 1988). The PEG, irrespective of its molecular weight, acts by volume exclusion resulting in osmotic forces driving membranes into close contact in dehydrated regions, which ultimately favours close-contact and agglutination of neighbouring cells (Lentz, 2007). The incoming chromosome, more than forcing its way into a single *Mycoplasma* cell, could be entrapped between several cells fusing together ultimately allowing the formation of syncytia (Labroussaa, Baby, Rodrigue, & Lartigue, 2019). This might lead to the formation of multi-nucleate enlarged cells, which can ultimately form individual cells carrying only one transplanted chromosome.

From a more technical point of view, a series of parameters were investigated to optimize the transplantation protocol, from the buffer used to wash *Mcap* cells up to the molecular weight and concentration of PEG used (Lartigue et al., 2009). The quantity of incoming chromosomal DNA has been thoroughly optimized. Originally, GT efficiencies followed a 'bell-shape' curve when increasing quantities of *Mmc* gDNA were used with an optimal quantity around 2 μg of chromosomal DNA (Lartigue et al., 2007). Interestingly, increasing DNA quantities were also tested for all the other transplanted species and a similar trend was observed for all of them, despite some obvious differences in GT efficiencies (Labroussaa et al., 2016). Considering that the transplanted genomes range between 0.8 and 1.2 Mbp in size, the use of $\sim 10^9$ chromosomes appear to be optimal for the GT process. The concentration of *Mcap* recipient cells was kept constant irrespectively of the mycoplasma species transplanted and averaged 6×10^9 and 1.2×10^{10} cells/transplantation for from-bacteria or from-yeast transplantations, respectively.

Attempts to generate transplants using the previously described GT protocol without the use of $CaCl_2$ and PEG consistently failed (Lartigue et al., 2007). The occurrence of GT in nature is therefore very unlikely and it is anticipated that the GT technique is strictly restricted to laboratory conditions. Despite its successful establishment for several *Mycoplasma* species belonging to the '*Spiroplasma* phylogenetic group', the molecular mechanisms involved in the GT process are still largely unknown. It could be speculated that genetic factors only present in this cluster of *Mollicutes* facilitated the establishment of the GT process. On the other hand, it is also possible that limiting factors, absent in the successfully transplanted genomes, prevent the process to be universally applied. The identification of such genetic factors, if they exist, would be key to extend the GT technique to other bacteria.

4.6 Limiting factors involved in the GT process

4.6.1 Restriction-modification (R-M) systems

Bacteria, including *Mollicutes*, have developed strategies to differentiate between self and foreign DNA to protect themselves against invading DNA such as phage DNA (Bernheim & Sorek, 2020). Among these strategies, restriction-modification (RM) systems are ubiquitous in *Mollicutes* (Roberts, Vincze, Posfai, & Macelis, 2015). Most RM systems consist of two enzymes, a restriction enzyme that recognizes and cleaves specific non-methylated sequences, and a methyltransferase that adds methyl groups to these specific DNA sequences to protect them from cleavage. Foreign non-methylated DNA, such as DNA of invading phages, will be restricted upon entry into the cell. The presence of active RM systems was shown to affect the capacity to transform many Gram-negative and Gram-positive bacteria such as *E. coli* or *Staphylococcus aureus* (Bickle & Kruger, 1993; Monk, Tree, Howden, Stinear, & Foster, 2015), among others. Consequently, numerous bacterial strains have been genetically modified to inactivate RM systems by deleting genes encoding the restriction enzyme. In the context of GT, the incoming donor genome needs to be protected from cleavage by the recipient cell upon entry. At first, the presence of RM systems in both *Mcap* and *Mmc* did not affect from-bacteria GT (Lartigue et al., 2007). In silico analyses using the REBASE database (Roberts et al., 2015) revealed that *Mcap* strain CK contained one type II RM system, whereas *Mmc* strain GM12 contained five type II and one type III systems (Algire, Montague, Vashee, Lartigue, & Merryman, 2012). However, the sole RM system found in the genome of *Mcap* is identical to one of the *Mmc* RM systems. When the *Mmc* incoming genome was isolated from bacteria to establish the original from-bacteria GT experiment, the latter was properly methylated and consequently not cleaved by the active RM system of the recipient *Mcap* cell (Lartigue et al., 2007). In contrast, the capacity of the yeast *S. cerevisiae* to methylate DNA is unclear. *S. cerevisiae* was shown to be capable of performing 5mC methylation but with 1–2 orders of magnitude lower than that in mammalian cells (Tang, Gao, Wang, Yuan, & Feng, 2012). When *Mmc* chromosomal DNA was isolated from yeast, chromosomes appeared to be non-methylated and sensitive to the *Mcap* RM system, resulting in the absence of transplants after from-yeast GT (Lartigue et al., 2009). Two strategies were developed to enable GT in this case. On one hand, *Mmc* genomes were methylated in vitro using either *Mcap/Mmc* cellular extracts or purified methyltransferases from the latter (Lartigue et al., 2009). These methylated chromosomes resulted in the generation of transplants using the wildtype *Mcap* strain CK as recipient cell. In parallel, the single *Mcap* RM system was inactivated by integrating a puromycin-resistance marker into the coding region of the restriction enzyme (i.e., MCAP_0050) to create the *Mcap* mutant strain *Mcap* RE(−), used ever since as the recipient cell for all successful GT experiments. The absence of a functional RM system in the recipient *Mcap* cell abolished the need to methylate the chromosomes isolated from yeast.

4.6.2 Phylogenetic distance of donor and recipient species

The phylogenetic distance between the donor and recipient genomes has been identified as an important factor for the successful establishment of GT (Labroussaa et al., 2016). The rationale surrounding this hypothesis is that the incoming donor genome needs to be compatible with the cellular machinery of the recipient cell to become the operating system of the latter. Indeed, the isolation of intact whole chromosomes includes the lysis of the *Mycoplasma* cells associated with a prolonged proteinase K treatment. When it enters the recipient cell during the GT process, the incoming donor genome is free of cellular factors or proteins and requires the cellular machinery of the recipient cell to get booted up. Key functions such as replication, transcription and translation of the genetic information are likely to be fundamental for the compatibility of incoming donor chromosome and recipient cell. *Mmc* and *Mcap* are closely related species and thus, their genomes are very similar. Indeed, more than 75% of the genome of *Mmc* can be easily mapped on the *Mcap* chromosome with more than 90% of identity at the nucleotide level. This core genome is likely to encode similar proteins involved in key cellular functions required for GT. The accessory *Mmc* genome contains regions rich in insertion sequence (IS) elements, which are not found in *Mcap*. In addition, the *Mmc/Mcap* pair shares >95% similarity at the core proteome level, used to determine the degree of relatedness between conserved proteins involved in key cellular functions (Labroussaa et al., 2016). In that context, seven different donor genomes with increasing phylogenetic distance from the recipient cell, all belonging to the 'Spiroplasma phylogenetic group,' were engineered and cloned into the yeast *S. cerevisiae*. GT was attempted for all of them (Fig. 3C) using both isolated chromosomes (frombacteria transplantation) and genomes cloned into yeast (from-yeast transplantation). A direct correlation was observed between the GT efficiency and the phylogenetic distance as *Mcap* closest relatives were back-transplanted with the highest efficiencies. A transplantation 'limit' was reached with *M. florum*, which has ~90% similarity to *Mcap* with respect to its core proteome (Fig. 3C). In contrast, the chromosome of the more distantly related *S. citri*, which shares less than 85% of similarity with *Mcap* at the core proteome level, could not be transplanted (Labroussaa et al., 2016).

4.6.3 Selection of the recipient cell

These two first factors, associated to the characteristics of the transplantation protocol itself, imply that the choice of the recipient cell is a key factor for GT. So far, *Mcap* is the only recipient cell that has been used to obtain *Mycoplasma* transplants. This might be explained by its close phylogenetic relatedness to other members of the '*M. mycoides* cluster'. As highlighted by the impossibility to back-transplant the genome of *S. citri*, increased phylogenetic distances are likely to limit or prevent the use of *Mcap* for the GT of more phylogenetically distant *Mollicutes* genomes. Therefore, alternative recipient strains need to be identified to extend GT. First, such strains would need to be easily transformed with foreign DNA. The establishment of efficient transformation protocols, especially based on cell fusion using PEG, is not trivial and often developed empirically. Such protocols have been reported for the porcine pathogens *Mesomycoplasma hyopneumoniae* (Trueeb, Gerber, Maes, Gharib, & Kuhnert, 2019) and *Mesomycoplasma hyorhinis* (Dybvig et al., 1998), *Mycoplasmoides gallisepticum* infecting chickens (Cao, Kapke, & Minion, 1994), and a handful of other *Mollicutes* species. However, no successful GT experiments have been reported using these species as recipient cells. Additionally, phylogenetic distances between the pair of bacteria used should be carefully considered. The phylogenetic distances between most 'closely related' species belonging to the *Mycoplasmoides* or *Metamycoplasma* clusters could be already problematic. Therefore, intra-species GT would be a good starting point to extend this technique to new species.

4.6.4 DNA recombination

A number of *Mollicutes* species are intrinsically capable of DNA recombination. Even if site-specific and illegitimate recombination have been observed in *Mollicutes*, HR is the one mechanism likely to negatively affect the GT. HR occurs between genomic regions sharing extensive stretches of nucleotide homology. As GT relies on the use of two phylogenetically closely related bacterial species, it is therefore expected that many conserved genes share high percentages of similarity that are likely to trigger HR events. A functional RecA protein was shown to be required for HR in *M. genitalium* among others, as a *recA* mutant of *M. genitalium* lost its ability to recombine homologous sequences (Burgos, Wood, Young, Glass, & Totten, 2012; Torres-Puig, Broto, Querol, Pinol, & Pich, 2015).

Even in the absence of a fully functional *recA* gene, some members of the former '*Spiroplasma* phylogenetic group' perform HR at very low frequencies, as exemplified for several *Spiroplasma* spp. (Duret, Danet, Garnier, & Renaudin, 1999; Marais, Bove, & Renaudin, 1996). Frequencies of HR in *Mmc* could be substantially increased by adding a heterologous *recA* gene originating from *E. coli* (Allam et al., 2010). The absence of efficient recombination machineries in *Mcap* and the rest of the '*M. mycoides* cluster' is now seen as a main advantage for GT. Recombination events were not reported between *Mmc* and *Mcap* chromosomes, even if only a limited number of transplants have been sequence analysed (Gibson et al., 2010; Lartigue et al., 2007, 2009).

4.6.5 Nucleases

Nucleases are enzymes hydrolyzing phosphodiester bonds between adjacent nucleotides and therefore have the capacity to degrade DNA or RNA (Minion et al., 1993). Several species of *Mollicutes* were shown to express nucleases either at the cell surface or as secreted proteins. These enzymes have been described to facilitate the recycling of DNA precursors or prevent their recognition by the host immune system. The hydrolysis of neutrophil extracellular traps (NETs) by specific extracellular nucleases was observed at least in vitro for *Mycoplasmopsis bovis* (Zhang et al., 2016) and *Mycoplasmoides pneumoniae* (Yamamoto, Kida, Sakamoto, & Kuwano, 2017) and therefore these nucleases represent candidate virulence factors. Active membrane-associated nucleases were found in these two pathogenic species but also in other species including *Mycoplasmopsis pulmonis* (Jarvill-Taylor, VanDyk, & Minion, 1999), *M. hyopneumoniae* (Schmidt, Browning, & Markham, 2007), *M. genitalium* (Li, Krishnan, Baseman, & Kannan, 2010), *M. gallisepticum* (Masukagami et al., 2013) and *Mycoplasmopsis synoviae* (Cizelj, Dusanic, Bencina, & Narat, 2016).

The expression of such nucleases by the recipient cell of GT would be detrimental for the establishment of GT, as they would have the capacity to degrade the donor chromosome upon entry. The detrimental effect of nucleases was previously demonstrated for the cloning of the chromosome of *A. laidlawii* strain PG-8A in yeast. In contrast to most *Mollicutes*, *A. laidlawii* uses the universal genetic code (Razin et al., 1998) and the expression of a surface-anchored endonuclease (i.e., ACL0117) in the eukaryotic host was shown to be toxic (Karas, Tagwerker, Yonemoto, Hutchison 3rd, & Smith, 2012). Its subsequent deletion resulted in the successful cloning of the chromosome as a YAC in yeast. Nuclease activity was so far not demonstrated for *Mcap* supporting its successful use as recipient cell for GT.

4.6.6 Toxin/antitoxin systems

Toxin-antitoxin (TA) systems are two-component modules found in many bacteria and were recently discovered in *Mollicutes* (Hill et al., 2021; Hutchison 3rd et al., 2019). They are classically composed of a toxin (mRNA or protein) and an antitoxin inhibiting the function of its cognate toxic partner (Jurenas, Fraikin, Goormaghtigh, & Van Melderen, 2022). The antitoxin degrades more rapidly upon exposure to different environmental stresses (Jurenas et al., 2022). If different TA systems are present in the donor and recipient cells, a toxin produced by the recipient cell could theoretically remain active even at residual level when the donor chromosome was transferred. Its activity would therefore affect the viability of the transplanted *Mycoplasma* cell. The capacity for a *Mycoplasma* toxin to trigger cell death in the absence of its cognate antitoxin was recently reported using *Mcap* as recipient of a toxin-encoding plasmid (Hill et al., 2021). Several TA systems were recently identified and characterized in the genomes of *Mollicutes* belonging to different phylogenetic groups. Additional candidate TA systems were detected in different members of the '*M. mycoides* cluster' but no functional TA system has been reported yet for *Mcap*.

4.6.7 Additional factors

Interestingly, none of the previous limiting factors could explain the repetitive failures to establish from-yeast GT using the genome of *Mmm* in *Mcap* RE(−). Despite being among the closest relatives to the GT-receptive *Mmc*, successful transplants were only obtained when using from-bacteria GT (Labroussaa et al., 2016). Unknown factors, probably specific to the genome of *Mmm*, are likely to be involved. The aberrant high numbers of IS elements on *Mmm*'s chromosome, representing 13% of its total genome size, could cause transcriptional or translational network problems in the transplanted cells due to the lack of adequate post-translational modifications in yeast, similar to what has been observed for RM systems. On the other hand, a total of 54 transposases of the IS*1634* family have been inactivated in the yeast-cloned chromosome of *Mmm* using a CRISPR-Cas Base Editor System, but the mutant chromosome was still refractory to GT from yeast using *Mcap* as recipient (Ipoutcha et al., 2022).

5 Future perspectives
5.1 GT and *Mollicutes*

Despite the great prospects of GT, its wide implementation has not been documented yet. The lack of transformation methods suitable for very large DNA fragments, together with the difficulties to find suitable recipient organisms capable of booting up the transplanted chromosomes, explains why the development of GT has been lagging. Furthermore, the exact molecular nature of the GT process is currently not fully understood, which limits the optimization and design of novel protocols for other bacterial species, even in the class *Mollicutes*.

Currently, only a few genomes belonging to *Mycoplasma* species of the 'Spiroplasma phylogenetic group' can be transplanted into *Mcap* using the standard protocol reported above. This fact suggests that the transfer of the chromosome from yeast to *Mcap* is possible from a technical point of view, but the main limiting factors reside in the capacity of the recipient cell to recognize and reboot the acquired chromosome defined by the phylogenetic distance. Phylogenetic distance is likely to dictate such concerted actions, but as seen in the case of *Mmm* this is not the only factor. The extension of GT to other *Mollicutes* will likely require the identification and testing of novel recipient cells in other phylogenetic groups. Upon selection, these recipient cells would likely require to be genetically modified to avoid undesired HR, the degradation of incoming genomes by nucleases and RM systems.

We predict the adaptation of the current GT protocol to other *Mollicutes* in the near and medium-term future, which in turn will increase our understanding of GT and foster its implementation in bacteria outside the *Mollicutes* class.

5.2 Adaptation of GT to bacterial species other than *Mollicutes*

The development of methodologies to clone, modify and maintain large DNA molecules in yeast triggered the cloning of viral and non-*Mollicutes* genomes by researchers. Many microbial and viral genomes are now available as YACs including several reduced or recoded versions of *E. coli* genomes (Karas et al., 2012; Lau et al., 2017; Noskov et al., 2012; Ostrov et al., 2016; Tagwerker et al., 2012; Vashee et al., 2020; Venetz et al., 2019; Zhou, Wu, Xue, & Qin, 2016). The main limitations to boot up other genomes are (i) the availability of transformation protocols enabling the transfer of entire intact genomes and (ii) recipient cells able to boot up the latter.

In *E. coli*, these boundaries have been partly overcome by sequential replacement of the genome—one part at a time—using Replicon Excision for enhanced genome engineering through programmed Recombination (REXER) followed by iterative conjugation (Fredens et al., 2019; Wang et al., 2016). REXER allows the assimilation of 100-kbp DNA fragments assembled in yeast into the *E. coli* genome and relies on CRISPR/Cas9 technology and lambda red recombination machinery. However, this strategy is limited to organisms in which REXER and conjugation can be applied. It requires multiple iterations which in turn can lead to the accumulation of off-target mutations. Recently, electroporation has been reported to enable the transformation of very large DNA molecules in *E. coli* (Yoneji, Fujita, Mukai, & Su'etsugu, 2021); such a protocol might be adopted for *Mollicutes*. A 1 Mbp bacterial artificial chromosome (BAC), representing one third of the whole genome, was swapped between two *E. coli* strains (Yoneji et al., 2021). This achievement was based on the supercoiling and repair of these BACs by enzymatic reaction (Fujita et al., 2022), suggesting that DNA topology is a key parameter for successful transfer of chromosomes between bacteria.

6 Ethical considerations

The discipline of synthetic biology is expanding rapidly, with novel methodologies being developed at an exponential rate, thus increasing the range of possibilities at the hand of scientists. Design and production of new drugs and vaccines, the creation of organisms capable of producing biofuels or aiding in the fight against climate change or waste recycling are just a few milestones on the horizon for synthetic biologists.

Like in other scientific disciplines the use of synthetic biology tools raises also ethical concerns. In particular, expansion of techniques like GT poses a more significant ethical concern commonly known as the *Dual-use Dilemma* (Miller & Selgelid, 2007). Misuse of knowledge regarding synthesis and design of novel biological agents can have calamitous consequences for all possible good outcomes of synthetic biology above-mentioned. Moreover, in contrast to other scientific disciplines like nuclear physics where new developments are kept highly classified to the open public, bioscientific advances are open and fully available to general audiences. Therefore, as more tools to design and create synthetic microorganisms become available, combined with the continuously decreasing cost of DNA synthesis, more safety regulations will need to be adopted. Synthetic organisms should be evaluated regarding their biosafety level according to their potential risk when used or upon accidental release. Consequently, national and international laws and regulations in the framework of biosafety and biosecurity have to be approved for synthetic organisms. Of particular importance is to prevent the generation and release of novel forms of pathogenic organisms inexistent in nature, to reinforce the biosafety by implementing biological kill switches and to increase bioethical discussions whether to pursue or openly disseminate certain kinds of knowledge (Douglas & Savulescu, 2010). However, some of these techniques have existed for more than a decade now and no such dual-use related issues have been reported to the best of our knowledge. Importantly, organisms controlled by synthetic genomes with an atypical genetic code restrict functionality of their synthetic genes in other microorganisms upon possible transfer events.

Acknowledgements

The study was supported by the Swiss National Science Foundation (grant number 310030_201152, www.snf.ch). We thank Peter Kuhnert for critical reading of the manuscript and his help in constructing the phylogenetic tree.

Appendix

The detailed protocol presented below is optimized for the preparation of intact donor chromosomes from either *M. mycoides* subsp. *capri* GM12 or yeast harbouring the GM12 chromosome to be transplanted into the recipient strain *M. capricolum* subsp. *capricolum* (*Mcap*) restriction-minus strain, namely *Mcap* RE(−). This protocol is based on previously published protocols (Labroussaa et al., 2016; Lartigue et al., 2007, 2009). Preparations of intact chromosomes derived from *Mollicutes* or yeast cultures are made in agarose plugs using the CHEF Mammalian Genomic DNA Plug Kit (BioRad). The use of a swing-out rotor is recommended for all centrifugation steps as it provides an optimal sedimentation of both mycoplasmas and yeast cells at the bottom of the conical centrifugation tubes. Following this protocol, transplantation rates (number of transplants/μg of gDNA) of 7.44×10^1 (± 9.66) and $2.41 \times 10^1 \pm 1.69 \times 10^1$ are generally obtained for from-bacteria and from-yeast transplantations, respectively.

A.1 Entrapping of intact donor chromosomes in agarose plugs

A.1.1 Isolation of intact Mmc *chromosomes from cultures*

a. Culture *Mmc* in 50 mL of SP5 medium supplemented with 5 μg/mL tetracycline and 10 μg/mL streptomycin[a] at 37 °C with no agitation until late logarithmic phase (indicated by pH 6.5–6.8). In general, 50 mL of culture are sufficient to prepare ~20 plugs at the optimal gDNA concentration.[b] If more plugs are needed, the volume of the culture has to be increased proportionally.

b. Add 100 μg/mL chloramphenicol and incubate at 37 °C for 90 min[c] without agitation. Chloramphenicol is added at sub-inhibitory concentration to ensure that DNA synthesis proceeds until completion of the current replication cycle.

c. Collect the *Mmc* cells by centrifugation at $5800 \times g$ for 15 min at 10 °C.

d. Remove the supernatant and resuspend the cells in 25 mL of cold resuspension buffer (10 mM Tris, 500 mM Sucrose, pH 6.5). Centrifuge at $5800 \times g$ for 15 min at 10 °C.

e. Repeat step d. one more time.

f. Remove the supernatant and resuspend the *Mmc* cells in 1 mL of resuspension buffer.

g. Incubate the cells at 50 °C for 10 min.

h. Mix the cells with 1 mL of 2% low-melting agarose prepared in TAE buffer (40 mM Tris base, 20 mM acetic acid, 1 mM EDTA), pre-incubated at 50 °C.

i. Cast plugs using 100 μL of the mixture per plug employing the mould provided in the CHEF Mammalian Genomic DNA Plug Kit (BioRad). Allow the agarose plugs to solidify at 4 °C for at least 20 min.

---

[a]Streptomycin at sub-inhibitory concentration is added to maintain the gDNA in compact form.
[b]A qPCR protocol was developed to assess the optimal gDNA concentration required for GT and ~2 μg of gDNA was determined to be the optimal quantity of gDNA to use for all species successfully transplanted so far.
[c]Incubation time corresponds approximatively to the doubling time of *Mmc* GM12 under the same culture conditions.

j. Remove the plugs carefully from the mould and transfer them into a 50-mL Falcon tube containing 3 mL of proteinase K buffer supplemented with 120 µL of proteinase K (provided in the CHEF Mammalian Genomic DNA Plug Kit) and incubate the plugs at 50 °C for 24 h.

k. Wash the plugs with 20 mL of 1× plug washing buffer (provided in the CHEF Mammalian Genomic DNA Plug Kit) for 1 h with slow agitation.

l. Repeat the step k. three more times. Plugs can then be stored in 1× plug washing buffer at 4 °C for several months.

A.1.2 Preparation of agarose plugs from yeast cultures containing modified donor chromosomes

The following protocol will be sufficient for the preparation of 5 agarose plugs. Volumes can be increased proportionally if more plugs are needed.

a. Culture *S. cerevisiae* containing the *Mmc* genome in 100 mL of SD-His medium (pH 5.8, Formedium™) at 30 °C under agitation (200 rpm) until it reaches an $OD_{600nm} \sim 2$.

b. Collect the yeast cells by centrifugation at $1750 \times g$ for 5 min at 4 °C. Discard the supernatant and resuspend the pellet in 10 mL of ice-cold 50 mM EDTA (pH 8). Incubate the cells on ice.

c. Determine the yeast cell concentration by counting the yeast cells using a Malassez counting chamber. The concentration of yeast cells obtained following this protocol ranges between 10^8 and 5×10^8 cells/mL of suspension.

d. Transfer the volume of yeast suspension needed to a new centrifugation tube. The desired concentration is $\sim 10^8$ yeast cells per plug. Collect the cells at $1750 \times g$ for 5 min at 4 °C.

e. Resuspend the pellet in 300 µL of cell suspension buffer (provided in the CHEF Mammalian Genomic DNA Plug Kit) and incubate the cells at 50 °C for 10 min.

f. Add 2.5 mg of Zymolyase® 100 T (AmsBio) to the yeast resuspension. Mix gently.

g. Add 300 µL of 2% low-melting agarose prepared in TAE buffer, pre-incubated at 50 °C. Mix gently.

h. Cast plugs using 100 µL of the mixture per plug employing the mould provided in the CHEF Mammalian Genomic DNA Plug Kit. Allow the agarose plugs to solidify at 4 °C for at least 20 min.

i. Remove the plugs carefully from the mould and transfer them into a 50-mL Falcon tube containing 2 mL of Lyticase buffer (10 mM Tris-HCl, pH 7.5; 50 mM EDTA, pH 8) supplemented with 10 mg of Zymolyase®-100 T. Incubate at 37 °C for 2 h.

j. Carefully remove the solution and wash the plugs twice with 25 mL of distilled sterile water.

k. Add 3 mL of proteinase K buffer supplemented with 120 µL of proteinase K (provided in the CHEF Mammalian Genomic DNA Plug Kit) and incubate the plugs at 50 °C for 24 h.

l. Remove the solution and repeat step k.

m. Wash the plugs with 20 mL of 1× plug washing buffer (provided in the CHEF Mammalian Genomic DNA Plug Kit) for 1 h with slow agitation.

n. Repeat step m. three more times. Plugs can then be stored in 1× plug washing buffer at 4 °C for several months.

A.2 Genome transplantation using Mcap RE(−) as a recipient cell

A.2.1 *Release of intact chromosomes from agarose plugs*

a. Gently take one agarose plug (containing *Mmc*- or yeast-derived chromosomes) and transfer it into a 1.5-mL Eppendorf tube with a small spatula.

b. Wash the plug twice with 1 mL of 0.1× plug washing buffer (1/10th dilution of the 1× plug washing buffer) for 30 min with gentle agitation.

c. Make sure to completely remove all the buffer and then add 10 μL of 10× β-agarase buffer (New England Biolabs).

d. Incubate at 42 °C for 5 min.

e. Incubate at 65 °C for 8 min to melt the agarose plug.

f. Incubate the mixture at 42 °C for 10 min and then add 3 μL of β-agarase (1 unit/μL; New England Biolabs) to the mixture. Incubate overnight[d] at 42 °C.

A.2.2 *Preparation of* Mcap *RE(−) recipient cells*

a. Start an overnight culture of *Mcap* RE(−) at 30 °C in SOB(+) medium until late logarithmic phase (∼pH 6.5). The volume of the culture depends on the number of experiments to be carried out. Six and 12 mL of culture are necessary to perform a single from-bacteria and from-yeast GT experiment, respectively.

Note: The determination of the initial *Mcap* RE(−) cell concentration is a prerequisite for the calculation of the transplantation efficiencies later on. To do so, perform a 10-fold serial dilution of the *Mcap* RE(−) culture and plate 500 μL of the appropriate dilutions (usually 10^{-6} and 10^{-7}) on SP5 plates (Labroussaa et al., 2016). Incubate the plates at 37 °C for 2 days and count the colony-forming units (CFUs) on each plate. The *Mcap* RE(−) cell concentration is calculated using the following formula: CFU/mL = number of CFUs counted × dilution factor × 2. A cell concentration of ∼10^9 CFU/mL is usually obtained.

b. Pellet *Mcap* RE(−) cells at 5800 × g for 15 min at 10 °C.

c. Resuspend the cells in wash buffer (10 mM Tris, 250 mM NaCl, pH 6.5) using half the volume of the initial culture volume. Centrifuge at 5800 × g for 15 min at 10 °C.

d. Discard the supernatant and resuspend the cell pellet (corresponding to 6 mL of culture) in 200 μL[e] of anhydrous 0.1 M CaCl₂.

e. Incubate on ice for 30 min.

---

[d]Shorter incubations are also possible (minimum 2 h).
[e]For from-yeast GT, 12 mL of culture will be ultimately resuspended in 400 μL of anhydrous 0.1 M CaCl₂).

A.2.3 Transplantation of donor chromosomes into Mcap RE(−) recipient cells

All volumes denoted below are only suitable for from-bacteria GT. When performing from-yeast GT, double all the indicated volumes.

a. Place 400 µL of SP5Δserum into a 15-mL Falcon tube. The SP5Δserum composition is similar to the one of the SP5 medium except that fetal bovine serum is omitted and 4.5 g/L of NaCl is added to maintain the isotonic nature of the medium.

b. Carefully add 50 µL of melted plugs.

c. Gently add the *Mcap* RE(−) competent cells to the mixture. Do not mix.

d. Add the same volume[f] of 2× Fusion buffer (20 mM Tris, 20 mM MgCl$_2$.6H$_2$O, 500 mM NaCl, 10% (*w/v*) PEG$_{6000}$, pH 6.5) to the reaction. Gently mix by rotation of the tube and incubate at 30 °C for 90 min without agitation.

e. Add 5 mL of SP5 and mix the solution by inverting the tube once.

f. Collect the *Mcap* RE(−) cells at 5800 × *g* for 15 min at 10 °C.

g. Resuspend the cells in 500 µL of SP5 medium and gently spread onto a SP5 plate supplemented with 5 µg/mL tetracycline. The use of a spreader is not recommended. Gently spread the cells evenly and completely over the agar surface by simply rotating and tilting the plates. For optimal from-yeast GT efficiencies, the cells can be spread on two separate plates. For detection of *lacZ*-encoded β-galactosidase expression in *Mcap* recipients, S5 plates supplemented with tetracycline and 5-bromo-4-chloro-3-indolyl-β-D-galactoside (X-Gal) can be used.

h. Allow the plates to dry for a few minutes and incubate for 2–4 days at 37 °C and protected from the light to limit antibiotic degradation.

A.2.4 Screening of the transplants

a. Pick 5–10 colonies using sterile tips or sterile Pasteur pipettes and transfer them to a 1.5-mL sterile Eppendorf tube containing 1 mL of SP5 supplemented with 5 µg/mL of tetracycline. Incubate at 37 °C without agitation until mid- to late-logarithmic phase (change from red to orange/yellow colour due to the acidification of the SP5 medium by *Mollicutes* associated to the presence of phenol red).

b. Transfer 5 µL of the culture into 995 µL of fresh SP5 medium (1/500th dilution) and incubate at 37 °C without agitation until mid- to late-logarithmic phase.

c. Repeat step b. one more time. These passaging steps are important to ensure that no residual *Mcap* cells are present in the cultures, which can interfere with the screening process later on.

d. Take 400 µL of the culture and freeze the remaining culture at −80 °C until further use.

---

[f]This corresponds to 650 µL for a from-bacteria GT and 1300 µL for a from-yeast GT.

e. Pellet the cells at $14,000 \times g$ for 2 min and resuspend the cells in 400 µL of phosphate-buffered saline (ThermoFisher Scientific).

f. Collect the cells at $14,000 \times g$ for 2 min and resuspend the cells in 200 µL of cell lysis buffer provided in the Wizard Genomic DNA Extraction Kit (Promega).

g. Incubate the cells at 95 °C for 10 min.

h. Collect the cells at $14,000 \times g$ for 2 min. Transfer 50 µL of the supernatant to a clean 1.5-mL Eppendorf reaction tube.

i. Use 1 µL as DNA template in a PCR reaction using primers specific to the donor genome (i.e., *Mmc*). A similar PCR reaction using primers specific to the recipient cell can be also performed to ensure the absence of the *Mcap* RE(−) genome.

j. Positives transplants can be then further analysed by multiplex PCR and Pulsed-Field Gel Electrophoresis (PFGE) to ensure the integrity of the whole chromosome. Ultimately, whole genome sequencing is performed to confirm the genotype of the obtained transplants and to identify any single nucleotide polymorphisms (SNPs) or transposed IS-elements that may have appeared during the GT process.

References

Algire, M. A., Montague, M. G., Vashee, S., Lartigue, C., & Merryman, C. (2012). A type III restriction-modification system in *Mycoplasma mycoides* subsp. *capri*. *Open Biology*, *2*(10), 120115. https://doi.org/10.1098/rsob.120115.

Allam, A. B., Reyes, L., Assad-Garcia, N., Glass, J. I., & Brown, M. B. (2010). Targeted homologous recombination in *Mycoplasma mycoides* subsp. *capri* is enhanced by inclusion of heterologous *recA*. *Applied and Environmental Microbiology*, *76*(20), 6951–6954. https://doi.org/10.1128/AEM.00056-10.

Baby, V., Labroussaa, F., Brodeur, J., Matteau, D., Gourgues, G., Lartigue, C., et al. (2018). Cloning and transplantation of the *Mesoplasma florum* genome. *ACS Synthetic Biology*, *7*(1), 209–217. https://doi.org/10.1021/acssynbio.7b00279.

Barroso, G., & Labarere, J. (1988). Chromosomal gene transfer in *Spiroplasma citri*. *Science*, *241*(4868), 959–961. https://doi.org/10.1126/science.3261453.

Benders, G. A., Noskov, V. N., Denisova, E. A., Lartigue, C., Gibson, D. G., Assad-Garcia, N., et al. (2010). Cloning whole bacterial genomes in yeast. *Nucleic Acids Research*, *38*(8), 2558–2569. https://doi.org/10.1093/nar/gkq119.

Bernheim, A., & Sorek, R. (2020). The pan-immune system of bacteria: Antiviral defence as a community resource. *Nature Reviews. Microbiology*, *18*(2), 113–119. https://doi.org/10.1038/s41579-019-0278-2.

Bickle, T. A., & Kruger, D. H. (1993). Biology of DNA restriction. *Microbiological Reviews*, *57*(2), 434–450. https://doi.org/10.1128/mr.57.2.434-450.1993.

Breton, M., Tardy, F., Dordet-Frisoni, E., Sagne, E., Mick, V., Renaudin, J., et al. (2012). Distribution and diversity of mycoplasma plasmids: Lessons from cryptic genetic elements. *BMC Microbiology*, *12*, 257. https://doi.org/10.1186/1471-2180-12-257.

Breuer, M., Earnest, T. M., Merryman, C., Wise, K. S., Sun, L., Lynott, M. R., et al. (2019). Essential metabolism for a minimal cell. *eLife*, *8*, e36842. https://doi.org/10.7554/eLife.36842.

Burgos, R., Wood, G. E., Young, L., Glass, J. I., & Totten, P. A. (2012). RecA mediates MgpB and MgpC phase and antigenic variation in *Mycoplasma genitalium*, but plays a minor role in DNA repair. *Molecular Microbiology, 85*(4), 669–683. https://doi.org/10.1111/j.1365-2958.2012.08130.x.

Cao, J., Kapke, P. A., & Minion, F. C. (1994). Transformation of *Mycoplasma gallisepticum* with Tn*916*, Tn*4001*, and integrative plasmid vectors. *Journal of Bacteriology, 176*(14), 4459–4462. https://doi.org/10.1128/jb.176.14.4459-4462.1994.

Chandran, S., Noskov, V. N., Segall-Shapiro, T. H., Ma, L., Whiteis, C., Lartigue, C., et al. (2014). TREC-IN: Gene knock-in genetic tool for genomes cloned in yeast. *BMC Genomics, 15*(1), 1180. https://doi.org/10.1186/1471-2164-15-1180.

Citti, C., Marechal-Drouard, L., Saillard, C., Weil, J. H., & Bove, J. M. (1992). *Spiroplasma citri* UGG and UGA tryptophan codons: Sequence of the two tryptophanyl-tRNAs and organization of the corresponding genes. *Journal of Bacteriology, 174*(20), 6471–6478. https://doi.org/10.1128/jb.174.20.6471-6478.1992.

Cizelj, I., Dusanic, D., Bencina, D., & Narat, M. (2016). Mycoplasma and host interaction: *In vitro* gene expression modulation in *Mycoplasma synoviae* and infected chicken chondrocytes. *Acta Veterinaria Hungarica, 64*(1), 26–37. https://doi.org/10.1556/004.2016.003.

DaMassa, A. J., Brooks, D. L., Adler, H. E., & Watt, D. E. (1983). Caprine mycoplasmosis: Acute pulmonary disease in newborn kids given *Mycoplasma capricolum* orally. *Australian Veterinary Journal, 60*(4), 125–126. http://www.ncbi.nlm.nih.gov/pubmed/6870714.

Djordjevic, S. R., Forbes, W. A., Forbes-Faulkner, J., Kuhnert, P., Hum, S., Hornitzky, M. A., et al. (2001). Genetic diversity among *Mycoplasma* species bovine group 7: Clonal isolates from an outbreak of polyarthritis, mastitis, and abortion in dairy cattle. *Electrophoresis, 22*(16), 3551–3561. https://doi.org/10.1002/1522-2683(200109)22:16<3551::AID-ELPS3551>3.0.CO;2-#.

Douglas, T., & Savulescu, J. (2010). Synthetic biology and the ethics of knowledge. *Journal of Medical Ethics, 36*(11), 687–693. https://doi.org/10.1136/jme.2010.038232.

Duret, S., Danet, J. L., Garnier, M., & Renaudin, J. (1999). Gene disruption through homologous recombination in *Spiroplasma citri*: An *scm1*-disrupted motility mutant is pathogenic. *Journal of Bacteriology, 181*(24), 7449–7456. https://doi.org/10.1128/JB.181.24.7449-7456.1999.

Dybvig, K., Sitaraman, R., & French, C. T. (1998). A family of phase-variable restriction enzymes with differing specificities generated by high-frequency gene rearrangements. *Proceedings of the National Academy of Sciences of the United States of America, 95*(23), 13923–13928. https://doi.org/10.1073/pnas.95.23.13923.

Fischer, A., Shapiro, B., Muriuki, C., Heller, M., Schnee, C., Bongcam-Rudloff, E., et al. (2012). The origin of the '*Mycoplasma mycoides* Cluster' coincides with domestication of ruminants. *PLoS One, 7*(4), e36150. https://doi.org/10.1371/journal.pone.0036150.

Fraser, C. M., Gocayne, J. D., White, O., Adams, M. D., Clayton, R. A., Fleischmann, R. D., et al. (1995). The minimal gene complement of *Mycoplasma genitalium*. *Science, 270*(5235), 397–403. https://doi.org/10.1126/science.270.5235.397.

Fredens, J., Wang, K., de la Torre, D., Funke, L. F. H., Robertson, W. E., Christova, Y., et al. (2019). Total synthesis of *Escherichia coli* with a recoded genome. *Nature, 569*(7757), 514–518. https://doi.org/10.1038/s41586-019-1192-5.

Fujita, H., Osaku, A., Sakane, Y., Yoshida, K., Yamada, K., Nara, S., et al. (2022). Enzymatic supercoiling of bacterial chromosomes facilitates genome manipulation. *ACS Synthetic Biology, 11*(9), 3088–3099. https://doi.org/10.1021/acssynbio.2c00353.

Gibson, D. G., Benders, G. A., Andrews-Pfannkoch, C., Denisova, E. A., Baden-Tillson, H., Zaveri, J., et al. (2008). Complete chemical synthesis, assembly, and cloning of a *Mycoplasma genitalium* genome. *Science, 319*(5867), 1215–1220. https://doi.org/10.1126/science.1151721.

Gibson, D. G., Glass, J. I., Lartigue, C., Noskov, V. N., Chuang, R. Y., Algire, M. A., et al. (2010). Creation of a bacterial cell controlled by a chemically synthesized genome. *Science, 329*(5987), 52–56. https://doi.org/10.1126/science.1190719.

Haas, D., Thamm, A. M., Sun, J., Huang, L., Sun, L., Beaudoin, G. A. W., et al. (2022). Metabolite damage and damage control in a minimal genome. *MBio, 13*(4), e0163022. https://doi.org/10.1128/mbio.01630-22.

Hanahan, D. (1983). Studies on transformation of *Escherichia coli* with plasmids. *Journal of Molecular Biology, 166*(4), 557–580. https://doi.org/10.1016/s0022-2836(83)80284-8.

Hill, V., Akarsu, H., Barbarroja, R. S., Cippa, V. L., Kuhnert, P., Heller, M., et al. (2021). Minimalistic mycoplasmas harbor different functional toxin-antitoxin systems. *PLoS Genetics, 17*(10), e1009365. https://doi.org/10.1371/journal.pgen.1009365.

Hossain, T., Deter, H. S., Peters, E. J., & Butzin, N. C. (2021). Antibiotic tolerance, persistence, and resistance of the evolved minimal cell, *Mycoplasma mycoides* JCVI-Syn3B. *iScience, 24*(5), 102391. https://doi.org/10.1016/j.isci.2021.102391.

Hutchison, C. A., 3rd, Chuang, R. Y., Noskov, V. N., Assad-Garcia, N., Deerinck, T. J., Ellisman, M. H., et al. (2016). Design and synthesis of a minimal bacterial genome. *Science, 351*(6280), aad6253. https://doi.org/10.1126/science.aad6253.

Hutchison, C. A., 3rd, Merryman, C., Sun, L., Assad-Garcia, N., Richter, R. A., Smith, H. O., et al. (2019). Polar effects of transposon insertion into a minimal bacterial genome. *Journal of Bacteriology, 201*(19), e00119–e00185. https://doi.org/10.1128/JB.00185-19.

Ipoutcha, T., Rideau, F., Gourgues, G., Arfi, Y., Lartigue, C., Blanchard, A., et al. (2022). Genome editing of veterinary relevant mycoplasmas using a CRISPR-Cas base editor system. *Applied and Environmental Microbiology, 88*(17), e0099622. https://doi.org/10.1128/aem.00996-22.

Janis, C., Bischof, D., Gourgues, G., Frey, J., Blanchard, A., & Sirand-Pugnet, P. (2008). Unmarked insertional mutagenesis in the bovine pathogen *Mycoplasma mycoides* subsp. *mycoides* SC: Characterization of a *lppQ* mutant. *Microbiology (Reading), 154*(Pt. 8), 2427–2436. https://doi.org/10.1099/mic.0.2008/017640-0.

Janis, C., Lartigue, C., Frey, J., Wroblewski, H., Thiaucourt, F., Blanchard, A., et al. (2005). Versatile use of *oriC* plasmids for functional genomics of *Mycoplasma capricolum* subsp. *capricolum*. *Applied and Environmental Microbiology, 71*(6), 2888–2893. https://doi.org/10.1128/AEM.71.6.2888-2893.2005.

Jarvill-Taylor, K. J., VanDyk, C., & Minion, F. C. (1999). Cloning of *mnuA*, a membrane nuclease gene of *Mycoplasma pulmonis*, and analysis of its expression in *Escherichia coli*. *Journal of Bacteriology, 181*(6), 1853–1860. https://doi.org/10.1128/JB.181.6.1853-1860.1999.

Jores, J., Baldwin, C., Blanchard, A., Browning, G. F., Colston, A., Gerdts, V., et al. (2020). Contagious bovine and caprine pleuropneumonia: A research community's recommendations for the development of better vaccines. *NPJ Vaccines, 5*, 66. https://doi.org/10.1038/s41541-020-00214-2.

Jores, J., Ma, L., Ssajjakambwe, P., Schieck, E., Liljander, A., Chandran, S., et al. (2019). Removal of a subset of non-essential genes fully attenuates a highly virulent *Mycoplasma* strain. *Frontiers in Microbiology, 10*, 664. https://doi.org/10.3389/fmicb.2019.00664.

Jores, J., Schieck, E., Liljander, A., Sacchini, F., Posthaus, H., Lartigue, C., et al. (2019). In vivo role of capsular polysaccharide in *Mycoplasma mycoides*. *The Journal of Infectious Diseases*, *219*(10), 1559–1563. https://doi.org/10.1093/infdis/jiy713.

Jurenas, D., Fraikin, N., Goormaghtigh, F., & Van Melderen, L. (2022). Biology and evolution of bacterial toxin-antitoxin systems. *Nature Reviews. Microbiology*, *20*(6), 335–350. https://doi.org/10.1038/s41579-021-00661-1.

Karas, B. J., Tagwerker, C., Yonemoto, I. T., Hutchison, C. A., 3rd, & Smith, H. O. (2012). Cloning the *Acholeplasma laidlawii* PG-8A genome in *Saccharomyces cerevisiae* as a yeast centromeric plasmid. *ACS Synthetic Biology*, *1*(1), 22–28. https://doi.org/10.1021/sb200013j.

Kent, B. N., Foecking, M. F., & Calcutt, M. J. (2012). Development of a novel plasmid as a shuttle vector for heterologous gene expression in *Mycoplasma yeatsii*. *Journal of Microbiological Methods*, *91*(1), 121–127. https://doi.org/10.1016/j.mimet.2012.07.018.

King, K. W., & Dybvig, K. (1992). Nucleotide sequence of *Mycoplasma mycoides* subspecies *mycoides* plasmid pKMK1. *Plasmid*, *28*(1), 86–91. http://www.ncbi.nlm.nih.gov/entrez/query.fcgi?cmd=Retrieve&db=PubMed&dopt=Citation&list_uids=1518915.

King, K. W., & Dybvig, K. (1994). Transformation of *Mycoplasma capricolum* and examination of DNA restriction modification in *M. capricolum* and *Mycoplasma mycoides* subsp. *mycoides*. *Plasmid*, *31*(3), 308–311. http://www.ncbi.nlm.nih.gov/entrez/query.fcgi?cmd=Retrieve&db=PubMed&dopt=Citation&list_uids=8058824.

Kitamura, K., Kaneko, T., & Yamamoto, Y. (1971). Lysis of viable yeast cells by enzymes of *Arthrobacter luteus*. *Archives of Biochemistry and Biophysics*, *145*(1), 402–404. https://doi.org/10.1016/0003-9861(71)90053-1.

Labroussaa, F., Baby, V., Rodrigue, S., & Lartigue, C. (2019). Whole genome transplantation: Bringing natural or synthetic bacterial genomes back to life. *Medical Science (Paris)*, *35*(10), 761–770. https://doi.org/10.1051/medsci/2019154. La transplantation de genomes—Redonner vie a des genomes bacteriens naturels ou synthetiques.

Labroussaa, F., Lebaudy, A., Baby, V., Gourgues, G., Matteau, D., Vashee, S., et al. (2016). Impact of donor-recipient phylogenetic distance on bacterial genome transplantation. *Nucleic Acids Research*, *44*(17), 8501–8511. https://doi.org/10.1093/nar/gkw688.

Labroussaa, F., Mehinagic, K., Cippa, V., Liniger, M., Akarsu, H., Ruggli, N., et al. (2021). In-yeast reconstruction of the African swine fever virus genome isolated from clinical samples. *STAR Protocol*, *2*(3), 100803. https://doi.org/10.1016/j.xpro.2021.100803.

Lartigue, C., Blanchard, A., Renaudin, J., Thiaucourt, F., & Sirand-Pugnet, P. (2003). Host specificity of mollicutes *oriC* plasmids: Functional analysis of replication origin. *Nucleic Acids Research*, *31*(22), 6610–6618. https://doi.org/10.1093/nar/gkg848.

Lartigue, C., Glass, J. I., Alperovich, N., Pieper, R., Parmar, P. P., Hutchison, C. A., 3rd, et al. (2007). Genome transplantation in bacteria: Changing one species to another. *Science*, *317*(5838), 632–638. https://doi.org/10.1126/science.1144622.

Lartigue, C., Lebaudy, A., Blanchard, A., El Yacoubi, B., Rose, S., Grosjean, H., et al. (2014). The flavoprotein Mcap0476 (RlmFO) catalyzes m5U1939 modification in *Mycoplasma capricolum* 23S rRNA. *Nucleic Acids Research*, *42*(12), 8073–8082. https://doi.org/10.1093/nar/gku518.

Lartigue, C., Vashee, S., Algire, M. A., Chuang, R. Y., Benders, G. A., Ma, L., et al. (2009). Creating bacterial strains from genomes that have been cloned and engineered in yeast. *Science*, *325*(5948), 1693–1696. https://doi.org/10.1126/science.1173759.

Lau, Y. H., Stirling, F., Kuo, J., Karrenbelt, M. A. P., Chan, Y. A., Riesselman, A., et al. (2017). Large-scale recoding of a bacterial genome by iterative recombineering of synthetic DNA. *Nucleic Acids Research*, *45*(11), 6971–6980. https://doi.org/10.1093/nar/gkx415.

Lentz, B. R. (2007). PEG as a tool to gain insight into membrane fusion. *European Biophysics Journal*, *36*(4–5), 315–326. https://doi.org/10.1007/s00249-006-0097-z.

Li, L., Krishnan, M., Baseman, J. B., & Kannan, T. R. (2010). Molecular cloning, expression, and characterization of a Ca2+−dependent, membrane-associated nuclease of *Mycoplasma genitalium*. *Journal of Bacteriology*, *192*(19), 4876–4884. https://doi.org/10.1128/JB.00401-10.

Manso-Silván, L., Vilei, E. M., Sachse, K., Djordjevic, S. P., Thiaucourt, F., & Frey, J. (2009). *Mycoplasma leachii* sp. nov. as a new species designation for *Mycoplasma* sp. bovine group 7 of Leach, and reclassification of *Mycoplasma mycoides* subsp. *mycoides* LC as a serovar of *Mycoplasma mycoides* subsp. *capri*. *International Journal of Systematic and Evolutionary Microbiology*, *59*(Pt 6), 1353–1358. https://doi.org/10.1099/ijs.0.005546-0.

Marais, A., Bove, J. M., & Renaudin, J. (1996). Characterization of the *recA* gene regions of *Spiroplasma citri* and *Spiroplasma melliferum*. *Journal of Bacteriology*, *178*(23), 7003–7009. https://doi.org/10.1128/jb.178.23.7003-7009.1996.

Masukagami, Y., Tivendale, K. A., Mardani, K., Ben-Barak, I., Markham, P. F., & Browning, G. F. (2013). The *Mycoplasma gallisepticum* virulence factor lipoprotein MslA is a novel polynucleotide binding protein. *Infection and Immunity*, *81*(9), 3220–3226. https://doi.org/10.1128/IAI.00365-13.

Miller, S., & Selgelid, M. J. (2007). Ethical and philosophical consideration of the dual-use dilemma in the biological sciences. *Science and Engineering Ethics*, *13*(4), 523–580. https://doi.org/10.1007/s11948-007-9043-4.

Minion, F. C., Jarvill-Taylor, K. J., Billings, D. E., & Tigges, E. (1993). Membrane-associated nuclease activities in mycoplasmas. *Journal of Bacteriology*, *175*(24), 7842–7847. https://doi.org/10.1128/jb.175.24.7842-7847.1993.

Monk, I. R., Tree, J. J., Howden, B. P., Stinear, T. P., & Foster, T. J. (2015). Complete bypass of restriction systems for major *Staphylococcus aureus* lineages. *MBio*, *6*(3), e00308–e00315. https://doi.org/10.1128/mBio.00308-15.

Murray, A. W., & Szostak, J. W. (1983). Construction of artificial chromosomes in yeast. *Nature*, *305*(5931), 189–193. https://doi.org/10.1038/305189a0.

Noskov, V. N., Karas, B. J., Young, L., Chuang, R. Y., Gibson, D. G., Lin, Y. C., et al. (2012). Assembly of large, high G+C bacterial DNA fragments in yeast. *ACS Synthetic Biology*, *1*(7), 267–273. https://doi.org/10.1021/sb3000194.

Noskov, V., Kouprina, N., Leem, S. H., Koriabine, M., Barrett, J. C., & Larionov, V. (2002). A genetic system for direct selection of gene-positive clones during recombinational cloning in yeast. *Nucleic Acids Research*, *30*(2), E8. https://doi.org/10.1093/nar/30.2.e8.

Noskov, V. N., Segall-Shapiro, T. H., & Chuang, R. Y. (2010). Tandem repeat coupled with endonuclease cleavage (TREC): A seamless modification tool for genome engineering in yeast. *Nucleic Acids Research*, *38*(8), 2570–2576. https://doi.org/10.1093/nar/gkq099.

Nottelet, P., Bataille, L., Gourgues, G., Anger, R., Lartigue, C., Sirand-Pugnet, P., et al. (2021). The mycoplasma surface proteins MIB and MIP promote the dissociation of the antibody-antigen interaction. *Science Advances*, *7*(10), eabf2403. https://doi.org/10.1126/sciadv.abf2403.

Oldfield, L. M., Grzesik, P., Voorhies, A. A., Alperovich, N., MacMath, D., Najera, C. D., et al. (2017). Genome-wide engineering of an infectious clone of herpes simplex virus type 1 using synthetic genomics assembly methods. *Proceedings of the National Academy of Sciences of the United States of America, 114*(42), E8885–E8894. https://doi.org/10.1073/pnas.1700534114.

Ostrov, N., Landon, M., Guell, M., Kuznetsov, G., Teramoto, J., Cervantes, N., et al. (2016). Design, synthesis, and testing toward a 57-codon genome. *Science, 353*(6301), 819–822. https://doi.org/10.1126/science.aaf3639.

Pelletier, J. F., Sun, L., Wise, K. S., Assad-Garcia, N., Karas, B. J., Deerinck, T. J., et al. (2021). Genetic requirements for cell division in a genomically minimal cell. *Cell, 184*(9), 2430–2440 e2416. https://doi.org/10.1016/j.cell.2021.03.008.

Razin, S. (1996). Mycoplasmas. In S. Baron (Ed.), *Medical microbiology* (4th ed.). https://www.ncbi.nlm.nih.gov/pubmed/21413254.

Razin, S., Yogev, D., & Naot, Y. (1998). Molecular biology and pathogenicity of mycoplasmas. *Microbiology and Molecular Biology Reviews, 62*(4), 1094–1156. https://doi.org/10.1128/MMBR.62.4.1094-1156.1998.

Roberts, R. J., Vincze, T., Posfai, J., & Macelis, D. (2015). REBASE—A database for DNA restriction and modification: enzymes, genes and genomes. *Nucleic Acids Research, 43*(Database issue), D298–D299. https://doi.org/10.1093/nar/gku1046.

Rogers, M. J., Simmons, J., Walker, R. T., Weisburg, W. G., Woese, C. R., Tanner, R. S., et al. (1985). Construction of the mycoplasma evolutionary tree from 5S rRNA sequence data. *Proceedings of the National Academy of Sciences of the United States of America, 82*(4), 1160–1164. https://doi.org/10.1073/pnas.82.4.1160.

Ruiz, E., Talenton, V., Dubrana, M. P., Guesdon, G., Lluch-Senar, M., Salin, F., et al. (2019). CReasPy-cloning: A method for simultaneous cloning and engineering of Megabase-sized genomes in yeast using the CRISPR-Cas9 system. *ACS Synthetic Biology, 8*(11), 2547–2557. https://doi.org/10.1021/acssynbio.9b00224.

Schieck, E., Lartigue, C., Frey, J., Vozza, N., Hegermann, J., Miller, R. A., et al. (2016). Galactofuranose in *Mycoplasma mycoides* is important for membrane integrity and conceals adhesins but does not contribute to serum resistance. *Molecular Microbiology, 99*(1), 55–70. https://doi.org/10.1111/mmi.13213.

Schmidt, J. A., Browning, G. F., & Markham, P. F. (2007). *Mycoplasma hyopneumoniae* mhp379 is a Ca2+−dependent, sugar-nonspecific exonuclease exposed on the cell surface. *Journal of Bacteriology, 189*(9), 3414–3424. https://doi.org/10.1128/JB.01835-06.

Tagwerker, C., Dupont, C. L., Karas, B. J., Ma, L., Chuang, R. Y., Benders, G. A., et al. (2012). Sequence analysis of a complete 1.66 Mb *Prochlorococcus marinus* MED4 genome cloned in yeast. *Nucleic Acids Research, 40*(20), 10375–10383. https://doi.org/10.1093/nar/gks823.

Talenton, V., Baby, V., Gourgues, G., Mouden, C., Claverol, S., Vashee, S., et al. (2022). Genome engineering of the fast-growing *Mycoplasma feriruminatoris* toward a live vaccine chassis. *ACS Synthetic Biology, 11*(5), 1919–1930. https://doi.org/10.1021/acssynbio.2c00062.

Tanaka, R., Muto, A., & Osawa, S. (1989). Nucleotide sequence of tryptophan tRNA gene in *Acholeplasma laidlawii*. *Nucleic Acids Research, 17*(14), 5842. https://doi.org/10.1093/nar/17.14.5842.

Tang, Y., Gao, X. D., Wang, Y., Yuan, B. F., & Feng, Y. Q. (2012). Widespread existence of cytosine methylation in yeast DNA measured by gas chromatography/mass spectrometry. *Analytical Chemistry, 84*(16), 7249–7255. https://doi.org/10.1021/ac301727c.

Thi Nhu Thao, T., Labroussaa, F., Ebert, N., V'Kovski, P., Stalder, H., Portmann, J., et al. (2020). Rapid reconstruction of SARS-CoV-2 using a synthetic genomics platform. *Nature, 582*(7813), 561–565. https://doi.org/10.1038/s41586-020-2294-9.

Thiaucourt, F., Manso-Silvan, L., Salah, W., Barbe, V., Vacherie, B., Jacob, D., et al. (2011). *Mycoplasma mycoides*, from "*mycoides* small colony" to "*capri*" A microevolutionary perspective. *BMC Genomics, 12*, 114. https://doi.org/10.1186/1471-2164-12-114.

Torres-Puig, S., Broto, A., Querol, E., Pinol, J., & Pich, O. Q. (2015). A novel sigma factor reveals a unique regulon controlling cell-specific recombination in *Mycoplasma genitalium. Nucleic Acids Research, 43*(10), 4923–4936. https://doi.org/10.1093/nar/gkv422.

Trueeb, B. S., Gerber, S., Maes, D., Gharib, W. H., & Kuhnert, P. (2019). Tn-sequencing of *Mycoplasma hyopneumoniae* and *Mycoplasma hyorhinis* mutant libraries reveals nonessential genes of porcine mycoplasmas differing in pathogenicity. *Veterinary Research, 50*(1), 55. https://doi.org/10.1186/s13567-019-0674-7.

Tsarmpopoulos, I., Gourgues, G., Blanchard, A., Vashee, S., Jores, J., Lartigue, C., et al. (2016). In-yeast engineering of a bacterial genome using CRISPR/Cas9. *ACS Synthetic Biology, 5*(1), 104–109. https://doi.org/10.1021/acssynbio.5b00196.

Vashee, S., Arfi, Y., & Lartigue, C. (2020). Budding yeast as a factory to engineer partial and complete microbial genomes. *Current Opinion in Systems Biology, 24*, 1–8. https://doi.org/10.1016/j.coisb.2020.09.003.

Vashee, S., Stockwell, T. B., Alperovich, N., Denisova, E. A., Gibson, D. G., Cady, K. C., et al. (2017). Cloning, assembly, and modification of the primary human cytomegalovirus isolate Toledo by yeast-based transformation-associated recombination. *mSphere, 2*(5), e00317–e00331. https://doi.org/10.1128/mSphereDirect.00331-17.

Venetz, J. E., Del Medico, L., Wolfle, A., Schachle, P., Bucher, Y., Appert, D., et al. (2019). Chemical synthesis rewriting of a bacterial genome to achieve design flexibility and biological functionality. *Proceedings of the National Academy of Sciences of the United States of America, 116*(16), 8070–8079. https://doi.org/10.1073/pnas.1818259116.

Venter, J. C., Glass, J. I., Hutchison, C. A., 3rd, & Vashee, S. (2022). Synthetic chromosomes, genomes, viruses, and cells. *Cell, 185*(15), 2708–2724. https://doi.org/10.1016/j.cell.2022.06.046.

Wang, K., Fredens, J., Brunner, S. F., Kim, S. H., Chia, T., & Chin, J. W. (2016). Defining synonymous codon compression schemes by genome recoding. *Nature, 539*(7627), 59–64. https://doi.org/10.1038/nature20124.

Woese, C. R., Maniloff, J., & Zablen, L. B. (1980). Phylogenetic analysis of the mycoplasmas. *Proceedings of the National Academy of Sciences of the United States of America, 77*(1), 494–498. https://doi.org/10.1073/pnas.77.1.494.

Yamamoto, T., Kida, Y., Sakamoto, Y., & Kuwano, K. (2017). Mpn491, a secreted nuclease of *Mycoplasma pneumoniae*, plays a critical role in evading killing by neutrophil extracellular traps. *Cellular Microbiology, 19*(3), e12666. https://doi.org/10.1111/cmi.12666.

Yoneji, T., Fujita, H., Mukai, T., & Su'etsugu, M. (2021). Grand scale genome manipulation via chromosome swapping in *Escherichia coli* programmed by three one megabase chromosomes. *Nucleic Acids Research, 49*(15), 8407–8418. https://doi.org/10.1093/nar/gkab298.

Zhang, H., Zhao, G., Guo, Y., Menghwar, H., Chen, Y., Chen, H., et al. (2016). *Mycoplasma bovis* MBOV_RS02825 encodes a secretory nuclease associated with cytotoxicity. *International Journal of Molecular Sciences, 17*(5), 628. https://doi.org/10.3390/ijms17050628.

Zhou, J., Wu, R., Xue, X., & Qin, Z. (2016). CasHRA (Cas9-facilitated homologous recombination assembly) method of constructing megabase-sized DNA. *Nucleic Acids Research, 44*(14), e124. https://doi.org/10.1093/nar/gkw475.

Recombineering and engineering

Genome engineering in bacteria: Current and prospective applications

Rubén D. Arroyo-Olarte[a,*], **Karla Daniela Rodríguez-Hernández**[b], **and Edgar Morales-Ríos**[c,*]

[a]*Unidad de Biomedicina, Facultad de Estudios Superiores-Iztacala, Universidad Nacional Autónoma de México, Tlalnepantla, México*
[b]*Instituto de Química, Universidad Nacional Autónoma de México, Mexico City, México*
[c]*Departamento de Bioquímica, Centro de Investigación y Estudios Avanzados del Instituto Politécnico Nacional (CINVESTAV), Mexico City, Mexico*
Corresponding authors: e-mail address: rubendao@gmail.com; edgar.morales@cinvestav.mx

Abbreviations

AB	antibiotic
ABr	antibiotic-resistance marker
C2c2	RNA-guided RNA targeting effector
Cas	CRISPR-associated genes and proteins
Cas9, Csn1	a CRISPR associated protein composed by two endonucleases domains that are programed by small RNA molecules
CRISPR	Clustered Regularly Interspaced Short Palindromic Repeats
Cmr-β	type III-B CRISPR-Cas complex
Cpf	endonuclease with catalytically active nuclease domains
crRNA	CRISPR RNA, small RNA molecule that forms part of the guide RNA and it is responsible for determining the specificity of the Cas nucleases, through homologous pairing with the intended target-site
Csm	type III-A CRISPR effector protein complex nuclease
CTc	chlortetracycline
DinG	damage inducible gene G with probable helicase activity
DSB	double-stranded break
dsDNA	double-stranded DNA
exo	α gene-encoded exonuclease of lambda red system
bet	β gene-encoded protein of lambda red system
gam	γ gene-encoded protein of lambda red system
G.O.I.	gene of interest
gRNA	guide RNA

Methods in Microbiology, Volume 52, ISSN 0580-9517, https://doi.org/10.1016/bs.mim.2023.01.003

HDR	homologous-directed repair, genetic repair of DNA strand breaks through homologous recombination
HNH	a Cas9 endonuclease domain named after catalytically active histidine and asparagine residues
HR	homologous recombination, type of genetic recombination where nucleotides are exchanged between two similar or identical strands of DNA
Merodiploid	partially diploid, haploid organism which has acquired a foreign copy of certain gene (e.g. through single homologous recombination with a plasmid) besides the endogenous wild-type gene
Ori	origin of replication
PAM	protospacer adjacent motif
REC	recognition lobe of Cas9, where the gRNA binds to
RecA	recombinase A
RecBCD	RecBC exonuclease
RecET	recombinase E+recombinase T
RT	DNA repair template
RuvC	a Cas9 endonuclease domain named after a DNA-repair protein in *Escherichia coli*
gRNA	guide RNA, RNA duplex of crRNA and tracrRNA
sgRNA	single-guide RNA, single RNA molecule that combines the crRNA and tracrRNA through a loop
ssDNA	single-stranded DNA
ssRNA	single-stranded RNA
Tc	tetracycline
tracrRNA	trans-activating CRISPR RNA, it combines with the crRNA to form the guide RNA. It is responsible for trans-activating the HNH and RuvC domains through its binding to the REC domain

1 Non-CRISPR methods for bacterial genome editing

1.1 Suicide plasmids

'Suicide' plasmids were developed in the 1980s (Selvaraj & Iyer, 1983). Suicide plasmids are circular DNA fragments that can replicate in one organism, the donor, but not in another, the recipient. These plasmids contain long 5′ and 3′ homologous sequences (about 1 kb) encompassing the desired insertion, deletion, or site-directed mutation, coupled to a marker, usually an antibiotic resistance cassette, and may harbour a transposon to facilitate their incorporation into the genome of the recipient strain. This is followed by antibiotic selection that will only permit the growth of colonies undergoing genome integration, as the plasmid cannot replicate in the recipient bacteria. (Fig. 1A, B). This strategy, however, is not very efficient, showing a high rate of false positives and often requiring two rounds of selection with different antibiotics to achieve edited colonies (Luo, He, Liu, & Hu, 2015; Wang, Li, et al., 2019).

Although developed in model organisms such as *E. coli*, currently suicide plasmids are widely used in other industrially relevant prokaryotic organisms where

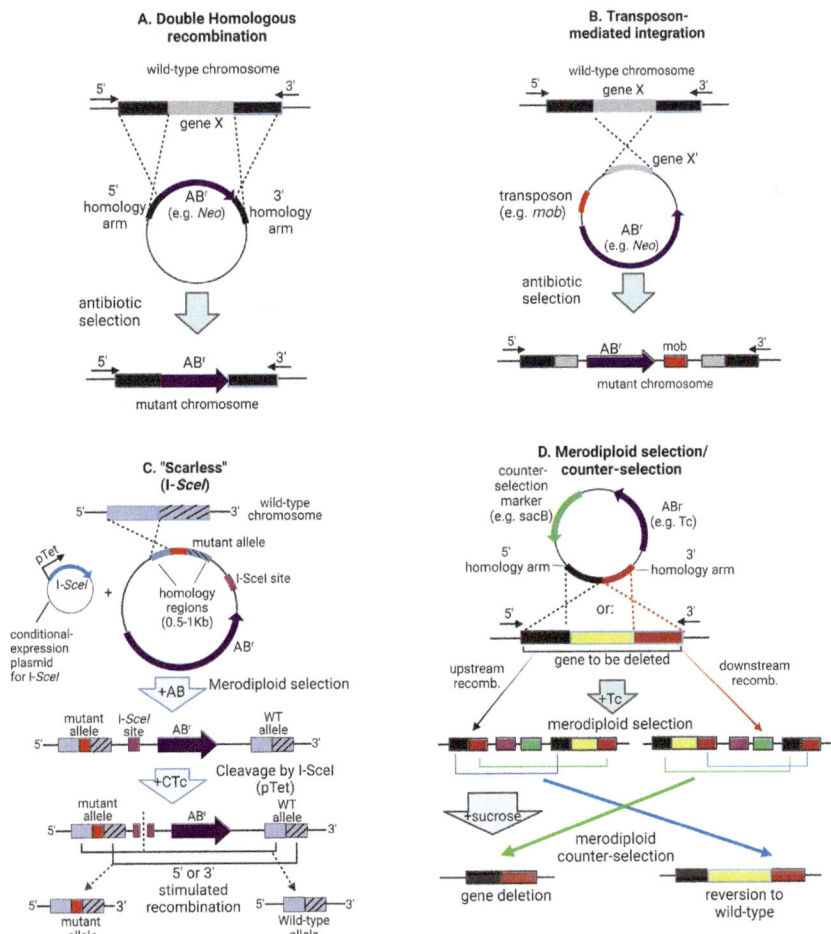

FIG. 1

Standard methods for genome editing in bacteria. Suicide plasmids. (A) The classic approach consists of transformation with a non-replicating plasmid (usually with a transposon element e.g. mob), which harbours a mutated recombination template and an antibiotic resistance marker (ABr). Antibiotic treatment will select only colonies that undergo homologous recombination to incorporate the antibiotic-resistance marker sequence encoded in the plasmid at the target locus (deleting gene X). (B) Integration of plasmid can be also aided by a plasmid-encoded transposon. In this case, a copy of the gene X (grey bar) is included in the plasmid to aid targeted insertion of the transposon and the antibiotic resistance marker (ABr) into the wild-type locus between the 5′ and 3′ flanking regions (black bars), leading to a disruption of gene X open-reading frame (C) In the 'scarless' variant a I-*Sce*I site is incorporated in the plasmid to be transformed in a I-*Sce*I expressing strain under an inducible promoter (pTet). After single homologous recombination and a first round

(Continued)

CRISPR-Cas9 is not yet as feasible and shows toxicity, e.g., *Corynebacterium glutamicum* (Jiang et al., 2017; Wang, Li, et al., 2019) or *Bifidobacterium* (Hoedt et al., 2021). Suicide plasmids have also been useful to generate site-directed mutants for protein purification to facilitate crystallization for structural studies, e.g., a C-terminal truncated, soluble cytochrome *c*1 in *Rhodobacter sphaeroides* (Konishi, van Doren, Kramer, Crofts, & Gennis, 1991; Page & Sockett, 1999).

Initially the use of suicide plasmids leaves behind a large antibiotic selection marker disrupting the gene of interest (Fig. 1B), however different alternatives have been developed to produce a markerless genotype, including the exploitation of I-*Sce*I or sucrose metabolism for counterselection of merodiploid co-integrates of the suicide plasmid and wild-type allele (Fig. 1C, D). I-*Sce*I is a homing endo-nuclease from *Saccharomyces cerevisiae* with a 18 bp asymmetric recognition sequence (TAGGGATAACAGGGTAAT), cleaving 3′-overhangs with a four base-pair length, which can facilitate homologous recombination (Choulika, Perrin, Dujon, & Nicolas, 1995; Niu, Tenney, Li, & Gimble, 2008). An I-*Sce*I site between the mutant allele and the antibiotic resistance marker, and the suicide plasmid is trans-formed into a bacterial strain with inducible I-*Sce*I expression. The suicide plasmid integrates into the gene of interest (G.O.I.) locus and is selected by an antibiotic marker at the non-permissive temperature for plasmid replication, leaving cells having both, plasmid mutant and wild-type alleles in their genome, also known as merodiploids (partially diploid). Afterwards, I-*Sce*I expression is induced to cleave the target gene locus, which is then repaired via native homologous recombination, provided there are large enough homology arms (>500 bp). This can result either in a reversion to the wild-type chromosome or in a markerless allele replacement (Fehér et al., 2007)

FIG. 1—Cont'd

of antibiotic treatment, cointegrating colonies harbouring the plasmid sequence with the mutant gene (mutation in red) and the wild-type allele at the target gene locus are selected. Addition of chlortetracycline (CTc) induces I-*Sce*I expression to cleave the target locus, which enhances homologous recombination to eliminate plasmid sequence resulting in either, reversion to wild-type or fixation of the mutant allele, depending on whether 5′ or 3′.
(D) Seamless counter-selection of merodiploids, which are co-integrates of wild-type gene (dark bar surrounded by yellow 5′ and red 3′ homology regions) and plasmid encoded mutant gene either by upstream or downstream single homologous recombination, can also be achieved by using a SacB marker which after selection with tetracycline (Tc) will render merodiploids sensitive to sucrose, whose function in this system is to be converted by *SacB* into levans, which causes cell death of merodiploids by over-accumulation of levans in the periplasmic space. Only colonies undergoing successful double homologous recombination to remove the plasmid counter-selection marker will survive, leading to either gene deletion or a reversion to the wild-type allele.

('Scarless' method, Fig. 1C). In theory, wild-type to mutant-allele colony ratio is expected to be 50:50; however, this will vary for each mutation. Short non-deleterious mutations are preferred over large deletions with low-efficiency repair (Chayot, Montagne, Mazel, & Ricchetti, 2010). In the cases of mutation leading to poor recombination or even to a small growth defect, a very large number of colonies will need to be screened. In a similar approach, merodiploid co-integrates can be discarded by means of a counter-selection marker encoded in the same plasmid such as *sacB*, which catalyses the conversion of sucrose to levans, a molecule that turns toxic upon accumulation in the periplasmic space of several bacteria, e.g., *Pseudomonas aeruginosa* (Huang & Wilks, 2017) (Fig. 1D).

1.2 ClosTron

Genome editing in clostridia has been hindered by the lack of efficient mutational tools for functional studies. The ClosTron method was developed using an endogenous intron with transposon activity (the ability to 'jump' and introduce into a genome either at a random or targeted position) that is a bacterial group II intron used as an insertional tool for gene inactivation (Heap et al., 2010; Heap, Pennington, Cartman, Carter, & Minton, 2007). Type II introns are broad host-range elements and as such their target specificity is mainly determined by homology between the intron RNA and the target site DNA sequences. Therefore, the re-targeting of these introns can be facilitated by altering the sequence of an intron RNA-encoding plasmid. The most widely used ClosTron system utilizes a broad host range Ll.LtrB intron of *Lactococcus lactis*. Intron target specificity is determined by a small region, so it is cost-effective to re-target an intron by sub-cloning a small DNA sequence (Fig. 2). Although it is still one of the most widely used genome engineering method in *Clostridium*, this tool is not free of some caveats (not present in CRISPR-based methods) such as the latent possibility of gene reversion due to the encoded-transposon activity, and its reliance on a precise algorithm to determine the best region within the transposon to incorporate the homology sequence for each G.O.I. (Wang, Hong, et al., 2018).

1.3 Lambda red

The lambda Red system is probably the most used approach for genome engineering in prokaryotes (Yu et al., 2000). This method was derived from lambda bacteriophage, and it is also known as 'recombineering' (recombination-mediated genetic engineering) (Sharan, Thomason, Kuznetsov, & Court, 2009). First, a strain with temperature-inducible expression of the Lambda Red proteins must be generated. The three main Lambda Red proteins are: α, β and γ: α has exonuclease activity

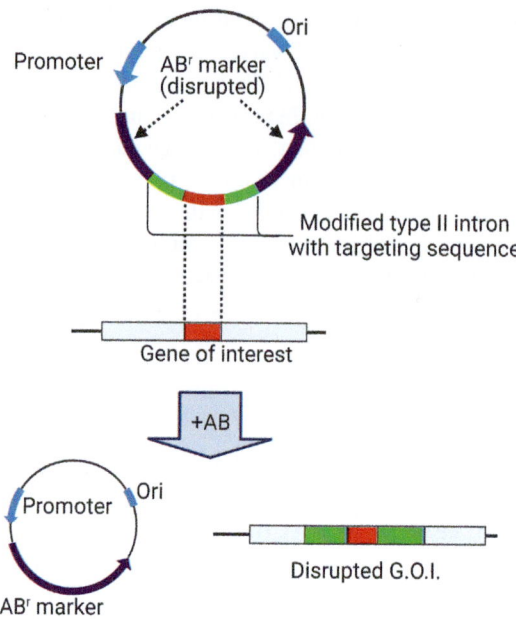

FIG. 2

ClosTron method. An antibiotic resistance cassette in a plasmid is disrupted by inserting a modified transposon gene (green) with a targeting sequence (red) homologous to the gene of interest. After transformation, the intron targets the gene of interest and disrupts it leaving behind a plasmid with a functional antibiotic resistance marker. Addition of antibiotic then enhances and simplifies the obtention of mutant colonies. AB: antibiotic, ABr: antibiotic resistance marker, G.O.I.: gene of interest, Ori: origin of replication.

(exo) and digests the 5′-ends of dsDNA; β (bet) binds to ssDNA and promotes strand annealing to the G.O.I; and γ (gam) binds to the bacterial RecBCD enzyme which degrades any linear DNA used as a template and inhibits its activities. These proteins induce a 'hyper-recombination' state in *E. coli* and other bacteria, in which recombination events between DNA species with as little as 35–50 bp of shared sequence occur at high frequency (Poteete, 2001; Sharan et al., 2009; Yu et al., 2000). The system itself is however selection-free and therefore is usually combined with the insertion of large antibiotic-resistance cassettes to improve the recovery of edited colonies (Sharan et al., 2009) (Fig. 3). In some cases, I-*Sce*I sites are also included in the targeting construct to proceed with a counter-selection step to eliminate the resistance marker by homologous recombination (Tas, Nguyen, Patel, Kim, & Kuhlman, 2015).

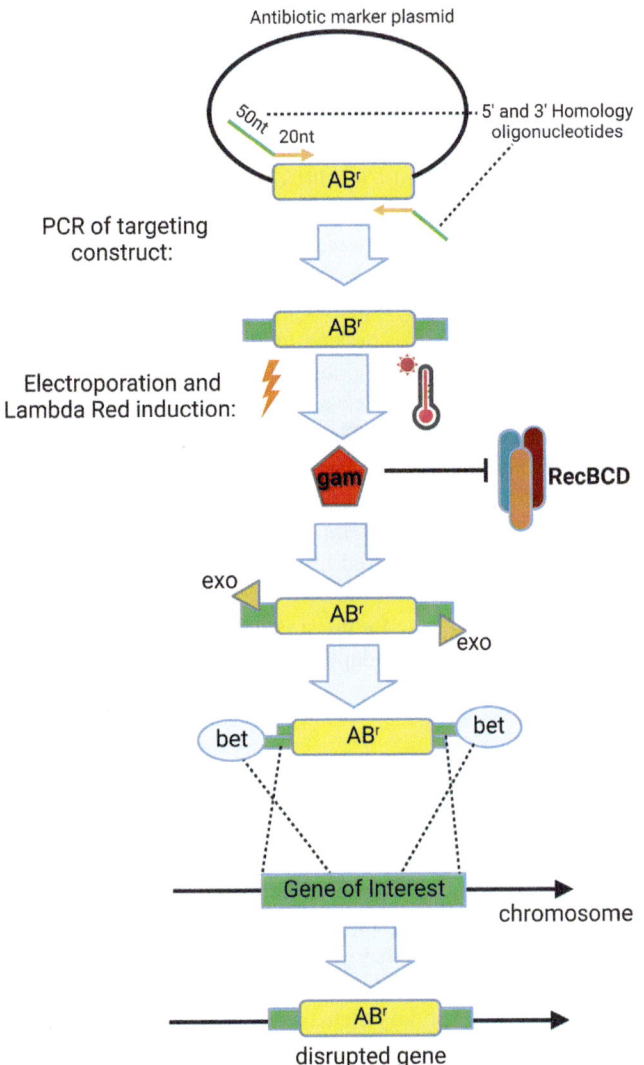

FIG. 3

Recombineering (lambda red system) for targeted gene disruption. A targeting construct is made by PCR of an antibiotic-encoding plasmid and oligonucleotides with 50 bp overhangs homologous to the 5′ and 3′ ends of the targeted gene. The PCR template is electroporated and expression of the lambda Red proteins is induced (e.g. via heat shock at 42 °C). Gam inhibits native RecBCD nuclease activity on linear DNA (protecting the targeting construct). Exo generates 3′ overhangs in the DNA linear template which will bind to bet protein to facilitate homologous recombination and integration, and disruption of the target gene (gene x). Edited colonies are then selected by antibiotic treatment.

2 CRISPR-Cas systems and mechanisms

CRISPR-Cas applications in genome editing for prokaryotes have revolutionized genetic engineering technology since it has enormous potential as antimicrobial agents, successfully eliminating virulence and resistance plasmids to design new drugs and therapies. The nuclease activity of CRISPR-Cas protein allows researchers to edit a genome with unprecedented ease, accuracy, and high throughput (Kamruzzaman, Yan, & Castro-Escarpulli, 2022). Here we describe the CRISPR-Cas systems, mechanism, properties, and their application.

According to the structure and function of Cas protein, the CRISPR/Cas systems can be categorized into two classes depending on the number of effector proteins or nucleases (class II for systems with one single nuclease protein, and class I with multi-subunit effector proteins), which are further subdivided into six types (type I–VI) (Makarova Kira et al., 2015). Class 1 includes type I, III, and IV, and class 2 includes type II, V, and VI (Fig. 4). The Cas operons of class 1 systems are highly diverse and consist of a gene encoding for the effector nuclease (types I and III) or helicase (type IV) (Chou-Zheng & Hatoum-Aslan, 2019; Liu & Doudna, 2020; Pinilla-Redondo et al., 2020), followed by the genes encoding the effector cascade complex and spacer adaptation. On the other hand, class 2 system operons have a simpler structure, including a gene for the single effector nuclease, and a set of genes responsible for the spacer (cRNA) adaptation (e.g. cas1, cas2, cas4) and/or maturation (tracrRNA).

2.1 Type I CRISPR-Cas systems

The CRISPR array is expressed as a premature CRISPR RNA (pre-crRNA), and the 3′-end of one repeat is processed by Cas6 (Cas5 in subtype I-C) to generate a mature crRNA with a 7–8 nt handle at the 5′-end of the flanking CRISPR repeat. This handle is followed by the CRISPR spacer sequence that defines the target of CRISPR-Cas immunity, and a processed repeat that generates a hairpin structure, representing the guide RNA of Type I systems. Often, the palindromic nature of CRISPR repeats leads to the formation of secondary structures, and the crRNA is processed at the 3′ bases of the hairpin. The crRNA is loaded into the Cascade where Cas5 interacts with the 5′ handles of the crRNA and the crRNA hairpin backbone interacts with various Cas7 subunits throughout the length of the guide. Noteworthy, the number of Cas7 subunits is correlated with the spacer length to adequately accommodate the crRNA (Cavanagh & Garrity, 2015; Van der Oost, Westra, Jackson, & Wiedenheft, 2014).

The Cascade complex, crRNA and Cas proteins, altogether constitute a double-stranded DNA recognition and targeting machinery that drives the sequence-specific cleavage of complementary DNA, based on nucleotide base-pairing between the spacer sequence of the crRNA and the complementary strand of the target DNA. The interaction between the crRNA: Cascade and the targeted DNA region loops

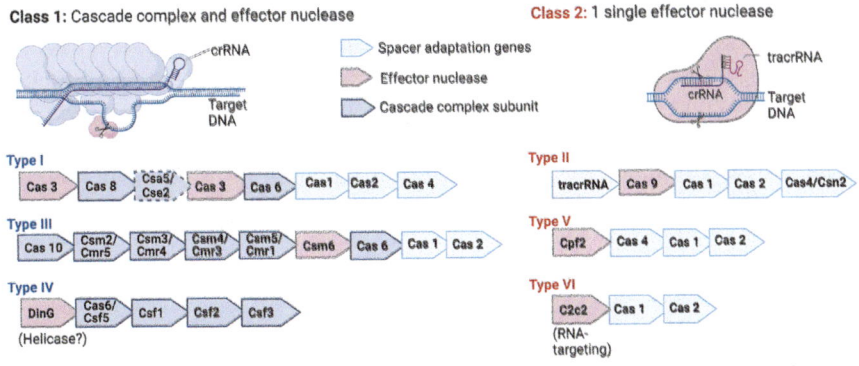

FIG. 4

The schematic representation of the most recent classification and functional modules of the CRISPR-Cas systems in prokaryotes. The effector complex is presented as follows: Class I is composed of multiple Cas effector proteins, principal effector proteins (pink), and additional cascade complex subunits (purple), there are also accessory genes and proteins involved in spacer adaptation (this is the process of crRNAs acquisition and maturation) are represented by light blue. Class II uses a single multidomain effector protein (pink) and also accessory genes and proteins to form a crRNA-binding complex (light blue). The components that are missing between subtypes are represented by dashed outlines. The conventional types I, II, and III, and types IV and V were added to classes 1 and 2, respectively. Types IV and V are those which do not have Cas1 and Cas2, necessary for the adaptation process, in the same CRISPR loci. Type VI was added most recently to class II. CRISPR-associated effector proteins: Cas. Csa, Cst, Csh, and Csm (subtype Cas systems in archaea). Csm, type III-A CRISPR effector protein complex nuclease; Cmr-β, type III-B CRISPR-Cas complex; Cpf, endonuclease with catalytically active nuclease domains; tracrRNA, trans-activating crRNA; C2c2 (RNA-guided-RNA-targeting effector) and DinG (Damage inducible gene G with probable helicase activity), are CRISPR-Cas proteins with potential gene knockdown capabilities.

out the non-targeted strand by generating an R-loop structure. Cascade binding to DNA is a prerequisite for subsequent Cas3 recruitment. Then, the signature Cas unwinds and degrades the non-target DNA strand in a 3′–5′ direction through ATP-dependent helicase and nuclease activities. This processive single-stranded DNA degradation typically causes cell death, especially in the absence of proper DNA repair mechanisms that help to overcome Cas damage.

2.2 Type II CRISPR-Cas

This system could prevent natural transformation in bacteria; however, the exact mechanism and whether other types of CRISPR systems also antagonize natural transformation is not known. Type II CRISPR nucleases such as Cas9 orthologs can be classified in three subgroups depending on their Cas operon architecture: IIA (cas9, cas1, cas2, cas4), IIB (cas9, cas1, cas2, Csn2) and IIC (cas9, cas1 and cas2 only). Structural comparisons reveal a relatively conserved catalytic core and a highly conserved arginine-rich bridge helix essential for R loop formation (DNA unwinding) and subsequent DNA cleavage. There is also a less conserved alpha-helical REC lobe essential for guide RNA binding and a divergent CTD that is responsible for both the PAM recognition and the guide RNA repeat–antirepeat heteroduplex binding. The divergent CTD domain may explain the differences in the PAM recognition sequence specific for each Cas9 ortholog. Despite this, the use of smaller Cas9 orthologs is highly valuable to ameliorate issues regarding large SpCas9 packing into vectors and difficulty to transform. For example, in *Trypanosoma cruzi*, due to a highly complex plasma-membrane glycocalyx, electroporation of large SpCas9/gRNA RNPs is not feasible; this issue has been addressed by using the smaller Cas9 ortholog from *Staphylococcus aureus* Cas9 (SaCas9, 123 kDa), with optimal results for gene knock-outs, gene deletions and endogenous gene-tagging. Other alternative Cas9 orthologs that have been used for genome editing in eukaryotes are from *Campylobacter jejunii* (CjCas9, 116 kDa), *Neisseria meningitides* (NmCas9, 124 kDa) and *Streptococcus thermophilus* (St1Cas9, 129 kDa and St3Cas9, 161 kDa) with more complex PAM sequence requirements. In the case of NmCas9 (PAM: NNNNGATT), a mismatch and indels study found an overall improvement over SpCas9. This indicates that a rare PAM sequence limits the number of off-targets for any given gRNA, providing a more specific genome-editing tool (Cavanagh & Garrity, 2015). Type II systems (e.g. Cas9) only depend on one effector nuclease which facilitates its heterologous expression, and are therefore the most popular tools for genome-editing.

2.3 Type III CRISPR-Cas systems

Type III CRISPR-Cas systems can target both RNA and single-stranded DNA and provide potent immunity against invaders, which is dependent on the target RNA transcription (Chou-Zheng & Hatoum-Aslan, 2019). The target RNA binding also activates the cyclic oligoadenylate (cOA) synthesis activity of the Cas10 subunit. Recent advances on cOA synthesis, cOA-activated effector protein, cOA signalling-mediated immunoprotection, cOA signalling inhibition, and possible crosstalk between cOA signalling and other cyclic oligonucleotide-mediated immunity have been discussed (Huang & Zhu, 2021).

2.4 Type IV CRISPR-Cas system

Primarily found on plasmids, is the least understood among the six CRISPR types. The lack of Cas nucleases, integrases, and other genetic features commonly found in most CRISPR systems has made it difficult to predict the mechanisms of action and biological functions of type IV CRISPR-Cas. Instead of Cas gene csf1, Cas-7 like gene csf2 was proposed to be employed to distinguish type IV from other types in the Class1 CRISPR-Cas group. Type IV-A systems protect bacteria from plasmids and phages, which needs DinG helicase along with other Cas proteins with an unknown mechanism of action. Recently identified type IV-C systems lack a Csf1 subunit and instead encode a Cas10-like subunit with an HD nuclease domain, while type IV-B systems lack a CRISPR locus and a crRNA processing enzyme and are associated with an ancillary gene identified as cysH-like. The mechanism of action and biological functions of type IV-B and -C are yet unknown (Van der Oost et al., 2014).

2.5 Type V CRISPR-Cas systems

These systems are distinguished by a single RNA-guided RuvC domain-containing effector, Cas12. Although effectors of subtypes V-A (Cas12a) and V-B (Cas12b) have been studied in detail, the distinct domain architectures and diverged RuvC sequences of uncharacterized Cas12 proteins suggest unexplored functional diversity. Here, we identify and characterize Cas12c, -g, -h, and -i. Cas12c, -h, and -i demonstrate RNA-guided double-stranded DNA (dsDNA) interference activity. Cas12i exhibits markedly different efficiencies of CRISPR RNA spacer complementary and noncomplementary strand cleavage resulting in predominant dsDNA nicking. Cas12g is an RNA-guided ribonuclease (RNase) with collateral RNase and single-strand DNase activities. Nowadays, there is functional diversity emerging along different routes of type V CRISPR-Cas evolution which expands the CRISPR toolbox (Tong et al., 2021).

2.6 Type VI CRISPR-Cas systems

This is the most recently discovered and exclusively targets single-stranded RNA (ssRNA). Four subtypes have been identified. The subtypes VI-A, VI-B, and VI-D have been functionally characterized along with respective effectors: Cas13a, Cas13b, and Cas13d. Functionally, all Cas13 effectors are crRNA-guided RNases with two distinct and independent catalytic centres. One catalytic centre processes pre-crRNA, and the other is formed by two R-X4-H motifs typical of higher eukaryotes and prokaryotes nucleotide-binding (HEPN) domains that mediate ssRNA cleavage. As opposed to Cas9 which specifically cleaves crRNA spacer-complementary dsDNA sequences (i.e. target DNA) in cis, the activated Cas13-crRNA interference complex cleaves nonspecifically both the crRNA-bound complementary ssRNA sequence (henceforth referred to as activator RNA) in cis and any other encountered RNA (both host and viral RNA) in trans (also known as collateral or bystander cleavage).

The cleavage preferentially occurs within structurally exposed regions of the RNA secondary structures, usually at uridine (U) or adenosine (A). Given their potent performance in mammalian cells, Cas13 effectors have already been used as tools for RNA manipulation, albeit their practical application is still in its infancy (Tong, Charusanti, Zhang, Weber, & Lee, 2015).

3 CRISPR as an adaptive immune mechanism in prokaryotes

CRISPR is the only known adaptive and hereditary immune response in prokaryotes. It is present in about 50% of bacteria and 90% of archaea. These CRISPR systems are mainly known as a new generation of genome editing tools that have specificity and efficiency thanks to the use of guide RNA to recognize unique sequences in the genome. This specificity of nucleic acid recognition is also the basis for adaptive immunity against viruses and other mobile genetic elements (MGEs) in bacteria (Koonin & Makarova, 2022).

The CRISPR immune response in prokaryotes consists of three stages:

3.1. *Adaptation:* This is the process of acquiring foreign DNA fragments (protospacers) that become spacers inserted between repeats in CRISPR arrays and that are subsequently used to produce guide RNAs that specifically target the related foreign nucleic acid.

3.2. *Interference:* In this step, the crRNAs are exploited as guides to recognize the target DNA or RNA which is then cleaved by the nuclease moiety of the CRISPR effector complex.

3.3. *Expression, maturation and biogenesis:* In this step, the long transcript of the pre-CRISPR crRNA matrix is processed to produce mature and functional crRNA.

Each of these stages in CRISPR immune function is mediated by a distinct set of Cas proteins and several accessory genes that comprise functional modules of CRISPR systems and are prone to rapid sequence evolution, rearrangements, and replacements. Some examples are the very conserved CRISPR-Cas systems of *Salmonella typhi*, which regulate the synthesis of major outer membrane proteins (OMP) OmpC, OmpF, OmpA, and quiescent OMP, OmpS2 by regulating the expression of the master porin regulator Omp and is also involved in the resistance to bile salts and in the formation of biofilms. CRISPR-Cas also acts by recompiling and storing genetic sequences from invader bacteriophages and noxious plasmids as spacers. These spacers are transcribed into crRNAs that bind to effector CRISPR nucleases (Cas proteins), which target specific complementary sequences, given they fulfil a specific PAM sequence requirement. It was previously discovered that crRNAs need to couple to an RNAse III-edited tracRNA before binding to the Cas nuclease (Deltcheva et al., 2011) (Fig. 5).

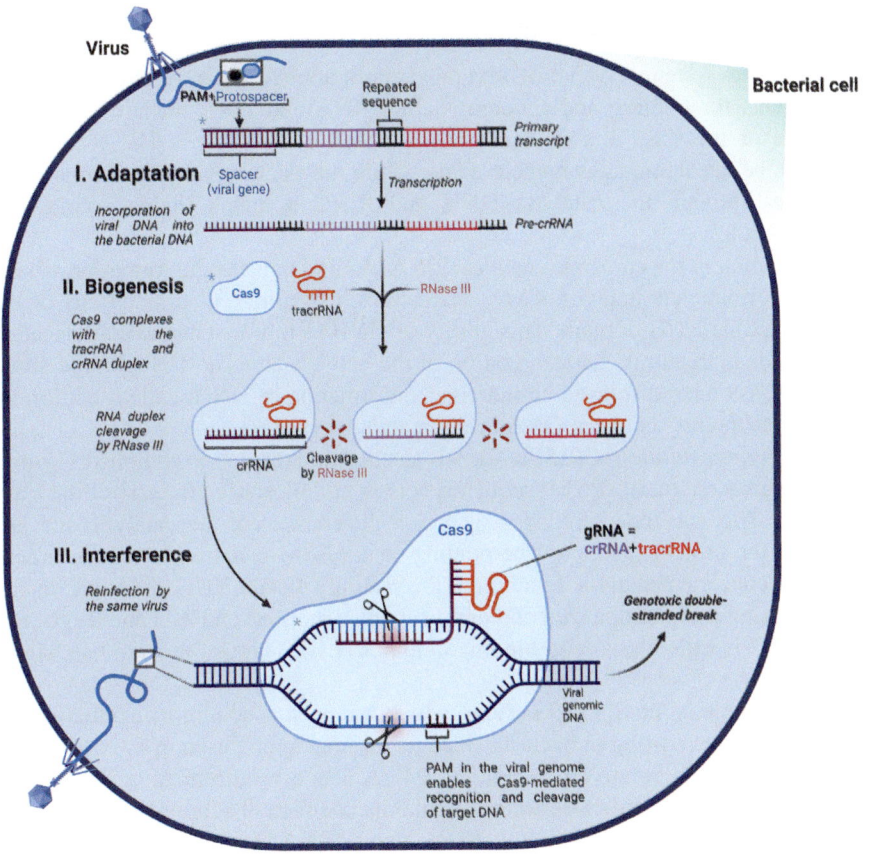

FIG. 5

The three stages of the CRISPR-Cas (type II) bacterial adaptive immune system.
(I) Adaptation, during this stage, viral DNA is injected into the bacterial genome through the phage to bacterial cell activates the adaptation module proteins (black box, PAM-protospacer), these proteins excise spacer-sized fragments of phage DNA (protospacers that are adjacent to a PAM in the viral genome) for incorporation into CRISPR loci. During CRISPR RNA biogenesis (II), CRISPR loci are transcribed and pre-crRNA is processed by a Cas9/RNaseIII complex at repeat sequences to generate mature crRNAs that couple to tracrRNA (gRNA). During the interference (III) process, individual gRNAs are bound by Cas protein effectors (e.g. Cas9). After a new phage infection with protospacer (and PAM) sequences matching a crRNA appears in the cell (lower right), specific Cas/gRNA complexes bind to viral DNA and cleave it.

4 Molecular mechanism

CRISPR-Cas systems consist of several conformational dynamic motions of Cas apoproteins for binding and accommodating RNA and DNA nucleic acids. The structure of the Cas9 endonuclease is composed of two lobes: The recognition (REC) lobe accommodates nucleic acids, while the nuclease lobe performs DNA cleavages through the catalytic HNH and RuvC domains (Saha, Arantes, & Palermo, 2022).

Formation of the catalytic complex: DNA binding induces the formation of several conformational states in Cas9, characterized by different orientations of the highly flexible HNH domain. This ability of HNH to rapidly change conformation has impeded structural characterization of the active state. The DNA-bound structures of Cas9 have displayed an inactive conformation of HNH, called 'conformational checkpoints' between DNA binding and cleavage.

Allosteric regulation: CRISPR-Cas9 is an allosteric regulation complex. The protospacer adjacent motif (PAM) sequence acts as an 'allosteric effector' of the Cas9 function. This substrate binding at a region different from the catalytic site and activates the protein function. The binding of the PAM recognition sequence activates the concerted catalytic function of the spatially distant HNH and RuvC nucleases inducing a population shift and highly coupled motions on HNH and RuvC; this 'cross-talk' is called look loop binding domains (L1/L2 loops), regarded as 'signal transducers'.

An example is the SpCas9 system, which has been used almost exclusively to perform genome editing in bacteria since its original application in *E. coli*. This is mostly due to its relatively simple PAM sequence requirement and its well-characterized crystal structure and molecular mechanism of action (Fig. 6). SpCas9 displays a striking conformational change upon gRNA binding. This, in turn, uncovers 2 endonuclease domains, RuvC cleaving the non-target DNA strand while the HNH cleaves the target DNA strand complementary to the gRNA. The spCas9 recognition mechanism is a critical determinant of initial target DNA binding and is a required element in subsequent strand separation and gRNA-target DNA hybridization on the PAM sequence (NGG). The critical residues of the PAM-binding domains (Toro and CTD) involved in the hydrogen bonding to the dinucleotide GG of the PAM sequence are R1333 and R1335. Interestingly, these steps can tolerate up to 5 base-pair mismatches between the target DNA and gRNA sequences depending on their position and distribution. Following general gRNA design guidelines, combined with bioinformatic tools to predict mismatches in a given target genome, these effects can be minimized (Fig. 6) (Anders, Niewoehner, Duerst, & Jinek, 2014).

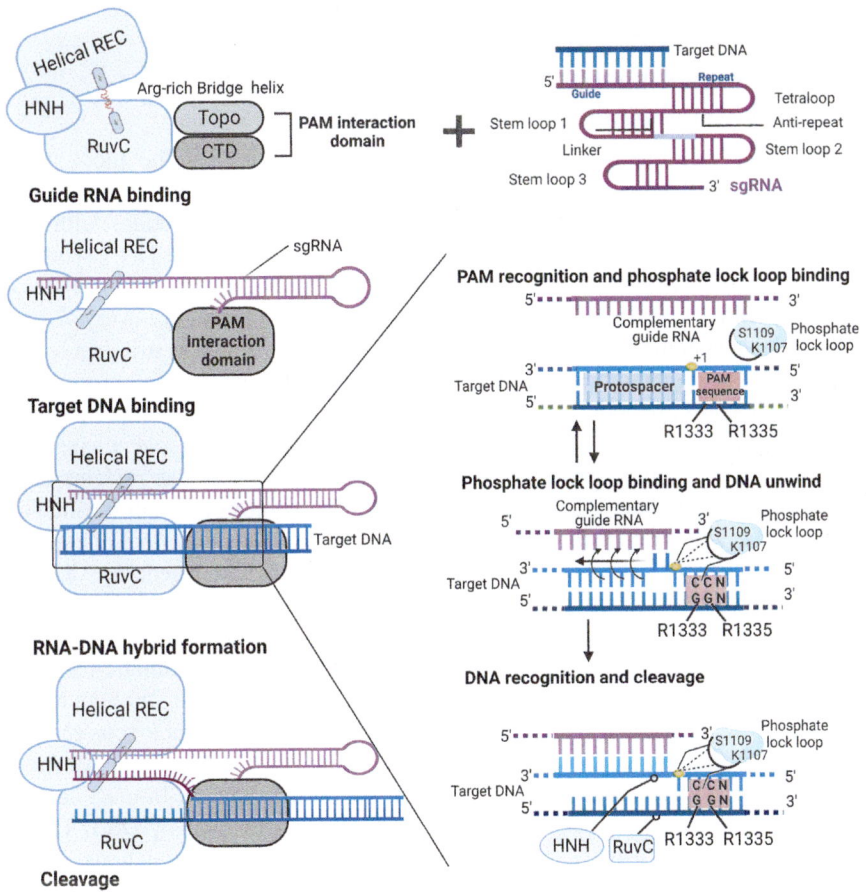

FIG. 6

Molecular mechanism of target DNA recognition and cleavage by Cas9. The guide RNA binding recognition and cleavage of dsDNA strictly require the presence of a protospacer adjacent motif (PAM). Cas9 results in the formation of a PAM binding site (PAM-interaction domain, conformed by TOPO and CTD subdomains, grey box). Cas9-sgRNA engages the PAM GG dinucleotide using Arg1333 and Arg1335, and positions the target DNA duplex such that the +1 phosphate (yellow circles) interacts with the phosphate lock loop, resulting in local strand separation immediately upstream of the PAM, the base-pairing between the displaced target DNA strands and the seed region of the guide RNA, promotes further stepwise strand displacement and propagation of the guide-target heteroduplex. The endonuclease domains of Cas9 (HNH and RuvC) generate site-specific breaks in the double-stranded DNA (dsDNA) target. The non-target DNA strand depends on the base-pair complementarity of the target DNA strand (tDNA) to the RNA guide template, SpCas9 adopts an overall bilobed architecture, in which the sgRNA: target DNA heteroduplex resides within a central channel between the α-helical recognition (REC). The REC lobe englobes two metal-ion-dependent nuclease domains, termed HNH and RuvC, which are responsible for cleaving the tDNA. sgRNA is depicted as pink strands, while target DNA is shown as blue strands.

5 CRISPR-based genome editing methods

5.1 DNA repair mechanisms and their application for CRISPR-mediated genome-editing in bacteria

The Cas9 protein from *Streptococcus pyogenes* (SpCas9) is the most widely employed CRISPR-based gene-editing tool. This is mostly due to its relatively common PAM sequence requirement: NGG (where N can be any nucleotide), with a theoretical frequency of once every 8 bp for a random double-strand DNA sequence. The actual frequency of the PAM motif will vary across genomes and is expected to be much rarer in AT rich genomes. For example, NGG occurs once every 42 bases in the human genome (Scherer, 2008). New Cas9 variants with less stringent or even absent PAM requirements have been generated, to expand the target site recognition of CRISPR-Cas9 methods (Walton, Christie, Whittaker, & Kleinstiver, 2020). The crRNA and tracrRNA are usually fused into a single-guide RNA molecule (sgRNA) with the same activity. An optional element of the CRISPR-Cas systems is the recombination template (RT), composed of 5′ and 3′ homology arms, the desired edit (insertion, deletion or specific mutation) and an internal sequence that disrupts the PAM target site, to prevent re-targeting of the successfully edited genome. Cleavage by CRISPR nucleases in the absence of successful recombination is often lethal in bacteria due to the formation of a double-strand break (DSB), and it serves as a powerful counterselection avoiding the necessity of inserting a large antibiotic-resistance cassette marker into the genome. In this way the DSB drives editing through homologous recombination (HR) or, more rarely in bacteria, via non-homologous end joining (NHEJ) (Shuman & Glickman, 2007). It is, therefore, the DNA repair systems of the host species/strain which actually perform the desired editing. In most bacterial organisms RecA-mediated HR is induced to repair DNA damage by DSB. This response however, is usually error-prone and inserts undesired mutations, mainly through the recruitment of the mutagenic DNA polymerase IV (PolIV) and inhibition of high-fidelity PolIII at the DSB site (Guirouilh-Barbat, Lambert, Bertrand, & Lopez, 2014; Pomerantz, Goodman, & O'Donnell, 2013; Rosenberg, Shee, Frisch, & Hastings, 2012). In most of the cases, where the CRISPR nucleases have been used to achieve highly efficient genome editing, particularly in *E. coli*, they are combined with an enhanced recombination system, e.g., the Lambda Red phage to promote homology-directed repair (HDR) (Jiang, Bikard, Cox, Zhang, & Marraffini, 2013).

The most common strategy used in *E. coli* and other model bacteria, uses a DNA linear template as well as a phage-derived recombinase to repair the DSB. In this approach (Fig. 7A), the first step (of three) is to include and induce the expression of the foreign recombinase in a plasmid followed by co-transformation with the recombination DNA template and the CRISPR plasmid (Cas9 + gRNA). The bottleneck here is the availability of a highly efficient recombinase to counter-select enough viable gene-edited colonies from DSB-killed non-edited colonies. In this regard, the original description of the use of SpCas9 for genome-editing in *E. coli* used the lambda Red phage recombinase system and linear double-stranded DNA as template to incorporate the desired edits (Jiang et al., 2013). This system has been used in

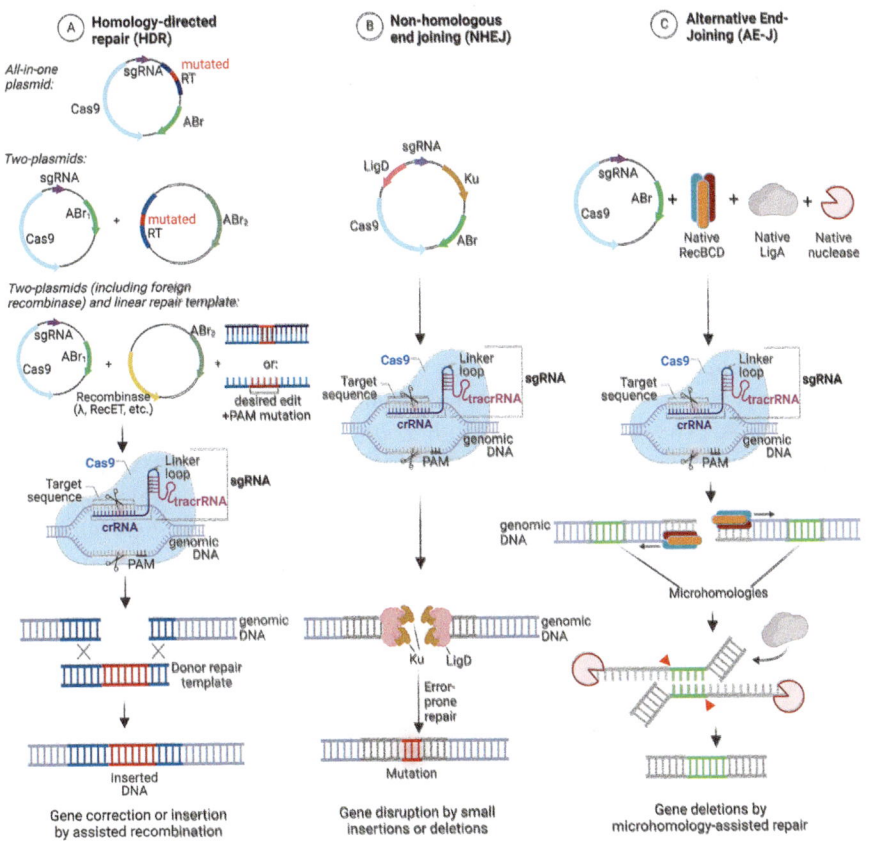

FIG. 7

Strategies used for CRISPR-Cas based genome editing in bacteria. (A) Editing via homologous recombination: Recombineering with a linear DNA template is followed by counterselection with CRISPR nucleases. A heterologous recombinase (e.g. λ red, RecT) is introduced via a plasmid (or phage) into the cell and co-transformed with the linear DNA template and CRISPR-nuclease plasmid with respective antibiotic-resistance marker (ABr). Genome editing may also be directed with a plasmid-encoded recombination template (RT) and endogenous or heterologous recombinase. The recombination template can be placed on the same plasmid encoding the CRISPR machinery for an all-in-one plasmid system, or it can be placed on a separate plasmid before transforming the CRISPR nuclease/gRNA plasmid. One-plasmid system is more streamlined, but due to its larger size it can be hard to transform, and cloning may not be possible if the gRNA can target the genome of the cloning strain. (B) Editing via the non-homologous end-joining (NHEJ) pathway. Depending on the strain, ku and/or ligD can be encoded on the CRISPR nuclease/gRNA plasmid and transformed into the strain.
(C) Alternative end joining (A-EJ) pathway can be found natively in many bacterial species with incomplete NHEJ. It does not require the introduction of foreign Ku or LigD, and instead relies in microhomology-assisted repair via native RecBCD, nucleases and LigA, leading to deletions of variable size (depending on the location of microhomologies) at the Cas9 cut site. For a more detailed insight on NHEJ and A-EJ mechanisms, the reader is advised to read (Chen, Zhang, Yeo, Bae, & Ji, 2017). All strategies require plasmid curing after nuclease targeting to isolate the mutant strain in order to avoid interference in pursuing downstream applications.

other species, mainly Proteobacteria (Aparicio, de Lorenzo, & Martínez-García, 2018; Oh & van Pijkeren, 2014; Wang, Wang, et al., 2018; Wu et al., 2017; Yan et al., 2017). Other heterologous recombination systems have recently been screened for their activity in different bacteria either alone or coupled to CRISPR-Cas. Among them, recombinase T (RecT), has been established successfully to enhance the CRISPR-Cas mediated genome editing in *Corynebacterium glutamicum* (Wang, Hu, et al., 2018), *Lactococcus lactis* (Guo, Xin, Zhang, Gu, & Kong, 2019), *Lactobacillus plantarum* and *Lactobacillus brevis* (Huang, Song, & Yang, 2019). RecT binds to ssDNA and protects it from degradation, fulfilling a similar function to gamma protein in the lambda Red system (Zhou et al., 2019).

Alternatively, the DNA repair template may be encoded in the same or different plasmid than SpCas9. In this case, foreign recombinases have been used (Jiang et al., 2015; Sun et al., 2018), though native recombination machinery may also be relied upon (Fig. 7A). This has been shown in *E. coli* with 1Kb homology arms in the recombination template plasmid (Cui & Bikard, 2016). In their work Vento, Crook, and Beisel (2019) described other bacteria where the native recombination machinery has been applied successfully with this approach, such as *Clostridium ljungdahlii* (Huang et al., 2016), *L. plantarum* (Leenay et al., 2019), *Pseudomonas putida* (Wirth, Kozaeva, & Nikel, 2020), *Streptomyces coelicolor* (Huang, Zheng, Jiang, Hu, & Lu, 2015) and *Staphylococcus aureus* (Chen et al., 2017). Using the native recombination machinery can simplify the system; however, in many species this machinery is either not reliable or efficient enough to achieve the desired edit.

Recombination template and/or machinery may also be omitted when relying on the non-homologous end-joining pathway to repair the CRISPR-Cas-directed double-strand break (Fig. 7C). However, very few bacterial species harbour a sufficiently active NHEJ machinery natively, therefore it must be usually heterologously encoded in the CRISPR plasmid. The NHEJ machinery in bacteria consists basically of two proteins: Ku and LigD. Ku binds to the cleaved DNA ends, while LigD joins them to seal the DNA together, often introducing non-specific mutations, insertions or deletions that render the gene non-functional (Fig. 7B). Similarly, the native alternative end-joining (A-EJ) pathway (also known as microhomology-mediated joining) can be exploited (Fig. 7C). This DNA repair pathway relies on microhomologies (1–9 nt) near the cut site by Cas9, which after resection of DNA ends by RecBCD are ligated by LigA, leaving behind deletions of variable size after repair (Finger-Bou, Orsi, van der Oost, & Staals, 2020). Native A-EJ has been combined with CRISPR-Cas9 in several species, including *E. coli* (Huang, Ding, et al., 2019), *S. coelicolor* (Tong et al., 2015) and *Pectobacterium atrosepticum* (Vercoe et al., 2013). Both strategies would not be useful to introduce specific mutations or insertions but would be effective for gene knockouts.

Overall, these strategies are not mutually exclusive and may be combined depending on the host species. In any case, they may have common drawbacks related to the continuous expression of a foreign Cas9 protein. SpCas9 overexpression can be highly cytotoxic in *E. coli* and many other bacteria (it will be explained within the next section) leading to little or no colonies, even when devoid of its nuclease activity (Cho et al., 2018; Misra et al., 2019).

5.2 **CRISPR-Cas9 difficulties and plasmid-based alternatives**

Overexpression of SpCas9 can have cytotoxic effects, which can be challenging for an efficient genome editing. In some species like *Corynebacterium glutamicum*, transformation of a SpCas9-encoding plasmid is not possible, even in the absence of gRNA, and no colonies can be obtained (Jiang et al., 2017). SpCas9 cytotoxicity can be due to residual, unspecific nuclease activity. One study however, showed that even overexpression of a nuclease-devoid SpCas9 (dCas9) leads to low colony count and to abnormal morphology, indicating instead a noxious effect for its PAM recognition and DNA binding activities across the genome (Cho et al., 2018). Strong effects on cell division and inner and outer membrane structure have also been shown, particularly in the absence of gRNA. On the other hand, in the cases where SpCas9 expression is well tolerated, other bottlenecks are efficient at DNA repair even in the presence of a recombination template (Cui & Bikard, 2016; Li et al., 2016). Another exception is, DNA repair mechanisms in prokaryotes are not as efficient as in eukaryotes, and this has been traditionally one of the bottlenecks for a wider application of CRISPR-Cas9 genome-editing tools in these organisms, even in the presence of a foreign recombination template. Therefore, these effects (toxicity of SpCas9, and inefficient DNA recombination mechanisms) are the main reasons that lead to no genome-edited colonies. In this regard, inducible promoters have been employed to drive SpCas9 expression (Fig. 8A). These promoters can be induced by the addition of a particular substance such as IPTG, which has been used for metabolic engineering in *E. coli* (Li et al., 2015). Other examples of inducible promoters used for SpCas9 and/or gRNA expression are dependent on tetracycline and its derivatives (pTet), mannose, nisin and arabinose, which have been employed in *E. coli* (Reisch & Prather, 2015), *Bacillus subtilis* (Altenbuchner, 2016), *Clostridium acetobutylicum* (Wasels, Jean-Marie, Collas, López-Contreras, & Lopes Ferreira, 2017) and *Lactococcus lactis* (Berlec, Škrlec, Kocjan, Olenic, & Štrukelj, 2018). These promoters, however, can have 'leaky expression' and elicit significant background SpCas9 activity at their 'off' state (Sun et al., 2019). One alternative still to be tested in bacteria, but quite successful in eukaryotes, showing little or no background SpCas9 activity, are the light-inducible systems (Nihongaki, Kawano, Nakajima, & Sato, 2015; Zhou et al., 2018), although their use requires specialized optical instruments.

The SpCas9-RuvC domain has been mutagenized (D10A) to function as a DNA nickase to produce single-strand breaks instead of the lethal DSB (Fig. 8B). The resulting nicking SpCas9 (nCas9) has been useful as a genome editing tool in cases where transformation with SpCas9 plasmids leads to no colonies, especially for large-scale genome deletions (Standage-Beier, Zhang, & Wang, 2015). Due to the non-lethal nature of single-strand breaks, nCas9 is not a counter-selection tool, leading often to poorly efficient genome editing (McAllister, Bouillaut, Kahn, Self, & Sorg, 2017; Song, Huang, Xiong, Ai, & Yang, 2017). Alternatively, two gRNAs targeting adjacent genome regions at opposing DNA strands can be used with nCas9 to generate a staggered double-strand break (Ran et al., 2013). This approach would be more specific and less susceptible to off-target edits, but also has more requirements including the presence of the PAM in both strands and in close proximity and finding two gRNAs with high activity targeting a short sequence of base pairs. In

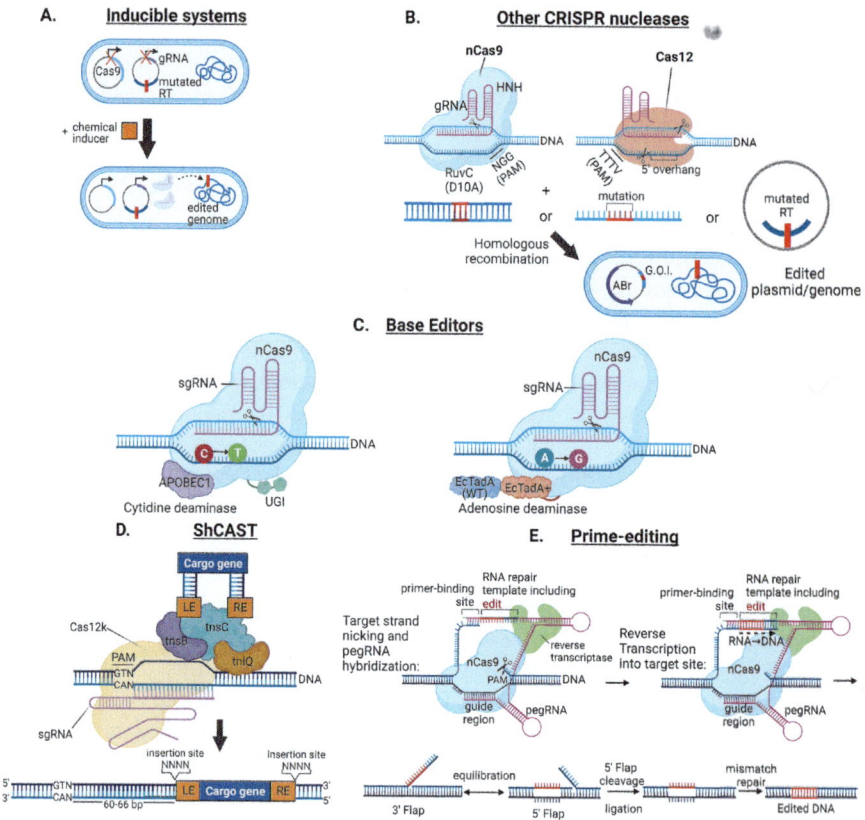

FIG. 8

Alternative strategies to circumvent SpCas9 cytotoxicity. (A) Use of inducible systems to express SpCas9. Via an inducible promoter, SpCas9 expression is strongly repressed without an inducer present (square) and only induced after exponential culture so that enough cells can survive and perform the genome edit aided by a mutated (red line) DNA repair template (RT). (B) Using less toxic nucleases to achieve editing. nCas9, which only cleaves one strand of DNA, and Cas12a (PAM: TTTV, where V is A or C or G), which makes two intercalated single-stranded breaks leaving a 5′ overhang, can be less toxic than SpCas9 to edit genome or plasmid-encoded gene of interest (G.O.I.). (C) SpCas9-derived base editors eliminate the double-stranded break requirement for genome editing. A translational fusion of dCas9 ('dead') or nCas9 (nickase), a cytidine (e.g. APOBEC1 in BE3) or adenosine (e.g. TadA−EcTadA+ in ABE2) deaminase domain, and a uracil DNA glycosylase inhibitor (UGI) is introduced on a plasmid into the cell. Upon nuclease binding and DNA strand unwinding, cytidines (or adenines) on the non-target strand within a defined window adjacent to the PAM (protospacer adjacent motif) are rapidly converted to uracils (or inosines), which are then processed as thymidines (or guanines) by DNA polymerase. (D) ShCAST insertion mechanism. A Tn7-like transposon from *Scytonema hoffmani* encodes transposases (tnsB, tnsC, tniQ), a nuclease deficient type V CRISPR protein (Cas12k) and guide RNA.

(Continued)

C. glutamicum, where neither exogenous SpCas9 nor nCas9 expression is possible due to toxicity and consequent plasmid loss, this issue has been solved by inserting the SpCas9 into the genome under the control of a native promoter. The resulting strain showed a low rate of escape colonies and a high gene-editing efficiency when transformed with a plasmid encoding a specific gRNA and recombination template (Cameron Coates et al., 2019).

The continuous research in the CRISPR gene-editing systems has created new CRISPR tools such as type II Cas9 ortholog, Nm2Cas9 and natural double-nicking CRISPR nucleases like the Type V-A Cas12a (also known as Cpf1) from *Francisella novicida* (FnCas12a) have also been characterized and studied as an alternative for genetic engineering.

Cas12a orthologs require one single gRNA and are usually smaller than SpCas9, recognizing a T-rich PAM and introducing a 5′ 5-nt overhang upon DNA cleavage (Zetsche et al., 2015) (Fig. 8B). As an alternative to Cas9, in *C. glutamicum*, a FnCas12a-encoding plasmid could be successfully transformed and used for genome editing (Jiang et al., 2017). Cas12a has since then been applied for genome editing in other bacteria, e.g., *Yersinia pestis* and *Mycobacterium smegmatis* (Yan et al., 2017). BhCas12b is another powerful CRISPR tool showing potential in gene editing more applied for human genome editing and cancer. Other CRISPR sub-types, such as Cas (12h), Cas (12i), and Cas (12g) have been identified from metagenome and are included in the type-V CRISPR system (Pinilla-Redondo et al., 2020). Some of these sub-types can be used in vitro as programmable endonucleases to cut ssDNA, dsDNA, ssRNA, or dsRNA. Cas12k was found as a RNA-guided site specific integration system in *E. coli* (Strecker et al., 2019), providing the potential to make CRISPR tools to produce precise targeted DNA insertion in bacteria and mammalian genomes.

There are, however, some aspects of this CRISPR nuclease that need to be characterized and further studied regarding its effects on genome editing. It has been found that once the Cas12a/gRNA complex cleaves its target DNA sequence, it

---

FIG. 8—Cont'd

This complex is combined with a cargo gene flanked by LE and RE elements. ShCAST is directed to the target locus and integrates the cargo gene 60–66 bp downstream of the PAM sequence, generating and insertion of the cargo gene flanked by the LE (transposon left-end) and RE (transposon right-end) elements, and a duplicated (4 bp) insertion site. (E) Prime-editing mechanism. Prime editors are fusion proteins composed of a Cas9 H840 nickase domain (nCas9), a reverse-transcriptase (RT) and a prime-editing guide RNA (pegRNA) containing an extended single guide RNA (sgRNA) with a primer binding site and a RNA repair template sequence containing the intended edit (red). After pegRNA-directed targeting, nicking of PAM strand and hybridization, reverse transcriptase (green) generates the edited DNA from the pegRNA sequence at the target site. Edited DNA equilibrates between 3′ and 5′ Flap states. Only when 5′ Flap is cleaved and ligated, is the desired edit incorporated into the PAM DNA strand. Finally, the DNA mismatch repair pathway also incorporates the desired edit in the anti-sense strand stabilizing genome editing.

remains active (contrary to Cas9, which is a single-turnover enzyme) and targets non-related sequences for cleavage. Although potentially troublesome for genome editing (possible off-targets), this feature has recently been applied for pathogen-infection diagnostics (e.g. SARS-Cov2) by cleaving fluorogenic DNA probes (Broughton et al., 2020).

One of the most recent advances in genome editing are the base editors, which specifically perform single-nucleotide edits without a double strand break or recombination template. The most extended base editor, BE3, is composed of a chimera of nCas9 or dCas9 to provide strong, specific gRNA-programmable DNA binding, and cytidine-deaminases, e.g., APOBEC1, to conduct C to T editing in the target gene (Komor, Kim, Packer, Zuris, & Liu, 2016) (Fig. 8C). Other variants of the system include an adenine-deaminase (A to G conversion) instead of a cytidine-deaminase (Gaudelli et al., 2017) or Cas12a instead of nCas9 (Li et al., 2018). The advantages of this system include its relative innocuity compared to Cas9-induced DSBs, and its independence from recombination machinery to introduce specific single-nucleotide mutations in a target gene. This system was initially developed in eukaryotes but is starting to be applied in some bacteria like *E. coli* (Banno, Nishida, Arazoe, Mitsunobu, & Kondo, 2018), *Klebsiella pneumonia* (Wang, Wang, et al., 2018), *Pseudomonas aeruginosa* (Chen et al., 2018), *Rhodobacter sphaeroides* (Luo et al., 2020) and *Staphylococcus aureus* (Zhang et al., 2020). Recently, however, a transcriptome-wide off-target RNA editing activity has been shown to be triggered by continuous expression of base editors, particularly those based on cytidine-deaminases in mammalian and plant cells (Xin, Wan, & Ping, 2019). Similarly, embryonic cells expressing base-editors show a higher-than-normal frequency of single-nucleotide polymorphisms (Zuo et al., 2019). These reports are consistent with the fact that cytidine-deaminases like APOBEC1 and APOBEC3G have a well-documented anti-DNA and anti-RNA virus replication activity, mainly through hyper-mutating viral genomes (Gee et al., 2011; Ikeda et al., 2008; Noguchi et al., 2005; Sawyer, Emerman, & Malik, 2004). Despite the encouraging results using base-editors in bacteria, more research is needed to address these possible caveats in prokaryotes.

Although base-editors are optimized for single-base edits, replacing larger stretches of genomic DNA by inserting sequences such as an epitope tag or a deletion usually requires a foreign DNA donor to repair a Cas9-induced DSB. A type V-K CRISPR-associated transposase (ShCAST) system avoids these requirements (Fig. 8D). This method is based on a naturally occurring Tn7-like transposon from *Scytonema hoffmani* which encodes besides its transposase genes, a nuclease-deficient Cas12k, tracRNA and 28–34 bp crRNAs (Strecker et al., 2019). ShCAST transposases, Cas12k and targeting sgRNAs are cloned into a helper plasmid, while cargo genes flanked by LE and RE elements to facilitate their insertion into a crRNA-targeted locus, are cloned into a donor plasmid. Integration is not 'scarless' as it also includes the LE and RE elements and a 5-bp duplication at the insertion site. The ShCAST system has shown up to 80% genome editing efficiency in several *E. coli* target loci without positive selection, highlighting its potential for genome engineering in prokaryotes (Strecker et al., 2019). A similar approach has been demonstrated in *E. coli* using the CAST locus from *Vibrio cholerae* (Klompe, Vo, Halpin-Healy, & Sternberg, 2019). Prime-editing is another recent DNA repair-free editing method. It combines an nCas9 and a reverse transcriptase that utilizes a pegRNA that works as both a guide RNA and as a reverse-transcriptase template to generate a desired DNA sequence that is integrated in the target locus (Anzalone et al., 2019). Prime-editing (Fig. 8E) shows higher or similar efficiency and fewer by products than homology-directed repair and induces much lower off-target editing than Cas9 nuclease at known Cas9 off-target sites in human cells (Anzalone et al., 2019). Prime editing has been successfully applied in mice (Liu et al., 2020) and plants (Lin et al., 2020), but its feasibility for bacterial genome editing still needs to be explored. Particularly, the large size of the prime-editing complex (about 7000 bp), may affect an efficient transformation and/or expression in bacteria.

6 Current applications of CRISPR-Cas for genome engineering in bacteria

A more laborious but hopefully much less deleterious way to use CRISPR-based genomic editing is to harness the endogenous CRISPR systems of bacteria. This would require however an extensive characterization of CRISPR loci and endogenous CRISPR nucleases for each species. In this regard, it was recently demonstrated that the endogenous Cas9 of *Mycoplasma gallisepticum* (MgaCas9) is active and can be used to perform genome editing in this species with low dependency on adjacent sequences (Mahdizadeh, Sansom, Lee, Browning, & Marenda, 2020). Despite being the most abundant CRISPR systems in prokaryotes, type I systems have not been used as often as type II systems (e.g. Cas9, Cas12) for genome engineering, owing to the relative difficulty of heterologous expression of the multicomponent Cascade complex (Cas1-2, Cas5-8, Cas11 and Cas3 as final endonuclease effector). Endogenous CRISPR type I systems would obviate this requirement. In *Clostridium difficile*, an endogenous CRISPR type I system has been characterized and redirected for Cas3-driven, DSB-induced auto-immunity control of this human pathogen (Maikova, Kreis, Boutserin, Severinov, & Soutourina, 2019). Another endogenous type I-A CRISPR system has also been exploited to facilitate genome editing by double-homologous recombination in *Heliobacterium modesticaldum* (Baker et al., 2019). Interestingly, Cas3 from *Pseudomonas aeruginosa* has been repurposed not only as an endogenous genome-editing tool but also as a heterologous editing tool more efficient than Cas9 for large deletions in *E. coli* and the plant pathogen *P. syringae* (Csörgő et al., 2020).

Table 1 shows a comprehensive list of the different applications of CRISPR-Cas system mediated genome editing in a wide array of bacterial species. As we can see, there has been a recent explosion of CRISPR-Cas methods, often combined with recombineering with variable host-dependent efficiencies.

Table 1 Strategies for CRISPR-mediated genome-editing in bacteria.

Strategies for editing	Strain	Results	Efficiency	References
Scarless Cas9 Assisted Recombineering (no-SCAR) λ-Red	*Escherichia coli MG1655*	This method does not leave recombinase recognition site scars, which can cause chromosomal instability and unwanted genomic rearrangements.	85–100%	Reisch and Prather (2015)
	pCas9cr4 pTET promoter			
Induction of a recombinase	*E. coli HME63*	Editing is facilitated by a co-selection of transformable cells and a small induction of recombination in the target site by Cas9 cleavage.	4.8×10^{5}/ 5.3×10^{2} CFU	Jiang et al. (2013)
	Streptococcus pneumoniae JEN53	Genome engineering works in highly recombinogenic bacteria.	10^{-1} CFU	Jiang et al. (2013)
	Corynebacterium glutamicum	Enables transformation to be simpler and more convenient than the two-plasmid-based CRISPR–Cas9 method.	2.1×103 CFU/µg	Wang, Hu, et al. (2018)
	Lactococcus lactis	Is highly efficient, time-saving and easy to use for introducing precise point mutations and seamlessly performing gene deletion and insertion.	87%	Guo et al. (2019)
	Lactobacillus plantarum WCFS1	Combination of RecE/T-assisted HDR and CRISPR–Cas9 targeted chromosomal DSBs offers a general and adaptable strategy to address the low HDR of *Lactobacillus* spp.	>89.4%	Huang, Song, and Yang (2019)
	Lactobacillus brevis ATCC367		83.3% (% colonies)	Huang, Song, and Yang (2019)

Encode DNA repair template in a plasmid	Clostridium ljungdahlii	More rapid, no added antibiotic resistance gene, scar-less and minimal polar effects.	<75%	Huang et al. (2016)
	L. plantarum NIZO2877	Uniquely capable of gene insertions. It showed vast differences for Cas9-mediated genome editing between methods and related strains.	10^2 CFU	Leenay et al. (2019)
	Pseudomonas putida KT2440	Adopted for counter selection of the correct mutants.	74.35%	Wirth et al. (2020)
	Streptomyces coelicolor	Improves the genome editing efficiency compared with the currently existing.	60–100%	Huang et al. (2015)
	Staphylococcus aureus RN4220	High editing efficiencies and easy use of a highly efficient transcription-inhibition system.	70–100%	Chen et al. (2017)
Inducible promoters	E. coli	Introduces various types of genomic modifications with near 100% editing efficiency and introduces three mutations simultaneously.	83%	Li et al. (2015)
	Bacillus subtilis	Shorter time to achieve the mutations. Sometimes it can be very laborious to find the corresponding mutant.	50%	Altenbuchner (2016)
	Clostridium acetobutylicum ATCC 824	Two-plasmid inducible CRISPR/Cas9 genome editing tool was successfully developed. This method enables the rapid introduction of marker-free genomic modification of any type, from the substitution of a few nucleotides to large deletions or insertions.	10^{-3} CFU/total colonies	Wasels et al. (2017)
	Lactococcus lactis dCas9	CRISPRi, is used in conjunction with a nisin-inducible promoter, for non-toxic, precise, targeted genome regulation and represents a valid alternative to RNAi.	50-fold mRNA downregulation	Berlec et al. (2018)
Nucleases of CRISPR-like DNA Nickase	C. glutamicum	Using either two plasmids or one plasmid consisting of FnCpf1, CRISPR RNA, and homologous arms.	86–100% for small changes	Jiang et al. (2017)
	Francisella novicida	CRISPR arrays are processed into mature crRNAs without the requirement of an additional trans-	25–100% in HEK293FT	Zetsche et al. (2015)

Continued

Table 1 Strategies for CRISPR-mediated genome-editing in bacteria.—cont'd

Strategies for editing	Strain	Results	Efficiency	References
		activating crRNA (tracrRNA) Cpf1-crRNA complexes efficiently cleave target DNA proceeded by a short T-rich protospacer-adjacent motif (PAM), in contrast to the G-rich PAM Cpf1 introduces a staggered DNA double-stranded break with a 4 or 5-nt 5'overhang.	80%	Yan et al. (2017)
	Mycobacterium smegmatis	CRISPR-Cas12a can efficiently introduce point mutations into PAM- and crRNA-targeting regions.		
	Yersinia pestis KIM 6+	CRISPR-Cas12a is a useful method for genetic manipulation of chromosomal and plasmid DNA.	81–83%	Banno et al. (2018)
Base editors	*E. coli*	The use of uracil DNA glycosylase inhibitor in combination with a degradation tag (LVA tag) resulted in a robustly high mutation efficiency, which allowed simultaneous multiplex editing of six different genes.	61.7–95.1%	Wang, Wang, et al. (2018)
	Klebsiella pneumonia	Development of a cytidine base-editing system, pBECKP, for precise C→T conversion by engineering the fusion of the cytidine deaminase APOBEC1 and a Cas9 nickase.	25–100%	Chen et al. (2018)
	Pseudomonas aeruginosa	Development of a genome editing method pCasPA/ pACRISPR by harnessing the CRISPR/Cas9 and the phage λ-Red recombination systems. The method allows for efficient and scarless genetic manipulation.	93–100%	Luo et al. (2020)
	Rhodobacter sphaeroides	CBEs (cytosine base editors) and ABEs (adenine base editors) serve as alternative methods for genetic manipulation of bacteria that are hard to directly edit by Cas9-sgRNA.	43–97%	Zhang et al. (2020)
	S. aureus	This method simplifies the genome editing process and achieves the conversion of adenine to guanine via	50–100%	

Endogenous CRISPR systems	*Clostridium difficile 630Δerm R20291*	an enzymatic deamination reaction and a subsequent DNA replication process rather than HDR. Repurposing of endogenous Type IB system coupled to a CRISPR mini-array plasmid to cause DSB-induced auto-immunity and to generate Δhfq mutant with plasmid-encoded HDR template.	30–100%	Maikova et al. (2019)
	Heliobacterium modesticaldum	Redeployment of endogenous type IA system, coupled to a HDR plasmid carrying a miniature CRISPR array, which targets sequences in pshA (downstream of a native PAM sequence) produced non-phototrophic transformants with clean replacements of the pshA gene.	80%	Baker et al. (2019)
	Mycoplasma gallisepticum S6	Use of endogenous MgaCas9 coupled to three constructs carrying different CRISPR arrays targeting regions in the ksgA gene. This leads to NHEJ-induced mutations (insertions and deletions) that prevent ribosomal methylation, which in turn confers resistance to the aminoglycoside antimicrobial kasugamycin, enabling the selection of mutants.	1.18×10^6 vs 2.47×10^8 CCU/mL (3 days cultures with vs without kasugamycin) 63–100% indel occurrence	Mahdizadeh et al. (2020)
	Pseudomonas aeruginosa	Repurposing and optimization of endogenous Type I CRISPR system (PaeCas3c) for genome engineering with a single crRNA and selecting only for survival after editing via native A-EJ. Self-targeting crRNAs lead to large genomic deletions (7–424 kb). When provided with an HDR template PaeCas3c promotes recombination compared to SpCas9.	A-EJ: • 20–40% of surviving colonies with native crRNAs. • 94–100% with modified-repeat crRNAs.HDR: • 22% for 249 kb deletion (vs 0% for SpCas9). • 61% for 56 kb deletion (vs 11% for SpCas9). • 100% for 0.17 kb (vs 78% for SpCas9).	Csörgő et al. (2020)

7 Alternative delivery methods for CRISPR-Cas: ribonucleoprotein complexes (RNPs) and lipid nanoparticles

Although several alternatives to Cas9 have been recently developed, an alternative method to plasmids is the delivery of pre-made Cas9/gRNA RNP complexes via electroporation (Fig. 9A) which is highly versatile, but depends on the host cell machinery to maintain an efficient, non-toxic expression of the Cas nuclease, Cas nickases, base-editors and gRNA. In theory, this approach is more laborious because it requires purifying active recombinant Cas9 protein from a heterologous system (mostly *E. coli*) and generating the sgRNA by in vitro transcription. Currently, however, both elements can be directly purchased from different vendors. The main advantage of this method is that it does not depend on the host transcription and translation machinery, allowing a direct assessment of the efficacy of the RNP preparation by in vitro nuclease assays. The RNP complex is degraded shortly after transfection, limiting the toxic effects of a continuous Cas9 expression. No cloning is required, therefore there is no restriction in the selection of gRNAs that may target a cloning strain genome. It also presents a more concise streamline than the plasmid methods, as no plasmid curing is required. This strategy has been used to efficiently target and edit eukaryote genomes, e.g., human, mouse, wheat and zebrafish (DeWitt, Corn, & Carroll, 2017; Kim, Kim, Cho, Kim, & Kim, 2014; Liang et al., 2017; Schumann et al., 2015). SpCas9 is a relatively large protein (160 kDa), which may limit the electroporation efficiency of the nuclease/gRNA complex.

On the other hand, bacteria with thick cell walls such as Gram-positive bacteria can be very difficult to transfect/electroporate. As an alternative, polymer-derivatized Cas9 has been developed (Kang et al., 2017). In this work, direct covalent modification of the protein with a cationic polymer (bPEI) was followed by complexation with a sgRNA to generate nanosized complexes (Fig. 9B). Treatment with Cr-nanocomplexes targeting antibiotic resistance inhibited bacterial cell growth on agar plates with oxacillin and demonstrated a higher genome-editing efficiency in methicillin-resistant *Staphylococcus aureus* (MRSA), compared to incubation with SpCas9/sgRNA RNP alone or combined with Lipofectamine, a traditional cationic lipid formulation which showed almost no effect on *S. aureus*. The removal of antibiotic resistance genes through this strategy could prove effective for the control of the rising problem of antibiotic resistance, while maintaining commensal bacteria in microbiota. Additionally, novel lipid nanoparticle formulations such as SORT (selective organ targeting) for Cas9 mRNA and sgRNA (Cheng et al., 2020), and polyethylene glycol phospholipid-modified cationic LNP for Cas9/sgRNA plasmid (Zhang et al., 2017) have shown a high efficiency in mammalian cells (Fig. 9C). However, it remains to be evaluated if they can also be redirected for genome editing in bacteria.

It has been shown that the Tyr66-His mutant (encoded by the single base substitution 196 T > C) shifts wild-type GFP absorption and emission towards the blue

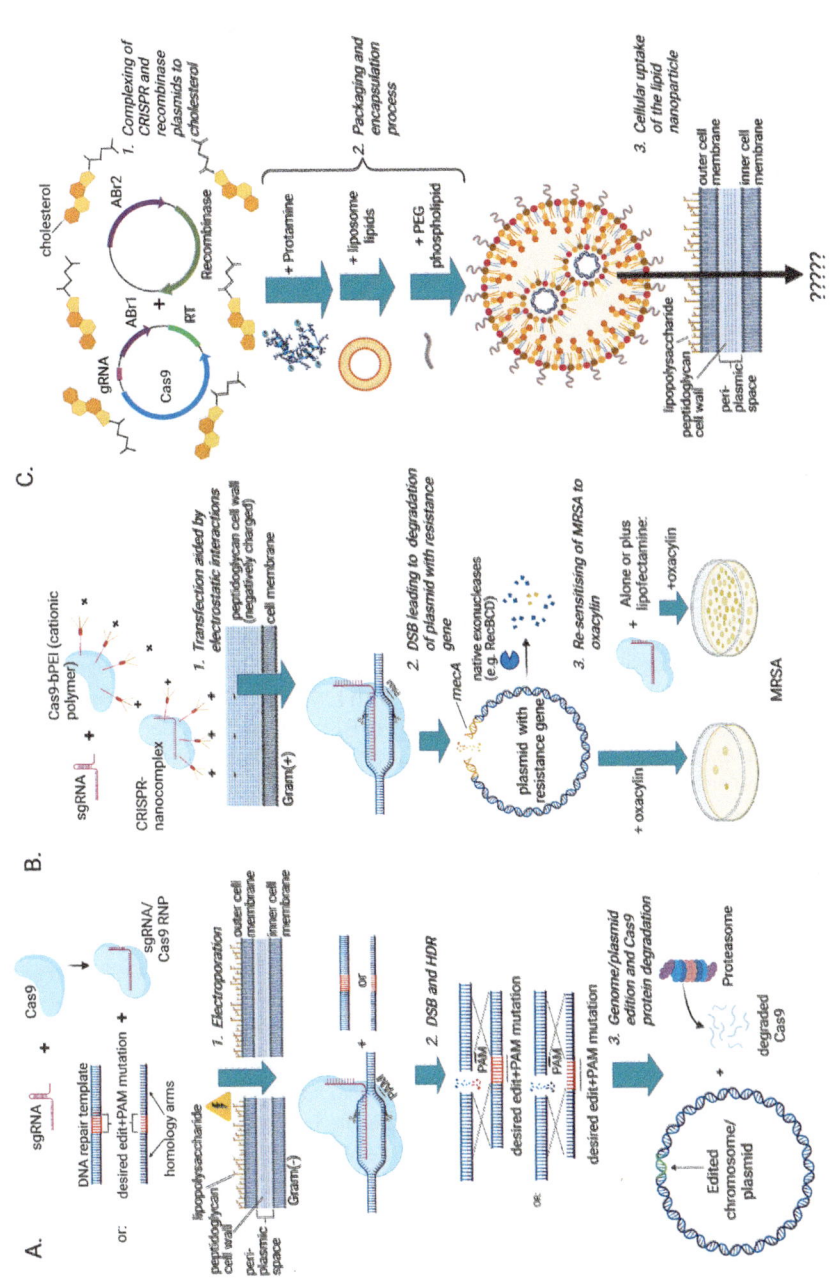

FIG. 9

See figure legend on next page.

spectrum, thus creating blue fluorescent protein (BFP) (Fig. 10A). As a proof of principle to evaluate the viability of RNP approach via electroporation, we performed an assay with Cas9/gRNA RNPs and a linear DNA repair template to induce specific mutations by homologous recombination in a GFP to BFP conversion assay in an *E. coli* strain harbouring a GFP encoding plasmid (Fig. 10B and C).

FIG. 9—Cont'd   Alternative delivery approaches for CRISPR-Cas mediated genome editing. (A) RNP electroporation: Recombinant CRISPR nuclease (e.g. Cas9) is combined with in vitro transcribed or synthetic sgRNA to form active Cas9/sgRNA RNP complexes. Electroporation (step 1) is usually used to form temporary holes in the bacterial cell wall to co-transform the RNPs with a linear single- or double-stranded recombination template harbouring the desired edit plus additional mutations at the PAM site to avoid Cas9/sgRNA targeting. Targeting to the desired locus occurs, DNA double-strand break is formed 2–3 bp upstream PAM sequence, which is repaired by double homologous recombination with the linear DNA template (step 2). Wild-type allele is replaced by the mutant allele, which is fixed in the target genome or plasmid. Cas9/sgRNA RNPs are maintained only transiently in the cell and are degraded shortly after gene edition by the cell protein-degradation machinery (step 3). This method does not require the introduction of antibiotic resistance markers or plasmid curing; however, its efficiency would be highly dependent on the transformation amenability and recombination machinery of the bacterial strain.
(B) Application of cationic polymer conjugation with Cas9/sgRNA to delete antibiotic-resistance genes in Gram-positive bacteria. Recombinant Cas9 is covalently linked to a cationic polymer (bPEI) followed by incubation with sgRNA to form CRISPR nanometric complexes. Electrostatic interactions between the positively-charged CRISPR nanocomplexes and the negatively-charged peptidoglycan cell wall facilitate binding and incorporation of Cr-nanocomplex into thick-cell walled Gram-positive bacteria (step 1). In this example, sgRNA targets incorporated Cas9 to the mecA gene, responsible for methicillin and oxacillin (oxa) resistance in *Staphylococcus aureus* (MRSA), leading to the loss of the resistance plasmid by native exonucleases such as RecBCD (Step 2). Due to the re-sensitizing to oxacylin, counterselection of MRSA is efficiently achieved compared to the incubation with RNP alone or combined with the cationic lipid lipofectamine (step 3).
(C) Formation of polyethylene glycol phospholipid-modified cationic lipid nano-particles (LNPs) for Cas9/sgRNA and recombinase plasmid delivery. The process begins with the complexing of the plasmids to cholesterol (step 1), followed by successive additions of protamine, liposome lipids and polyethylene glycol (step 2). Their application has not yet been proved to cross bacterial cell walls for an efficient cellular uptake of the LNPs (step 3).

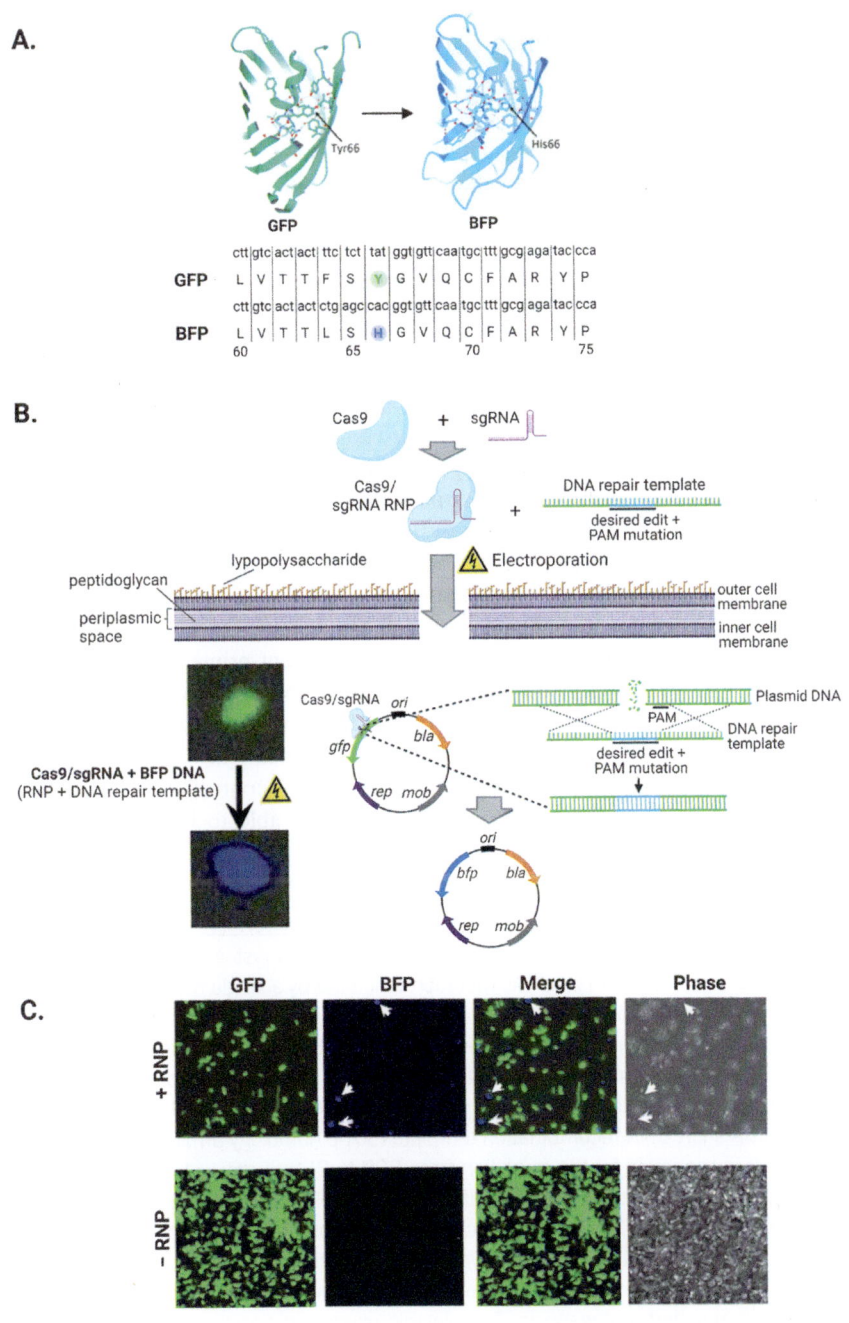

FIG. 10

Assay for conversion of the green fluorescent protein (GFP) to the blue fluorescent protein (BFP). (A) Crystallographic structures of the GFB and BFP shows the amino acid residues

(Continued)

8 Summary, conclusions and future directions

CRISPR-Cas technology has revolutionized genome editing and has become the predominant approach in eukaryotic organisms. Its application in prokaryotes has been slower but is quickly being adapted to several bacteria of industrial and biomedical importance. Several challenges need to be addressed for a widespread application of this revolutionary technology in bacteria (Table 2).

In particular, if it is to be superior to current methods based on suicide plasmids and recombineering that keep being adapted and improved in bacteria (Volke, Friis, Wirth, Turlin, & Nikel, 2020), efficiency needs to be adjusted regarding available transformation tools and genetic accessibility for each species. The greatest disadvantage of CRISPR-editing tools in bacteria is so far the cytotoxicity induced by a continuous expression of foreign CRISPR-nucleases. CRISPR-Cas9 is mostly utilized as a counter-selection mechanism against colonies that do not undergo a desired genomic edit, rather than as an actual genome editing tool in bacteria. Several alternatives have been developed including inducible-promoters, nCas9, dCas9, Cas12a and base-editors with different degrees of success depending on the bacterial strain or species. A complex chimeric effector combining engineered dCas9 without PAM binding activity coupled to other inducible DNA binding proteins such as PhlF domain has also been developed, with reduced toxicity in *E. coli* (Zhang & Voigt, 2018). Additionally, the development of highly efficient prime editors that do not require a DNA repair template or DSB still needs to be explored in bacteria.

However, the continuous expression of any foreign protein with DNA binding/editing activity seems to be particularly toxic for many bacteria. The natural function of CRISPR as an adaptive immune system is highly controlled in prokaryotes. Ultimately, more research to fully understand and be able to harness endogenous

FIG. 10—Cont'd

close to the tyrosine 66 in GFP's and the histidine 66 in BFP's crystallographic structures and amino acid sequences. The Tyr66-His mutation (encoded by a single nucleotide substitution at 196 T > C) has been shown to alter wild-type GFP uptake and blue spectrum luminescence, resulting in the blue fluorescent protein (BFP). (B) Electroporation (represented by the bolt-induced pore in the cell wall and bacterial membrane as the double parallel-arranged purple lines) of GFP-expressing bacteria with the Cas9 (blue) and sgRNA (pink) RNP. A short DNA recombination template (30 bp homology arms in green) encoding the mutation responsible for the transition from GFP to BFP (Tyr66 → His) and a PAM (protospacer motif) mutation to avoid self-targeting (in blue), was also used to turn blue RNP-transformed cells. C. Cells with successful reparation produced fluorescence in colour blue. Blue-fluorescent cells can be observed after the electroporation. Although the observed efficiency was low (<5%), these results are a proof of principle for the use of RNP in bacteria for in vivo site directed-mutagenesis. The plasmid used in this application contains the following elements: *bfp*: blue fluorescent protein, *bla*: beta-lactamase, *gfp*: green fluorescent protein, *mob*: mobilization gene, *ori*: origin of replication, *rep*: replication gene.

Table 2 Advantages and disadvantages of most commonly used genome-editing methodologies in bacteria.

Method	Advantages	Disadvantages
'Suicide' plasmids	– Low cost. – Does not require specialized strains. – Useful for large genomic deletions or targeted gene disruption.	– Low efficiency. – High rate of false positives. – Often requires several rounds of antibiotic selection. – Long homology flanking regions (~1 kb) to the desired edit need to be cloned.
'Recombineering' (Lambda Red, RecE/T)	– Low cost. – Highly efficient, particularly for small-scale edits. – Utilizes DNA templates with only short regions of homology (50 bp) to promote gene edits by homologous recombination.	– Requires development of specialized strains with controlled foreign recombinase expression. – Usually requires counter-selection steps to eliminate antibiotic resistance markers from the genome.
ClosTron method (Retrotransposition-Activated Marker)	– Can be programmed by designing a 344 bp region homologous to the target gene. – Broad-host range of Ll.LtrB intron theoretically allows its use in any bacterial species.	– So far only tested in members of the Clostridium (Clostridioides) genus. – Requires extensive cloning or expensive out-sourced synthesis of modified targeting intron. – The application of the method is straightforward only for targeted gene disruption.
CRISPR-Cas (Plasmid-encoded)	– Low cost. – Can be combined with recombineering for an enhanced efficiency. – Highly customizable. – Double strand breaks induce cell death in non-edited cells diminishing background (false positive colonies). – Highly versatile genome editing from large genome deletions/insertions to single base mutations.	– High cytotoxicity of Cas9 expression can alter morphology and survival even when devoid of nuclease activity due to steric hindrance posed by Cas9 PAM binding and subsequent DNA unwinding activity along the genome. – Induction of off-target effects (undesired genome edits) due to non-specific DNA cleaving, particularly after prolonged Cas9/gRNA expression.
CRISPR-Cas (Endogenous systems)	– Do not necessitate the expression of a foreign CRISPR nuclease. – Highly programmable by altering the homology repair template and the CRISPR array sequence.	– Requires extensive characterization of the endogenous CRISPR system (nucleases, PAM requirement, efficiency, etc.) and DNA repair pathways (e.g. NHEJ) for each particular species/strain.

CRISPR loci (spacers and Cas proteins) for genome editing would be in principle the most effective way to avoid foreign CRISPR systems in bacteria. This approach would require a case-by-case scenario of efficiency and tuning for each native CRISPR effector. Nevertheless, more recent studies using the native CRISPR machinery (type I or type II) have been reported with high efficiencies (Baker et al., 2019; Csörgő et al., 2020; Mahdizadeh et al., 2020; Maikova et al., 2019) (Table 1), indicating that this may be the future for biomedical and industrially relevant bacterial species with endogenous CRISPR systems. Continuous research on endogenous CRISPR systems also helps to create and diversify the strategies for heterologous genome editing.

In eukaryotic organisms, the ribonucleoprotein (RNP) format with foreign but well-characterized Cas enzymes, such as SpCas9, has shown higher efficiency and much lower cytotoxic and off-target effects compared to plasmids. Further research would show if this strategy could have similar benefits for genome editing in bacteria with the available transformation methods (e.g. electroporation). The RNP approach is by no means limited to SpCas9, as it has been tested successfully, mostly in eukaryotic organisms, with other natural Cas9 orthologs such as SaCas9 and CjCas9 with more complex PAM requirements that are smaller and easier to transfect than SpCas9. Also, several Cas9 orthologs have been engineered and promise higher efficiency, specificity, and broader PAM requirements. These novel alternatives expand the available toolbox that should be explored in bacteria to enhance the potential of CRISPR-mediated genome editing in these relevant organisms.

Ultimately genome editing would allow the creation of synthetic genomes combining a wide array of genes, metabolic pathways and even full chromosomes (Wang, de la Torre, Robertson, & Chin, 2019) from different organisms to optimize the production of relevant metabolites, e.g., natural products (Xu, Liu, Du, Ledesma-Amaro, & Liu, 2020).

Acknowledgements

R.D.A.-O. received a post-doctoral fellowship from CONACYT (Consejo Nacional de Ciencia y Tecnología), application number 2466755.

References

Altenbuchner, J. (2016). Editing of the Bacillus subtilis genome by the CRISPR-Cas9 system. *Applied and Environmental Microbiology, 82*(17), 5421–5427. https://doi.org/10.1128/AEM.01453-16.

Anders, C., Niewoehner, O., Duerst, A., & Jinek, M. (2014). Structural basis of PAM-dependent target DNA recognition by the Cas9 endonuclease. *Nature, 513*(7519), 569–573. https://doi.org/10.1038/nature13579.

Anzalone, A. V., Randolph, P. B., Davis, J. R., Sousa, A. A., Koblan, L. W., Levy, J. M., et al. (2019). Search-and-replace genome editing without double-strand breaks or donor DNA. *Nature, 576*(7785), 149–157. https://doi.org/10.1038/s41586-019-1711-4.

Aparicio, T., de Lorenzo, V., & Martínez-García, E. (2018). CRISPR/Cas9-based counterselection boosts recombineering efficiency in Pseudomonas putida. *Biotechnology Journal*, *13*(5), e1700161. https://doi.org/10.1002/biot.201700161.

Baker, P. L., Orf, G. S., Kevershan, K., Pyne, M. E., Bicer, T., & Redding, K. E. (2019). Using the endogenous CRISPR-Cas system of *Heliobacterium modesticaldum* to delete the photochemical reaction center core subunit gene. *Applied and Environmental Microbiology*, *85*(23), e01644–19. https://doi.org/10.1128/AEM.01644-19.

Banno, S., Nishida, K., Arazoe, T., Mitsunobu, H., & Kondo, A. (2018). Deaminase-mediated multiplex genome editing in Escherichia coli. *Nature Microbiology*, *3*(4), 423–429. https://doi.org/10.1038/s41564-017-0102-6.

Berlec, A., Škrlec, K., Kocjan, J., Olenic, M., & Štrukelj, B. (2018). Single plasmid systems for inducible dual protein expression and for CRISPR-Cas9/CRISPRi gene regulation in lactic acid bacterium Lactococcus lactis. *Scientific Reports*, *8*(1), 1009. https://doi.org/10.1038/s41598-018-19402-1.

Broughton, J. P., Deng, X., Yu, G., Fasching, C. L., Servellita, V., Singh, J., et al. (2020). CRISPR–Cas12-based detection of SARS-CoV-2. *Nature Biotechnology*, *38*(7), 870–874. https://doi.org/10.1038/s41587-020-0513-4.

Cameron Coates, R., Blaskowski, S., Szyjka, S., van Rossum, H. M., Vallandingham, J., Patel, K., et al. (2019). Systematic investigation of CRISPR–Cas9 configurations for flexible and efficient genome editing in Corynebacterium glutamicum NRRL-B11474. *Journal of Industrial Microbiology and Biotechnology*, *46*(2), 187–201. https://doi.org/10.1007/s10295-018-2112-7.

Cavanagh, P., & Garrity, A. (2015). *CRISPR mechanism, DNA binding and cleavage.* https://sites.tufts.edu/crispr/crispr-mechanism/rna-guide/.

Chayot, R., Montagne, B., Mazel, D., & Ricchetti, M. (2010). An end-joining repair mechanism in Escherichia coli. *Proceedings of the National Academy of Sciences of the United States of America*, *107*(5), 2141–2146. https://doi.org/10.1073/pnas.0906355107.

Chen, W., Zhang, Y., Yeo, W.-S., Bae, T., & Ji, Q. (2017). Rapid and efficient genome editing in *Staphylococcus aureus* by using an engineered CRISPR/Cas9 system. *Journal of the American Chemical Society*, *139*(10), 3790–3795. https://doi.org/10.1021/jacs.6b13317.

Chen, W., Zhang, Y., Zhang, Y., Pi, Y., Gu, T., Song, L., et al. (2018). CRISPR/Cas9-based genome editing in Pseudomonas aeruginosa and cytidine deaminase-mediated base editing in Pseudomonas species. *IScience*, *6*, 222–231. https://doi.org/10.1016/j.isci.2018.07.024.

Cheng, Q., Wei, T., Farbiak, L., Johnson, L. T., Dilliard, S. A., & Siegwart, D. J. (2020). Selective organ targeting (SORT) nanoparticles for tissue-specific mRNA delivery and CRISPR–Cas gene editing. *Nature Nanotechnology*, *15*(4), 313–320. https://doi.org/10.1038/s41565-020-0669-6.

Cho, S., Choe, D., Lee, E., Kim, S. C., Palsson, B., & Cho, B. K. (2018). High-level dCas9 expression induces abnormal cell morphology in Escherichia coli. *ACS Synthetic Biology*, *7*(4), 1085–1094. https://doi.org/10.1021/acssynbio.7b00462.

Choulika, A., Perrin, A., Dujon, B., & Nicolas, J. F. (1995). Induction of homologous recombination in mammalian chromosomes by using the I-SceI system of Saccharomyces cerevisiae. *Molecular and Cellular Biology*, *15*(4), 1968–1973. https://doi.org/10.1128/MCB.15.4.1968.

Chou-Zheng, L., & Hatoum-Aslan, A. (2019). A type III-A CRISPR-Cas system employs degradosome nucleases to ensure robust immunity. *eLife*, *8*, e45393. https://doi.org/10.7554/ELIFE.45393.

Csörgő, B., León, L. M., Chau-Ly, I. J., Vasquez-Rifo, A., Berry, J. D., Mahendra, C., et al. (2020). A compact Cascade–Cas3 system for targeted genome engineering. *Nature Methods*, *17*(12), 1183–1190. https://doi.org/10.1038/s41592-020-00980-w.

Cui, L., & Bikard, D. (2016). Consequences of Cas9 cleavage in the chromosome of Escherichia coli. *Nucleic Acids Research*, *44*(9), 4243–4251. https://doi.org/10.1093/nar/gkw223.

Deltcheva, E., Chylinski, K., Sharma, C. M., Gonzales, K., Chao, Y., Pirzada, Z. A., et al. (2011). CRISPR RNA maturation by trans-encoded small RNA and host factor RNase III. *Nature*, *471*(7340), 602. https://doi.org/10.1038/NATURE09886.

DeWitt, M. A., Corn, J. E., & Carroll, D. (2017). Genome editing via delivery of Cas9 ribonucleoprotein. *Methods*, *121–122*, 9–15. https://doi.org/10.1016/j.ymeth.2017.04.003.

Fehér, T., Karcagi, I., Gyorfy, Z., Umenhoffer, K., Csörgo, B., & Pósfai, G. (2007). Scarless engineering of the Escherichia coli genome. *Methods in Molecular Biology*, *416*, 251–259. https://doi.org/10.1007/978-1-59745-321-9_16.

Finger-Bou, M., Orsi, E., van der Oost, J., & Staals, R. H. J. (2020). CRISPR with a happy ending: Non-templated DNA repair for prokaryotic genome engineering. *Biotechnology Journal*, *15*(7), 1900404. https://doi.org/10.1002/biot.201900404.

Gaudelli, N. M., Komor, A. C., Rees, H. A., Packer, M. S., Badran, A. H., Bryson, D. I., et al. (2017). Programmable base editing of A•T to G•C in genomic DNA without DNA cleavage. *Nature*, *551*(7681), 464–471. https://doi.org/10.1038/nature24644.

Gee, P., Ando, Y., Kitayama, H., Yamamoto, S. P., Kanemura, Y., Ebina, H., et al. (2011). APOBEC1-mediated editing and attenuation of herpes simplex virus 1 DNA indicate that neurons have an antiviral role during herpes simplex encephalitis. *Journal of Virology*, *85*(19), 9726–9736. https://doi.org/10.1128/JVI.05288-11.

Guirouilh-Barbat, J., Lambert, S., Bertrand, P., & Lopez, B. S. (2014). Is homologous recombination really an error-free process? *Frontiers in Genetics*, *5*, 175. https://doi.org/10.3389/fgene.2014.00175.

Guo, T., Xin, Y., Zhang, Y., Gu, X., & Kong, J. (2019). A rapid and versatile tool for genomic engineering in Lactococcus lactis. *Microbial Cell Factories*, *18*(1), 22. https://doi.org/10.1186/s12934-019-1075-3.

Heap, J. T., Kuehne, S. A., Ehsaan, M., Cartman, S. T., Cooksley, C. M., Scott, J. C., et al. (2010). The ClosTron: Mutagenesis in Clostridium refined and streamlined. *Journal of Microbiological Methods*, *80*(1), 49–55. https://doi.org/10.1016/j.mimet.2009.10.018.

Heap, J. T., Pennington, O. J., Cartman, S. T., Carter, G. P., & Minton, N. P. (2007). The ClosTron: A universal gene knock-out system for the genus Clostridium. *Journal of Microbiological Methods*, *70*(3), 452–464. https://doi.org/10.1016/j.mimet.2007.05.021.

Hoedt, E. C., Bottacini, F., Cash, N., Bongers, R. S., van Limpt, K., Ben Amor, K., et al. (2021). Broad purpose vector for site-directed insertional mutagenesis in Bifidobacterium breve. *Frontiers in Microbiology*, *12*, 636822. https://doi.org/10.3389/FMICB.2021.636822.

Huang, F., & Zhu, B. (2021). The cyclic oligoadenylate signaling pathway of type III CRISPR-Cas systems. *Frontiers in Microbiology*, *11*, 602789. https://doi.org/10.3389/FMICB.2020.602789.

Huang, H., Chai, C., Li, N., Rowe, P., Minton, N. P., Yang, S., et al. (2016). CRISPR/Cas9-based efficient genome editing in Clostridium ljungdahlii, an autotrophic gas-fermenting bacterium. *ACS Synthetic Biology*, *5*(12), 1355–1361. https://doi.org/10.1021/acssynbio.6b00044.

Huang, C., Ding, T., Wang, J., Wang, X., Guo, L., Wang, J., et al. (2019). CRISPR-Cas9-assisted native end-joining editing offers a simple strategy for efficient genetic engineering in Escherichia coli. *Applied Microbiology and Biotechnology*, *103*(20), 8497–8509. https://doi.org/10.1007/s00253-019-10104-w.

Huang, H., Song, X., & Yang, S. (2019). Development of a RecE/T-assisted CRISPR–Cas9 toolbox for Lactobacillus. *Biotechnology Journal*, *14*(7), 1800690. https://doi.org/10.1002/biot.201800690.

Huang, W., & Wilks, A. (2017). A rapid seamless method for gene knockout in Pseudomonas aeruginosa. *BMC Microbiology*, *17*(1), 1–8. https://doi.org/10.1186/S12866-017-1112-5/FIGURES/6.

Huang, H., Zheng, G., Jiang, W., Hu, H., & Lu, Y. (2015). One-step high-efficiency CRISPR/Cas9-mediated genome editing in Streptomyces. *Acta Biochimica et Biophysica Sinica*, *47*(4), 231–243. https://doi.org/10.1093/abbs/gmv007.

Ikeda, T., Ohsugi, T., Kimura, T., Matsushita, S., Maeda, Y., Harada, S., et al. (2008). The antiretroviral potency of APOBEC1 deaminase from small animal species. *Nucleic Acids Research*, *36*(21), 6859–6871. https://doi.org/10.1093/nar/gkn802.

Jiang, W., Bikard, D., Cox, D., Zhang, F., & Marraffini, L. A. (2013). RNA-guided editing of bacterial genomes using CRISPR-Cas systems. *Nature Biotechnology*, *31*(3), 233–239. https://doi.org/10.1038/nbt.2508.

Jiang, Y., Chen, B., Duan, C., Sun, B., Yang, J., & Yang, S. (2015). Multigene editing in the Escherichia coli genome via the CRISPR-Cas9 system. *Applied and Environmental Microbiology*, *81*(7), 2506–2514. https://doi.org/10.1128/AEM.04023-14.

Jiang, Y., Qian, F., Yang, J., Liu, Y., Dong, F., Xu, C., et al. (2017). CRISPR-Cpf1 assisted genome editing of Corynebacterium glutamicum. *Nature Communications*, *8*, 15179. https://doi.org/10.1038/ncomms15179.

Kamruzzaman, M., Yan, A., & Castro-Escarpulli, G. (2022). Editorial: CRISPR-Cas Systems in Bacteria and Archaea. *Frontiers in Microbiology*, *1098*, 887778. https://doi.org/10.3389/FMICB.2022.887778.

Kang, Y. K., Kwon, K., Ryu, J. S., Lee, H. N., Park, C., & Chung, H. J. (2017). Nonviral genome editing based on a polymer-derivatized CRISPR nanocomplex for targeting bacterial pathogens and antibiotic resistance. *Bioconjugate Chemistry*, *28*(4), 957–967. https://doi.org/10.1021/acs.bioconjchem.6b00676.

Kim, S., Kim, D., Cho, S. W., Kim, J., & Kim, J.-S. (2014). Highly efficient RNA-guided genome editing in human cells via delivery of purified Cas9 ribonucleoproteins. *Genome Research*, *24*(6), 1012–1019. https://doi.org/10.1101/gr.171322.113.

Klompe, S. E., Vo, P. L. H., Halpin-Healy, T. S., & Sternberg, S. H. (2019). Transposon-encoded CRISPR–Cas systems direct RNA-guided DNA integration. *Nature*, *571*(7764), 219–225. https://doi.org/10.1038/s41586-019-1323-z.

Komor, A. C., Kim, Y. B., Packer, M. S., Zuris, J. A., & Liu, D. R. (2016). Programmable editing of a target base in genomic DNA without double-stranded DNA cleavage. *Nature*, *533*(7603), 420–424. https://doi.org/10.1038/nature17946.

Konishi, K., van Doren, S. R., Kramer, D. M., Crofts, A. R., & Gennis, R. B. (1991). Preparation and characterization of the water-soluble heme-binding domain of cytochrome c1 from the Rhodobacter sphaeroides bc1 complex. *Journal of Biological Chemistry*, *266*(22), 14270–14276. https://doi.org/10.1016/S0021-9258(18)98678-3.

Koonin, E. V., & Makarova, K. S. (2022). Evolutionary plasticity and functional versatility of CRISPR systems. *PLoS Biology*, *20*(1), e3001481. https://doi.org/10.1371/JOURNAL.PBIO.3001481.

Leenay, R. T., Vento, J. M., Shah, M., Martino, M. E., Leulier, F., & Beisel, C. L. (2019). Genome editing with CRISPR-Cas9 in Lactobacillus plantarum revealed that editing outcomes can vary across strains and between methods. *Biotechnology Journal*, *14*(3), e1700583. https://doi.org/10.1002/biot.201700583.

Li, Q., Chen, J., Minton, N. P., Zhang, Y., Wen, Z., Liu, J., et al. (2016). CRISPR-based genome editing and expression control systems in Clostridium acetobutylicum and Clostridium beijerinckii. *Biotechnology Journal*, *11*(7), 961–972. https://doi.org/10.1002/biot.201600053.

Li, Y., Lin, Z., Huang, C., Zhang, Y., Wang, Z., Tang, Y., et al. (2015). Metabolic engineering of Escherichia coli using CRISPR-Cas9 meditated genome editing. *Metabolic Engineering*, *31*, 13–21. https://doi.org/10.1016/j.ymben.2015.06.006.

Li, X., Wang, Y., Liu, Y., Yang, B., Wang, X., Wei, J., et al. (2018). Base editing with a Cpf1−cytidine deaminase fusion. *Nature Biotechnology*, *36*(4), 324–327. https://doi.org/10.1038/nbt.4102.

Liang, Z., Chen, K., Li, T., Zhang, Y., Wang, Y., Zhao, Q., et al. (2017). Efficient DNA-free genome editing of bread wheat using CRISPR/Cas9 ribonucleoprotein complexes. *Nature Communications*, *8*, 14261. https://doi.org/10.1038/ncomms14261.

Lin, Q., Zong, Y., Xue, C., Wang, S., Jin, S., Zhu, Z., et al. (2020). Prime genome editing in rice and wheat. *Nature Biotechnology*, *38*(5), 582–585. https://doi.org/10.1038/s41587-020-0455-x.

Liu, T. Y., & Doudna, J. A. (2020). Chemistry of class 1 CRISPR-Cas effectors: Binding, editing, and regulation. *The Journal of Biological Chemistry*, *295*(42), 14473–14487. https://doi.org/10.1074/JBC.REV120.007034.

Liu, Y., Li, X., He, S., Huang, S., Li, C., Chen, Y., et al. (2020). Efficient generation of mouse models with the prime editing system. *Cell Discovery*, *6*(1), 27. https://doi.org/10.1038/s41421-020-0165-z.

Luo, Y., Ge, M., Wang, B., Sun, C., Wang, J., Dong, Y., et al. (2020). CRISPR/Cas9-deaminase enables robust base editing in Rhodobacter sphaeroides 2.4.1. *Microbial Cell Factories*, *19*(1), 93. https://doi.org/10.1186/s12934-020-01345-w.

Luo, P., He, X., Liu, Q., & Hu, C. (2015). Developing universal genetic tools for rapid and efficient deletion mutation in Vibrio species based on suicide T-vectors carrying a novel counterselectable marker, vmi480. *PLoS One*, *10*(12), e0144465. https://doi.org/10.1371/journal.pone.0144465.

Mahdizadeh, S., Sansom, F. M., Lee, S.-W., Browning, G. F., & Marenda, M. S. (2020). Targeted mutagenesis of mycoplasma gallisepticum using its endogenous CRISPR/Cas system [article]. *Veterinary Microbiology*, *250*, 108868. https://doi.org/10.1016/j.vetmic.2020.108868.

Maikova, A., Kreis, V., Boutserin, A., Severinov, K., & Soutourina, O. (2019). Using an endogenous CRISPR-Cas system for genome editing in the human pathogen Clostridium difficile [JOUR]. *Applied and Environmental Microbiology*, *85*(20), e01416–e01419. https://doi.org/10.1128/AEM.01416-19.

Makarova Kira, S., Wolf Yuri, I., Alkhnbashi Omer, S., Fabrizio, C., Shah Shiraz, A., Saunders Sita, J., et al. (2015). An updated evolutionary classification of CRISPR-Cas systems. *Nature Reviews. Microbiology*, *13*(11), 722–736.

McAllister, K. N., Bouillaut, L., Kahn, J. N., Self, W. T., & Sorg, J. A. (2017). Using CRISPR-Cas9-mediated genome editing to generate C. difficile mutants defective in selenoproteins synthesis. *Scientific Reports*, *7*, 14672. https://doi.org/10.1038/s41598-017-15236-5.

Misra, C. S., Bindal, G., Sodani, M., Wadhawan, S., Kulkarni, S., Gautam, S., et al. (2019). Determination of Cas9/dCas9 associated toxicity in microbes. *BioRxiv*, *848135*, 1–28. https://doi.org/10.1101/848135.

Nihongaki, Y., Kawano, F., Nakajima, T., & Sato, M. (2015). Photoactivatable CRISPR-Cas9 for optogenetic genome editing. *Nature Biotechnology, 33*(7), 755–760. https://doi.org/10.1038/nbt.3245.

Niu, Y., Tenney, K., Li, H., & Gimble, F. S. (2008). Engineering variants of the I-SceI homing endonuclease with strand-specific and site-specific DNA nicking activity. *Journal of Molecular Biology, 382*(1), 188. https://doi.org/10.1016/J.JMB.2008.07.010.

Noguchi, C., Ishino, H., Tsuge, M., Fujimoto, Y., Imamura, M., Takahashi, S., et al. (2005). G to A hypermutation of hepatitis B virus. *Hepatology, 41*(3), 626–633. https://doi.org/10.1002/hep.20580.

Oh, J. H., & van Pijkeren, J. P. (2014). CRISPR-Cas9-assisted recombineering in Lactobacillus reuteri. *Nucleic Acids Research, 42*(17), e131. https://doi.org/10.1093/nar/gku623.

Page, M. D., & Sockett, R. E. (1999). 13 molecular genetic methods in Paracoccus and Rhodobacter with particular reference to the analysis of respiration and photosynthesis. *Methods in Microbiology, 29*(C), 427–466. https://doi.org/10.1016/S0580-9517(08)70124-7.

Pinilla-Redondo, R., Mayo-Muñoz, D., Russel, J., Garrett, R. A., Randau, L., Sørensen, S. J., et al. (2020). Type IV CRISPR-Cas systems are highly diverse and involved in competition between plasmids. *Nucleic Acids Research, 48*(4), 2000–2012. https://doi.org/10.1093/NAR/GKZ1197.

Pomerantz, R. T., Goodman, M. F., & O'Donnell, M. E. (2013). DNA polymerases are error-prone at RecA-mediated recombination intermediates. *Cell Cycle, 12*(16), 2558–2563. https://doi.org/10.4161/cc.25691.

Poteete, A. R. (2001). What makes the bacteriophage λ red system useful for genetic engineering: Molecular mechanism and biological function. *FEMS Microbiology Letters, 201*(1), 9–14. https://doi.org/10.1111/j.1574-6968.2001.tb10725.x.

Ran, F. A., Hsu, P. D., Lin, C.-Y., Gootenberg, J. S., Konermann, S., Trevino, A. E., et al. (2013). Double nicking by RNA-guided CRISPR Cas9 for enhanced genome editing specificity. *Cell, 154*(6), 1380–1389. https://doi.org/10.1016/j.cell.2013.08.021.

Reisch, C. R., & Prather, K. L. J. (2015). The no-SCAR (scarless Cas9 assisted recombineering) system for genome editing in Escherichia coli. *Scientific Reports, 5*, 15096. https://doi.org/10.1038/srep15096.

Rosenberg, S. M., Shee, C., Frisch, R. L., & Hastings, P. J. (2012). Stress-induced mutation via DNA breaks in Escherichia coli: A molecular mechanism with implications for evolution and medicine. *BioEssays, 34*(10), 885–892. https://doi.org/10.1002/bies.201200050.

Saha, A., Arantes, P. R., & Palermo, G. (2022). Dynamics and mechanisms of CRISPR-Cas9 through the lens of computational methods. *Current Opinion in Structural Biology, 75*, 102400. https://doi.org/10.1016/J.SBI.2022.102400.

Sawyer, S. L., Emerman, M., & Malik, H. S. (2004). Ancient adaptive evolution of the primate antiviral DNA-editing enzyme APOBEC3G. *PLoS Biology, 2*(9), e275. https://doi.org/10.1371/journal.pbio.0020275.

Scherer, S. (2008). *A short guide to the human genome*. Cold Spring Harbor Laboratory Press.

Schumann, K., Lin, S., Boyer, E., Simeonov, D. R., Subramaniam, M., Gate, R. E., et al. (2015). Generation of knock-in primary human T cells using Cas9 ribonucleoproteins. *Proceedings of the National Academy of Sciences of the United States of America, 112*(33), 10437–10442. https://doi.org/10.1073/pnas.1512503112.

Selvaraj, G., & Iyer, V. N. (1983). Suicide plasmid vehicles for insertion mutagenesis in Rhizobium meliloti and related bacteria. *Journal of Bacteriology, 156*(3), 1292–1300. https://doi.org/10.1128/jb.156.3.1292-1300.1983.

Sharan, S. K., Thomason, L. C., Kuznetsov, S. G., & Court, D. L. (2009). Recombineering: A homologous recombination-based method of genetic engineering. *Nature Protocols*, *4*(2), 206–223. https://doi.org/10.1038/nprot.2008.227.

Shuman, S., & Glickman, M. S. (2007). Bacterial DNA repair by non-homologous end joining. In. *Nature Reviews. Microbiology*, *5*(11), 852–861. https://doi.org/10.1038/nrmicro1768.

Song, X., Huang, H., Xiong, Z., Ai, L., & Yang, S. (2017). CRISPR-Cas9D10A nickase-assisted genome editing in Lactobacillus casei. *Applied and Environmental Microbiology*, *83*(22), e01259–17. https://doi.org/10.1128/AEM.01259-17.

Standage-Beier, K., Zhang, Q., & Wang, X. (2015). Targeted large-scale deletion of bacterial genomes using CRISPR-Nickases. *ACS Synthetic Biology*, *4*(11), 1217–1225. https://doi.org/10.1021/acssynbio.5b00132.

Strecker, J., Ladha, A., Gardner, Z., Schmid-Burgk, J. L., Makarova, K. S., Koonin, E. V., et al. (2019). RNA-guided DNA insertion with CRISPR-associated transposases. *Science*, *365*(6448), 48–53. https://doi.org/10.1126/science.aax9181.

Sun, N., Petiwala, S., Wang, R., Lu, C., Hu, M., Ghosh, S., et al. (2019). Development of drug-inducible CRISPR-Cas9 systems for large-scale functional screening. *BMC Genomics*, *20*(1), 225. https://doi.org/10.1186/s12864-019-5601-9.

Sun, J., Wang, Q., Jiang, Y., Wen, Z., Yang, L., Wu, J., et al. (2018). Genome editing and transcriptional repression in Pseudomonas putida KT2440 via the type II CRISPR system. *Microbial Cell Factories*, *17*(1), 41. https://doi.org/10.1186/s12934-018-0887-x.

Tas, H., Nguyen, C. T., Patel, R., Kim, N. H., & Kuhlman, T. E. (2015). An integrated system for precise genome modification in Escherichia coli. *PLoS One*, *10*(9), e0136963. https://doi.org/10.1371/journal.pone.0136963.

Tong, Y., Charusanti, P., Zhang, L., Weber, T., & Lee, S. Y. (2015). CRISPR-Cas9 based engineering of actinomycetal genomes. *ACS Synthetic Biology*, *4*(9), 1020–1029. https://doi.org/10.1021/acssynbio.5b00038.

Tong, B., Dong, H., Cui, Y., Jiang, P., Jin, Z., & Zhang, D. (2021). The versatile type V CRISPR effectors and their application prospects. *Frontiers in Cell and Developmental Biology*, *8*, 1835. https://doi.org/10.3389/FCELL.2020.622103/BIBTEX.

Van der Oost, J., Westra, E. R., Jackson, R. N., & Wiedenheft, B. (2014). Unravelling the structural and mechanistic basis of CRISPR-Cas systems. *Nature Reviews. Microbiology*, *12*(7), 479–492.

Vento, J. M., Crook, N., & Beisel, C. L. (2019). Barriers to genome editing with CRISPR in bacteria. *Journal of Industrial Microbiology & Biotechnology*, *46*(9–10), 1327–1341. https://doi.org/10.1007/s10295-019-02195-1.

Vercoe, R. B., Chang, J. T., Dy, R. L., Taylor, C., Gristwood, T., Clulow, J. S., et al. (2013). Cytotoxic chromosomal targeting by CRISPR/Cas systems can reshape bacterial genomes and expel or remodel Pathogenicity Islands. *PLoS Genetics*, *9*(4), e100345. https://doi.org/10.1371/journal.pgen.1003454.

Volke, D. C., Friis, L., Wirth, N. T., Turlin, J., & Nikel, P. I. (2020). Synthetic control of plasmid replication enables target- and self-curing of vectors and expedites genome engineering of Pseudomonas putida. *Metabolic Engineering Communications*, *10*, e00126. https://doi.org/10.1016/J.MEC.2020.E00126.

Walton, R. T., Christie, K. A., Whittaker, M. N., & Kleinstiver, B. P. (2020). Unconstrained genome targeting with near-PAMless engineered CRISPR-Cas9 variants. *Science*, *368*(6488), 290–296. https://doi.org/10.1126/science.aba8853.

Wang, K., de la Torre, D., Robertson, W. E., & Chin, J. W. (2019). Programmed chromosome fission and fusion enable precise large-scale genome rearrangement and assembly. *Science (New York, N.Y.)*, *365*(6456), 922–926. https://doi.org/10.1126/science.aay0737.

Wang, S., Hong, W., Dong, S., Zhang, Z. T., Zhang, J., Wang, L., et al. (2018). Genome engineering of Clostridium difficile using the CRISPR-Cas9 system. *Clinical Microbiology and Infection, 24*(10), 1095–1099. https://doi.org/10.1016/J.CMI.2018.03.026.

Wang, B., Hu, Q., Zhang, Y., Shi, R., Chai, X., Liu, Z., et al. (2018). A RecET-assisted CRISPR-Cas9 genome editing in Corynebacterium glutamicum. *Microbial Cell Factories, 17*(1), 63. https://doi.org/10.1186/s12934-018-0910-2.

Wang, T., Li, Y., Li, J., Zhang, D., Cai, N., Zhao, G., et al. (2019). An update of the suicide plasmid-mediated genome editing system in *Corynebacterium glutamicum. Microbial Biotechnology, 12*(5), 907–919. https://doi.org/10.1111/1751-7915.13444.

Wang, Y., Wang, S., Chen, W., Song, L., Zhang, Y., Shen, Z., et al. (2018). CRISPR-Cas9 and CRISPR-assisted cytidine deaminase enable precise and efficient genome editing in Klebsiella pneumoniae. *Applied and Environmental Microbiology, 84*(23), e01834–18. https://doi.org/10.1128/AEM.01834-18.

Wasels, F., Jean-Marie, J., Collas, F., López-Contreras, A. M., & Lopes Ferreira, N. (2017). A two-plasmid inducible CRISPR/Cas9 genome editing tool for Clostridium acetobutylicum. *Journal of Microbiological Methods, 140*, 5–11. https://doi.org/10.1016/j.mimet.2017.06.010.

Wirth, N. T., Kozaeva, E., & Nikel, P. I. (2020). Accelerated genome engineering of *Pseudomonas putida* by I- *Sce* I—mediated recombination and CRISPR -Cas9 counterselection. *Microbial Biotechnology, 13*(1), 233–249. https://doi.org/10.1111/1751-7915.13396.

Wu, Y., Hao, Y., Wei, X., Shen, Q., Ding, X., Wang, L., et al. (2017). Impairment of NADH dehydrogenase and regulation of anaerobic metabolism by the small RNA RyhB and NadE for improved biohydrogen production in Enterobacter aerogenes. *Biotechnolgy for Biofuels, 10*, 248. https://doi.org/10.1186/s13068-017-0938-2.

Xin, H., Wan, T., & Ping, Y. (2019). Off-targeting of base editors: BE3 but not ABE induces substantial off-target single nucleotide variants. *Signal Transduction and Targeted Therapy, 4*(1), 9. https://doi.org/10.1038/s41392-019-0044-y.

Xu, X., Liu, Y., Du, G., Ledesma-Amaro, R., & Liu, L. (2020). Microbial chassis development for natural product biosynthesis. *Trends in Biotechnology, 38*(7), 779–796. https://doi.org/10.1016/j.tibtech.2020.01.002.

Yan, M. Y., Yan, H. Q., Ren, G. X., Zhao, J. P., Guo, X. P., & Sun, Y. C. (2017). CRISPR-Cas12a-assisted recombineering in bacteria. *Applied and Environmental Microbiology, 83*(17), e00947–17. https://doi.org/10.1128/AEM.00947-17.

Yu, D., Ellis, H. M., Lee, E.-C., Jenkins, N. A., Copeland, N. G., & Court, D. L. (2000). An efficient recombination system for chromosome engineering in Escherichia coli. *Proceedings of the National Academy of Sciences of United States of America, 97*(11), 5978–5983. https://doi.org/10.1073/pnas.100127597.

Zetsche, B., Gootenberg, J. S., Abudayyeh, O. O., Slaymaker, I. M., Makarova, K. S., Essletzbichler, P., et al. (2015). Cpf1 is a single RNA-guided endonuclease of a class 2 CRISPR-Cas system. *Cell, 163*(3), 759–771. https://doi.org/10.1016/j.cell.2015.09.038.

Zhang, S., & Voigt, C. A. (2018). Engineered dCas9 with reduced toxicity in bacteria: Implications for genetic circuit design. *Nucleic Acids Research, 46*(20), 11115. https://doi.org/10.1093/NAR/GKY884.

Zhang, L., Wang, P., Feng, Q., Wang, N., Chen, Z., Huang, Y., et al. (2017). Lipid nanoparticle-mediated efficient delivery of CRISPR/Cas9 for tumor therapy. *NPG Asia Materials, 9*(10), e441. https://doi.org/10.1038/am.2017.185.

Zhang, Y., Zhang, H., Wang, Z., Wu, Z., Wang, Y., Tang, N., et al. (2020). Programmable adenine deamination in bacteria using a Cas9-adenine-deaminase fusion. *Chemical Science*, *11*(6), 1657–1664. https://doi.org/10.1039/c9sc03784e.

Zhou, D., Jiang, Z., Pang, Q., Zhu, Y., Wang, Q., & Qi, Q. (2019). CRISPR/Cas9-assisted seamless genome editing in Lactobacillus plantarum and its application in N-acetylglucosamine production. *Applied and Environmental Microbiology*, *85*(21), e01367–19. https://doi.org/10.1128/AEM.01367-19.

Zhou, X. X., Zou, X., Chung, H. K., Gao, Y., Liu, Y., Qi, L. S., et al. (2018). A single-chain photoswitchable CRISPR-Cas9 architecture for light-inducible gene editing and transcription. *ACS Chemical Biology*, *13*(2), 443–448. https://doi.org/10.1021/acschembio.7b00603.

Zuo, E., Sun, Y., Wei, W., Yuan, T., Ying, W., Sun, H., et al. (2019). Cytosine base editor generates substantial off-target single-nucleotide variants in mouse embryos. *Science (New York, N.Y.)*, *364*(6437), 289–292. https://doi.org/10.1126/science.aav9973.

Towards a circular bioeconomy: Engineering biology for effective assimilation of cellulosic biomass

Marcos Valenzuela-Ortega[a],
Florentina Winkelmann[a], and Christopher E. French[a,b,*]

[a]*School of Biological Sciences, University of Edinburgh, Edinburgh, United Kingdom*
[b]*Joint Research Centre for Engineering Biology, Zhejiang University, Haining, Zhejiang, China*
[*]*Corresponding author: e-mail address: c.french@ed.ac.uk*

Abbreviations

AA	auxiliary activity
ADH	alcohol dehydrogenase
AZCL-HE-cellulose	Azurine-cross-linked hydroxyethylcellulose
BG	β-glucosidase
CAZY	carbohydrate-active enzymes database
CAZYme	carbohydrate active enzyme
CBD	cellulose-binding domain
CBH	cellobiohydrolase
CBM	carbohydrate-binding module
CBP	consolidated bioprocessing
CMC	carboxymethyl cellulose
CPEC	circular polymerase extension cloning
DNS	3,5-dinitrosalicylic acid
DoE	design of experiments
EG	endoglucanase
ER	endoplasmic reticulum
EX	endoxylanase
FP	filter paper
GFP	green fluorescent protein
GH	glycosyl hydrolase
GPI	glycosylphosphatidylinositol

Methods in Microbiology, Volume 52, ISSN 0580-9517, https://doi.org/10.1016/bs.mim.2023.01.004

HTS	high throughput screening
LPMO	lytic polysaccharide monooxygenase
OD	optical density
PASC	phosphoric acid swollen cellulose
PL	polysaccharide lyase
RBS	ribosome binding site
SEVA	standard European vector architecture
SLiCE	seamless ligation cloning extract
SN	supernatant
SP	secretion signal peptide
SSCF	simultaneous saccharification and cofermentation
SSF	simultaneous saccharification and fermentation

1 Introduction

There is currently a great deal of industrial and governmental interest in improving the sustainability of our society, especially through replacement of fossil carbon (coal, oil and gas) by renewable resources. For the manufacture of the many chemicals our society requires, this means either synthesis from CO_2 captured from the atmosphere, or synthesis from plant or algal biomass which has been generated by biological CO_2 fixation. Many important chemicals are currently manufactured biologically or from biological precursors, and many more could be, with improved process economics. However, such processes must be operated on an extremely large scale to be of any significance, and the most accessible forms of biological carbon available on large scales are sucrose, starch and vegetable oils, which also have critically important roles in the human food supply. Expansion of non-food uses for such materials is unthinkable, especially as the world requires larger amounts of food to feed a growing population. Hence, for use of renewable materials to be important in achieving sustainability, it is essential to move away from the use of starch and oils towards the use of non-edible biomass materials. These are essentially composed of polysaccharides (especially cellulose, hemicellulose and pectin in land plants, as well as others such as ulvan, carrageenan and agarose in algae) and lignin. Since these are structural materials in biology, they are much more difficult to process than starch or oils, which serve energy storage roles and as such are relatively easy to break down. This tension defines the essential problem of large-scale conversion of renewable biomass to fuels and feedstock chemicals without interference with the human food supply: how can the reduced carbon of structural biomass components be economically converted to useful products? In principle this can be accomplished by either biological or thermochemical processes. In this chapter, we will focus on biological processes (though in an industrial context, these often also require thermochemical pre-processing, known as pre-treatment). We should note that, to the best of our knowledge, this problem has not been solved in any satisfactory way (for a recent review, see Valenzuela-Ortega and French (2019)), despite its critical importance to the circular economy. Here we will present a summary of approaches that have been reported and new methods which may offer a path forward.

Many important products are manufactured from biological materials, often through the use of microorganisms. Some well-known examples include ethanol (*Saccharomyces cerevisiae*), butanol (*Clostridium acetobutylicum*), citric acid (*Aspergillus niger*), lactic acid (*Lactobacillus* spp.), glutamic acid (*Corynebacterium glutamicum*) and polyhydroxybutyrate (*Ralstonia* spp.) as well as important therapeutic products such as antibiotics. These processes have traditionally used wildtype or mutant strains of microorganisms, in highly engineered processes using the tools of chemical engineering. New tools developed for engineering biology have allowed remarkable progress in engineering microorganisms so as to improve process economics, as well as engineering new pathways in well-behaved industrial microorganisms (especially *S. cerevisiae*) to produce new chemicals such as isobutanol (Das, Patra, & Ghosh, 2020), farnesene, limonene, bisabolene (Walls & Rios-Solis, 2020) and many others, with excellent yields and titres. However, such processes are invariably based on sucrose or glucose (derived from starch) and attempts to base them directly on cellulosic biomass have given rather low yields and titres (for example: Bokinsky et al., 2011; Frederix et al., 2016).

In principle there are several ways forward. The most obvious would be to improve the current chemical and enzymic processes used to convert cellulosic materials to soluble sugars (especially D-glucose, D-xylose and L-arabinose) which can be recovered and used as the basis for growth media for production strains. Generally, these processes consist of pre-treatment (for example, steam explosion or ammonia fibre expansion) to disrupt lignin, followed by saccharification with exogenous enzyme blends mainly derived from *Trichoderma*. New pre-treatment approaches are a good starting point, since current methods require harsh chemicals or large amounts of energy and generate growth inhibitors; for example, acid hydrolysis can lead to furfural and 5-hydroxymethyl furfural from pentoses and hexoses, respectively (Hassan, Williams, & Jaiswal, 2018). In addition to improving pre-treatment of the feedstock and valorisation of lignin, new process variants also aim to reduce the cost of the whole transformation procedure. One such approach is *Simultaneous Saccharification and Fermentation (SSF)*, which combines enzymatic hydrolysis with microbial fermentation. This strategy increases cellulase activity by continuously removing hydrolysis products, which would otherwise inhibit the enzymes. A progression of this approach is *Simultaneous Saccharification and Co-Fermentation (SSCF)*, where non-glucose sugars from hemicellulose are also fermented. Currently these processes remain rather expensive, with knock-on effects on process economics. An alternative, which has attracted much research attention for several decades, is the prospect of engineering a single strain which can directly degrade biomass materials, producing its own degradative enzymes, and then use the resulting sugars to produce the desired product. This is known as *consolidated bioprocessing (CBP)* (Banner, Toogood, & Scrutton, 2021; Lynd, Weimer, van Zyl, & Pretorius, 2002; Salehi Jouzani & Taherzadeh, 2015). Since many natural microorganisms are capable of effective breakdown of cellulosic materials under suitable conditions, this does not seem an unreasonable aim. CBP might offer considerably simpler processes and improved economics, but requires either engineering a native cellulose-degrading host strain to produce desired products in high yields and titres, or engineering a tractable industrial strain (especially *Escherichia coli, Bacillus subtilis* or *S. cerevisiae*) to degrade and assimilate carbon from cellulosic biomass. Both approaches have proven rather challenging and neither has been particularly successful, to judge by reports in the literature (Valenzuela-Ortega & French, 2019).

Synthetic biology (engineering biology) can be defined as a multidisciplinary approach to science or engineering based on the generation of living systems (or systems based on biological components) with new-to-nature properties. When synthetic biology is applied to industrial biotechnology, it takes a top-down approach where an existing microorganism (chassis or host) is manipulated, altering its biochemical behaviour to generate the desired product from the available substrate. This is achieved via multiple means: modification of genetic elements native to the host, introducing metabolic pathways or functions present in other organisms or entirely new to nature (Peretó, 2020), and sometimes changes to the biochemical conditions beyond those encoded genetically (Sadler, 2020). Synthetic biology approaches the engineering of the biological host through the *"design-build-test-learn"* (DBTL) research cycle (Fig. 1).

In principle, such tools and approaches should be highly suited to the development of organisms for CBP (Banner et al., 2021; French, 2009). In this chapter we will consider methods that have been applied to these purposes, limiting factors, and possible solutions which may be explored in the future.

1.1 Defining the problem: The structure of biomass

The food components of plants (sucrose, starch and oils) are easily processed by enzymes as part of their role in energy storage *in planta*, and of course, this is the basis for their use in food. The non-food components are structural and are designed by evolution to be extremely difficult to break down. The most widely available types of non-food biomass are the inedible parts of crop plants, or non-food crops specifically grown as a source of biomass. The carbon in these is mainly in the form of plant cell walls, which have a characteristic structure, though with considerable variety in different taxonomic groups. The main structural element is *cellulose*, a long polymer of D-glucose units linked by β-(1,4)-glycosidic bonds. This bonding causes each residue to be rotated by 180° relative to its neighbours. This configuration allows strong intrachain hydrogen bonds, tending to hold the chain in a rigid extended conformation exposing hydrophobic surfaces, and also allows formation of interchain hydrogen bonds stacking multiple chains together in an ordered array. This is very different from the loose configuration adopted by glucan chains with the α-(1,4) and α-(1,6) bonds of amylose and amylopectin (starch), and as a result, cellulose is very insoluble in water and most solvents, though it can be dissolved in certain ionic liquids and other complex solvents, and this can be useful in processing and fractionation of

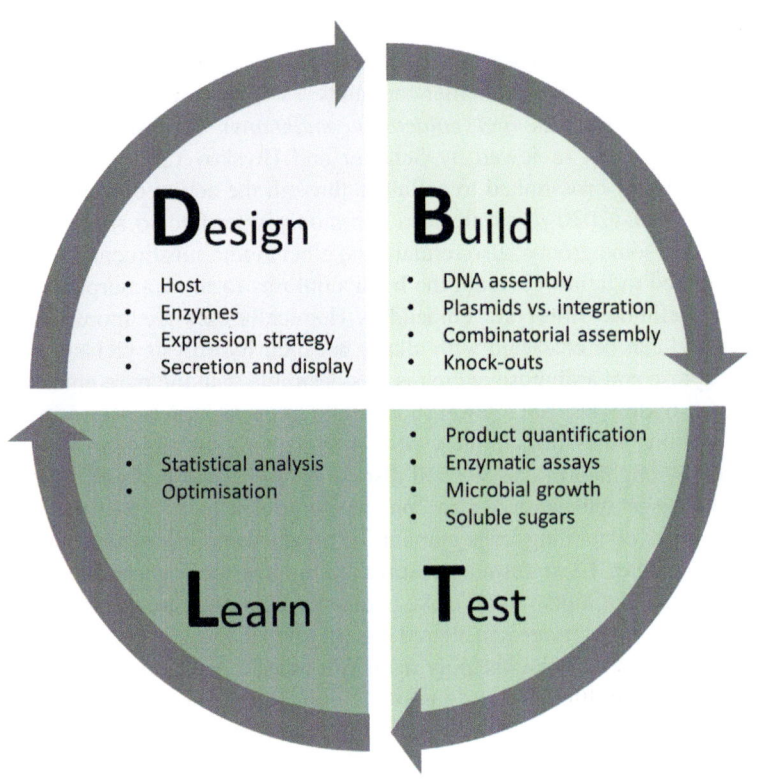

FIG. 1

The design, build, test, learn cycle. Development of an effective strain for cellulosic biomass degradation is likely to undergo several iterations of this cycle.

biomass (Morais et al., 2020; Satari, Karimi, & Kumar, 2019). Even cellobiose and cello-oligosaccharides are much less water-soluble than maltose and maltodextrins. In cellulose, individual cellulose chains, with a DP (degree of polymerization) of several thousand, are arrayed in groups of approximately 24 (Haigler & Roberts, 2019) by the action of the cellulose synthase complexes which produce and secrete them, forming microfibrils in which the chains are ordered and aligned. These join together to form fibrils, which are the structural basis of the plant cell wall. Some regions of the fibrils have a very ordered structure and are known as crystalline regions; others are less ordered, known as amorphous regions, and are generally more accessible to enzymes and other treatments.

In native biomass, cellulose fibrils are further surrounded by various other polysaccharides generically known as *hemicelluloses*. These generally have much shorter chains, often branched, and vary in type in different plants. Common types include *xylans*, with a backbone of β-(1,4)-linked D-xylose and side chains of glucuronic acid (*glucuronoxylans*) or arabinose (*arabinoxylans*) or both (*glucuronoarabinoxylans*), and *xyloglucans*, with a backbone of β-(1,4)-linked D-glucose and D-xylose with D-xylose sidechains, with smaller amounts of mannose-containing polymers (*mannans, glucomannans and galactoglucomannans*). The structures of hemicelluloses have been reviewed by Scheller and Ulvskov (2010). Hemicelluloses may be covalently crosslinked to cellulose through the action of transglycosylases (Herburger et al., 2020), and are also commonly esterified to some degree with acetate, and in some groups, also ferulate and other acidic substituents, and may also be cross-linked to lignin. Overall, the hemicelluloses form an amorphous matrix in which the cellulose fibres are embedded. Hemicelluloses are more soluble than cellulose and can be extracted with alkali, but their hydrolysis yields a mixture of glucose, xylose and arabinose, which is less desirable than the pure glucose yielded by cellulose hydrolysis, since fewer industrially important microorganisms have the native capability to assimilate these pentose sugars, though this capacity has been introduced by engineering, as we will discuss below. Many types of biomass, such as fruit and vegetable wastes, also contain significant amounts of *pectins*, acidic polysaccharides containing large amounts of D-galacturonic acid along with L-arabinose and D-xylose. These form a gel surrounding other polysaccharide components (Jayani, Saxena, & Gupta, 2005). Like hemicelluloses, they are easily extracted but their hydrolysis products are considered less useful than those of cellulose. Finally, in most types of plant material, the polysaccharides are embedded in a matrix of *lignin*, a three-dimensional polymer of aromatic monomers, predominantly coniferyl, sinapyl and paracoumaryl alcohols, linked randomly by ether bonds introduced by radical polymerization (Atiwesh, Parrish, Banoub, & Le, 2022). Lignin may also be covalently linked to polysaccharides via ferulic acid ester bonds. Disruption of lignin is required to expose polysaccharide chains for attack, and this is the principal reason why expensive pre-treatment is required. Of energy crops and other relevant biomass sources, hemicellulose represents 16–33% of the dry weight, lignin 9-32%, and cellulose 37–52% (Ragauskas et al., 2014).

Algal biomass is also a potentially attractive feedstock as there is a great deal of scope for expansion of the supply by harvest as well as onshore and offshore farming. Algae, especially marine algae, grow rapidly and do not require farmland or fresh water (Zaky, Carter, Meng, & French, 2021) and their rapid growth can draw down CO_2 from the atmosphere in globally significant quantities (Yong, Thien, Rupert, & Rodrigues, 2022). Algae generally contain cellulose and hemicellulose but not lignin, which may reduce the costs of pre-treatment. In addition, they contain characteristic acidic and sulphated polysaccharides which vary according to the group. The main groups of macroalgae are the green, red and brown seaweeds (Chlorophyta, Rhodophyta and Phaeophyta). Among other distinctive polysaccharides, Chlorophyta contain *ulvan*, a polymer of L-idose and L-iduronic acid; Rhodophyta contain *agar* and *carrageenan*, sulphated polymers of galactose and anhydrogalactose; and Phaeophyta contain *alginate*, a polymer of D-mannuronic and L-guluronic acids (Bäumgen, Dutschei, & Bornscheuer, 2021; Li, He, Liang, Peng, & Hu, 2022). The unusual sugars in algal polysaccharides may pose problems for assimilation by commercially useful organisms, but in principle these could be overcome in the same way as has been accomplished for xylose and arabinose, as described below.

1.2 Defining the scope of solutions: Enzymic depolymerization of biomass

Despite the recalcitrance of cellulosic biomass, many microorganisms in nature are capable of effectively degrading it and assimilating the resulting products, and in warm and damp environments, such as the rumen of ruminant mammals and the floors of tropical forests, biomass is rapidly degraded. The organisms responsible include a variety of bacteria and fungi belonging to several groups. Degradation is accomplished by means of secreted enzymes known generically as cellulases and hemicellulases. This process has been studied for decades and a paradigmatic understanding of the process was developed (Lynd et al., 2002), based mainly on the cellulase systems of the ascomycetous fungus *Trichoderma reesei*, which was isolated in New Guinea during the second World War, and has become the parent of most commercial cellulase-producing strains due to its production and secretion of abundant cellulases (Bischof, Ramoni, & Seiboth, 2016; Gupta et al., 2016). The paradigmatic process proceeds as follows (Fig. 2) (Lynd et al., 2002). Initially hemicelluloses are degraded by various types of hemicellulase, including *endoxylanases* (which cleave internal xylose-xylose bonds in xylan chains), *exoxylanases* (which cleave off terminal xylose residues), arabinanases, and *acetyl esterases* which remove ester-linked substituents. Pectins are degraded by hydrolytic *polygalacturonases*

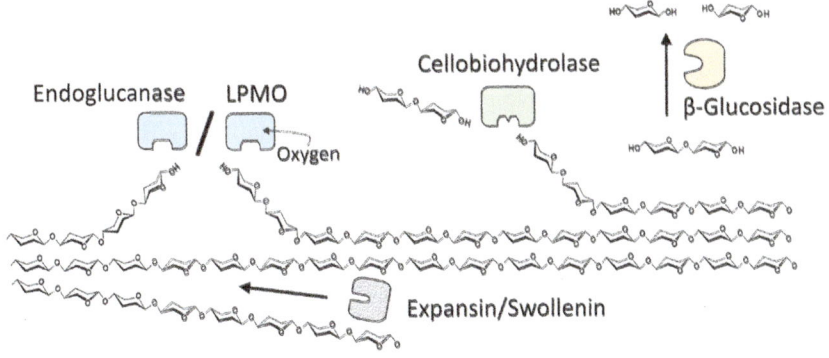

FIG. 2

Schematic of cellulose breakdown. Expansins/swollenins open crystalline microfibres of cellulose increasing access to hydrolytic enzymes such as endoglucanase and oxidative enzymes such as Lytic Polysaccharide Monooxygenases (LPMO) which attack fibres in the middle of cellulose chains. Cellobiohydrolases release cellobiose units from either reducing or non-reducing ends. Lastly, β-glucosidases cleave cellobiose to form glucose monomers.

as well as by *pectic lyases*, which cleave uronic acid-linked bonds to generate products with unsaturated bonds (Jayani et al., 2005). Acidic algal polymers are similarly degraded by lyases such as *alginate lyase* and *ulvan lyase*, and also require removal of sulphate esters by *sulphatases* (Bäumgen et al., 2021; Li et al., 2022). Thus the structural load-bearing cellulose fibres are revealed and made accessible, although only at their surfaces. *Endoglucanases* (EC3.2.1.4) now bind to these surfaces and cleave random internal bonds revealing a reducing (C1) and non-reducing (C4) end. These exposed ends are then attacked by *exoglucanases*, also known as *cellobiohydrolases*, which are generally processive, moving along the chains and releasing successive molecules of cellobiose. The exoglucanases come in two types: those which attack the reducing end (EC3.2.1.176), and those which attack the non-reducing ends (EC3.2.1.91); it is postulated that they may work in pairs with complementary activities (Gilkes et al., 1997). In addition to their catalytic domains, which are classified into families of *Glycosyl Hydrolases* (*GH*), both endoglucanases and exoglucanases generally possess separate *cellulose-binding domains* (*CBD*), known more generally as *carbohydrate-binding modules* (*CBM*), which are attached to the catalytic GH domains by flexible linkers and are believed to aid in binding of the insoluble substrate (Boraston, Bolam, Gilbert, & Davies, 2004). Finally, the soluble product cellobiose may be assimilated directly in some organisms; in others it is hydrolysed to glucose by secreted *β-glucosidases* (EC3.2.1.21). This is necessary to prevent product inhibition of the exoglucanases.

T. reesei has been useful for research and industrial production of cellulase because it secretes all the enzymes required to hydrolyse cellulose completely to glucose; however, the mechanisms to digest cellulose are more diverse in different organisms. Cellodextrin glucohydrolases (also named *cellodextrinases*, EC 3.2.1.74) can release glucose directly from cellooligosaccharides (Liu, Bevan, & Zhang, 2010). The bacterium *Cytophaga hutchinsonii* glides along cellulose fibres, degrading it with enzymes attached to the cell-wall or secreted to the periplasm. *C. hutchinsonii* does not possess any identifiable exoglucanase; instead, it uses endoglucanases and cellodextrin glucohydrolases such as Chu2268 (Zhu & McBride, 2017). Other organisms are capable of importing and degrading cellooligosaccharides or cellobiose intracellularly, either by hydrolysis or by phosphorolysis (Parisutham, Chandran, Mukhopadhyay, Lee, & Keasling, 2017).

An additional complication is that in some organisms (paradigmatically, the anaerobic Gram-positive bacteria such as *Clostridium* spp.), the various enzyme activities are not secreted free into the medium, but rather remain bound to the cell surface in complex structures known as *cellulosomes*. In the cellulosome of the model organism *Clostridium thermocellum* (Barth et al., 2018), cellulases have a *cohesin* domain that binds to one of the multiple *dockerin* domains of a *scaffoldin* protein, which has a cohesin domain that attaches the complex to the bacterial cell wall (Fig. 3). Since the scaffoldin also contains carbohydrate-binding modules, this obviates the requirement for CBMs on each enzyme. Co-localisation of enzymes and proximity to the cell wall makes cellulosomes very efficient at biomass degradation and absorption of resulting soluble sugars. It is conjectured, with some supporting

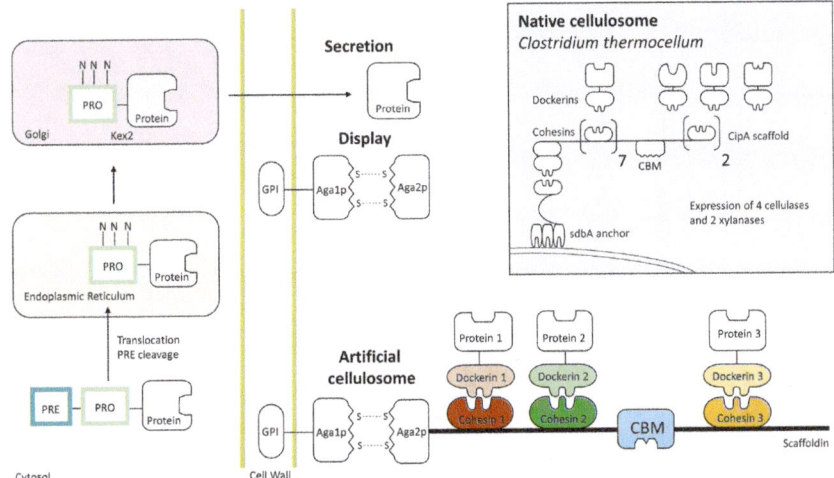

FIG. 3

Schematic diagram of cellulosomes and *S. cerevisiae* secretion and surface display. For secretion the protein with the secretion signal (PRE and PRO regions) is processed through the Endoplasmic Reticulum, where it is glycosylated (N), and the Golgi, where the PRO region is cleaved. If the protein of interest is fused with Aga2p this can bind Aga1p (inserted into the cell wall via a GPI anchor) for display. The protein can alternatively be fused to dockerins to be able to bind cohesins expressed together with a carbohydrate binding module (CBM) in scaffoldin which is fused to Aga2p to form an artificial cellulosome. An example of a native cellulosome from *C. thermocellum* is shown in the grey box.

Adapted from Chang, J. J., Anandharaj, M., Ho, C. Y., Tsuge, K., Tsai, T. Y., Ke, H. M., et al. (2018). Biomimetic strategy for constructing Clostridium thermocellum cellulosomal operons in Bacillus subtilis. Biotechnology for Biofuels and Bioproducts, 11, 157. doi:10.1186/s13068-018-1151-7.

evidence, that the physical proximity of the different activities associated with the cellulosome aids in their synergistic activity, and that their proximity to the cell results in a greater recovery of the soluble sugars thus released. This may be assumed to be particularly important in anaerobic contexts, where the energy available from each sugar residue assimilated is much less than in aerobic environments.

An extremely useful resource in planning CBP experiments is CAZy, the database of Carbohydrate Active Enzymes (CAZymes) (Cantarel et al., 2009). The enzymes described above mainly fall into the classification of *Glycosyl Hydrolases (GH)*, and polysaccharide Lyases (PL), with many also having CBM domains. At the time of writing, CAZy includes 173 families of GH, 42 families of PL, and 93 families of CBM, though most of the best studied enzymes in cellulosic biomass degradation fall into a handful of these families such as GH3, GH6 and GH7 (see Table 1).

Table 1 Examples of enzymes used in recent genome engineering experiments for consolidated bioprocessing.

Enzyme	Source	Type	Structure	References
EGI (Cel7B)	*Trichoderma reesei*	Endoglucanase	GH7-CBM1	Guo, Duquesne, Bozonnet, Cioci, et al. (2017), Guo, Duquesne, Bozonnet, Nicaud, et al. (2017)
EGII (Cel5A)	*Trichoderma reesei*	Endoglucanase	CBM1-GH5	Nakatani, Yamada, Ogino, and Kondo (2013)
CBHI (Cel7A)	*Trichoderma reesei*	Exoglucanase (cellobiohydrolase)	GH7-CBM1	Guo, Duquesne, Bozonnet, Cioci, et al. (2017)
CBHII (Cel6A)	*Trichoderma reesei*	Exoglucanase (cellobiohydrolase)	CBM1-GH6	Nakatani et al. (2013); Guo, Duquesne, Bozonnet, Cioci, et al. (2017); Guo, Duquesne, Bozonnet, Nicaud, et al. (2017)
CBHI	*Talaromyces (Rasamsonia) emersonii*	Exoglucanase (cellobiohydrolase)	GH7	Liu et al. (2015)
CBHI	*Neurospora crassa*	Exoglucanase (cellobiohydrolase)	GH7-CBM1	Guo, Duquesne, Bozonnet, Cioci, et al. (2017); Guo, Duquesne, Bozonnet, Nicaud, et al. (2017)
BGLI (XP_006964076)	*Trichoderma reesei*	β-Glucosidase	GH3	Kickenweiz, Glieder, and Wu (2018)
XP_001398816	*Aspergillus niger*	β-Glucosidase	GH3-FN3	Kickenweiz et al. (2018)
BGLI	*Aspergillus aculeatus*	β-Glucosidase	GH3-FN3	Nakatani et al. (2013)
BGLI	*Saccharomyces fibuligera*	β-Glucosidase	GH3-FN3	Tang et al. (2018)
CelA	*Clostridium thermocellum*	Endoglucanase	GH8-dockerin1	Chang et al. (2018); Tozakidis, Brossette, Lenz, Maas, and Jose (2016)
CelR	*Clostridium thermocellum*	Endoglucanase	GH9-CBM3-dockerin1	Chang et al. (2018)
CelK	*Clostridium thermocellum*	Exoglucanase (cellobiohydrolase)	CBM4-GH9-dockerin1	Chang et al. (2018); Daas et al. (2018); Tozakidis et al. (2016)
CelS	*Clostridium thermocellum*	Exoglucanase (cellobiohydrolase)	Gh48-dockerin1	Chang et al. (2018); Daas et al. (2018)
BglA	*Clostridium thermocellum*	β-Glucosidase	GH1	Tozakidis et al. (2016)
CenA	*Cellulomonas fimi*	Endoglucanase	CBM2-GH6	Hetzler, Bröker, and Steinbüchel (2013);Lakhundi et al. (2017)

Name	Organism	Function	Domain structure	Reference
CenB	Cellulomonas fimi	Endoglucanase	GH9-CBM3-FN3-CBM2	Hetzler et al. (2013)
CenC	Cellulomonas fimi	Endoglucanase	CBM4/9-GH9	Hetzler et al. (2013)
CbhA	Cellulomonas fimi	Exoglucanase (cellobiohydrolase)	GH6-FN3-CBM2	Hetzler et al. (2013)
Cex	Cellulomonas fimi	Exoglucanase/xylanase	CBM2-GH10	Lakhundi et al. (2017)
Cel	Bacillus sp. D04	Endoglucanase/exoglucanase	GH5-CBM3	Bokinsky et al. (2011)
XynB	Clostridium stercorarium	Xylanase	GH10	Bokinsky et al. (2011)
Cel3A	Cellvibrio japonicus	β-Glucosidase	GH3	Bokinsky et al. (2011)
Cel3B	Cellvibrio japonicus	β-Glucosidase	GH3	Bokinsky et al. (2011)
Gly43F	Cellvibrio japonicus	β-Xylosidase	GH43	Bokinsky et al. (2011)
BglC	Thermobifida fusca	β-Glucosidase	GH1	Gao, Luan, Wang, Liang, and Qi (2015)
Cel6A	Thermobifida fusca	Endoglucanase	GH6	Hetzler et al. (2013)
Cel-CD	Bacillus sp. Z16	Endoglucanase	GH5-CBM17-CBM28	Gao, Luan, et al. (2015)
SWO1	Trichoderma reesei	Swollenin	CBM1-expansin	Guo, Duquesne, Bozonnet, Nicaud, et al. (2017); Nakatani et al. (2013)
Aoelp1	Aspergillus oryzae	Expansin-like protein	(unclear)	Nakatani et al. (2013)
LPMOA	Trichoderma reesei	LPMO	GH61-CBM1	Guo, Duquesne, Bozonnet, Nicaud, et al. (2017)
GH61a	Thermoascus aurantiacus	LPMO	GH61	Liang, Si, Ang, and Zhao (2014)

GH, glycosyl hydrolase domain; CBM, carbohydrate binding module; FN, fibronectin-like domain (function unknown).

In the above description we have not mentioned lignin. In the case of lignified materials, this must initially be degraded to expose the polysaccharides. In industrial processes this necessity may be obviated by physico-chemical pre-treatment, but in nature lignin is degrade enzymically. In the case of fungi, especially the white-rot fungi, lignin is degraded by oxidative enzymes such as *lignin peroxidase, manganese peroxidase*, and *"versatile peroxidase"*, as well as *laccases* (Atiwesh et al., 2022; Silva, Ticona, Hamann, Quirino, & Noronha, 2021). These enzymes are classified by CAZy as families *AA* (*Auxiliary Activities*) AA1 and AA2. Bioprocesses based on such enzymes may also lead to production of valuable compounds from lignin (Weng, Peng, & Han, 2021); however, lignin degrading enzymes have not generally been included in the construction of organisms for consolidated bioprocessing.

The above description has also been supplemented recently by the discovery of several auxiliary components whose importance is still under assessment. In particular we may consider the *expansins* (found in plants, as well as some bacteria, such as EXLX1/YoaJ of *Bacillus subtilis*) and *swollenins* (found in fungi), which are related to GH45 glycosyl hydrolases (Cosgrove, 2017; Martinez-Anaya, 2016) but lack critical catalytic residues; they are believed to function in reducing the crystallinity of cellulose, converting ordered regions to amorphous regions and thus enhancing degradation (with the plant expansins presumably playing a role in growth). It has also been reported that CBM domains of cellulases may have amorphogenic activity (Bernardes et al., 2019). Another major discovery is several new classes of oxidative enzymes known collectively as *Lytic Polysaccharide Monooxygenases* (*LPMO*). These were initially considered to be glycosyl hydrolases (family GH61) or carbohydrate-binding modules (CBM33) but have now been demonstrated to have oxidative activity, using oxygen, or preferably hydrogen peroxide, to cleave polysaccharide chains (Bissaro, Várnai, Røhr, & Eijsink, 2018). LPMO are copper-containing oxidoreductases that cleave cellulase by oxidation of either the C1 or C4 carbon of a β-1,4-glycosidic bond, which then spontaneously breaks. In addition to O_2, the LPMO catalytic cycle requires a reductant, which may include lignin degradation products (Brenelli, Squina, Felby, & Cannella, 2018) or electrons received from reductant oxidases such as cellobiose dehydrogenase (Loose et al., 2016). Alternatively, a recent study has also shown that these enzymes might replace O_2 and the additional reductant with H_2O_2, just requiring reductant to reduce Cu^{2+} to Cu^{1+} (Kuusk et al., 2019; Müller, Chylenski, Bissaro, Eijsink, & Horn, 2018). LPMO act in synergy with other cellulases increasing the overall hydrolysis (Müller et al., 2018); however, the need for oxygen might limit their activity to aerobic conditions. LPMO have now been reclassified in the CAZy database as Auxiliary Activities, with GH61 becoming AA9 and CBM33 becoming AA10; several more families of LPMO have also been discovered (families AA11 and AA13 to AA17).

2 Design

From the above it can be seen that engineering a non-cellulolytic organism to degrade cellulose in a pure culture is not a trivial challenge, and may require expression of many different components. We will now move on to consider these design issues in more detail.

2.1 Design—Choosing a host

The "Design" step comprises selecting a relevant microorganism or "chassis" and planning which set of genetic modifications are required to achieve the desired function and how to incorporate them. Previous knowledge (such as genomics and biochemical data) and modelling (systems biology) allow prediction of which (genetic) modifications are likely to result in the desired behaviour. However, because of the complexity of living systems, it is not possible to predict the effect of specific genetic modifications (Chao, Mishra, Si, & Zhao, 2017; Del Vecchio, 2015; Gyorgy et al., 2015). The complexity of interactions will increase with the complexity of the genetic construct introduced into the host, and consequently, multiple iterations of the research cycle might be necessary to achieve the desired effect in an optimal way (Naseri & Koffas, 2020).

The first step in the design process is to choose your host organism, also known as a chassis. For CBP, in principle, there are three possibilities:

1. Choose a cellulose degrading host and introduce a pathway for forming the desired product. Unfortunately, most effective cellulose-degrading organisms are rather challenging to engineer and are not necessarily well suited to current large scale industrial processes. Ultimately this may well be the best choice, but since the focus of this chapter is on engineering genomes for the degradation and assimilation of biomass, we will not consider it further here.
2. Choose a host which natively produces a useful product at high yields and titres, and which is suitable for large scale processes, and engineer it to degrade biomass and assimilate the components.
3. Choose a host which is genetically tractable and suitable for large scale processes, and engineer it to both degrade/assimilate biomass and produce a useful product.

In practice there is a considerable overlap between options 2 and 3, since industrially useful organisms have often been intensively studied and genetic modification systems have been developed for them to improve production; conversely, well studied laboratory organisms have often been applied to develop new industrial processes. By far the most important organism in this context is the yeast *Saccharomyces cerevisiae*, which is used industrially on an enormous scale for the manufacture of ethanol and bread, as well as other products such as flavourings derived from yeast extract, and more recently, recombinant protein therapeutics. *S. cerevisiae* is a robust industrial host and methods for very large scale cultivation are well developed. It is also Generally Regarded As Safe for food-related uses. Due to its ubiquity, *S. cerevisiae* has also become an important laboratory model organism, and methods for its genetic modification have been developed, up to the current Sc2.0 project which aims to synthesise its entire genome (Pretorius & Boeke, 2018). *S. cerevisiae* has also been engineered to produce many other products at high yield and titre, including farnesene, squalene, bisabolene, and isobutanol, and is the focus of many similar current projects to produce molecules such as cannabidiol and opiates (Jensen & Keasling, 2015). *S. cerevisiae* cannot natively assimilate cellobiose, arabinose or xylose, but these capabilities have been introduced by genetic modification

(Endalur Gopinarayanan & Nair, 2019) with the aim of generating strains which can be used for cellulosic ethanol production on an industrial scale (Kickenweiz et al., 2018). Other yeasts have also been used for biomass degradation experiments, including *Pichia pastoris* (*Komagataella phaffii*) (Kickenweiz et al., 2018; Shin et al., 2015) and the oil-producing yeast *Yarrowia lipolytica* (Guo, Duquesne, Bozonnet, Cioci, et al., 2017; Wei et al., 2014).

Among bacteria, the most frequently used host is *Escherichia coli. E. coli* is a very important laboratory organism, with many genetic tools available, and is almost always the first choice for genetic modification experiments; even where it is not the ultimate host, intermediate genetic constructs are almost always made in *E. coli. E. coli* has also been genetically modified to produce many other products such as ethanol, acetone, butanol, isobutanol, fatty acid ethyl esters (microdiesel), and many others (Zhao, Zhang, & Li, 2019). *E. coli* is also capable of anaerobic growth, and natively assimilates hemicellulose sugars such as arabinose, xylose and mannose as well as glucose, but not cellobiose, though this capacity is easy to add either by genetic modification or mutation (Kachroo, Kancherla, Singh, Varshney, & Mahadevan, 2007). Close relatives of *E. coli* such as *Citrobacter freundii* (Lakhundi et al., 2017) and *Klebsiella oxytoca* (Zhou & Ingram, 2001), which can use the same genetic tools but natively assimilate cellobiose, have also been proposed for such use. Another interesting Gram negative host is *Pseudomonas putida*, which is less closely related to *E. coli*, less versatile in use of sugars, and does not grow well in anaerobic conditions, but has superior tolerance to inhibition by toxic chemicals, which may be advantageous in some applications (Tozakidis et al., 2016).

Another important bacterial host is the model Gram positive organism *Bacillus subtilis*. Genetic tools are also well developed for *B. subtilis*, and it is used commercially in the manufacture of enzymes (proteases and amylases). Compared to *E. coli*, *B. subtilis* has the advantage of natively secreting enzymes much better, and having a cell envelope more similar to that of the highly effective cellulose-degrading cellulosome-producing organisms such as *Clostridium thermocellum* and *Clostridium cellulolyticum*, and can be engineered to produce synthetic cellulosomes (Chang et al., 2018) as described below. *B. subtilis* is capable of assimilating most biomass-derived sugars, though it is not capable of good anaerobic growth under normal conditions. *B. subtilis* strains are also used in the manufacture of foodstuffs such as natto, and as such are Generally Regarded As Safe, like *S. cerevisisae*. Other Grampositive bacterial hosts which have been investigated for CBP include *Geobacillus thermoglucosidasius* (Daas et al., 2018), a thermophilic relative of *B. subtilis*, as well as other industrially useful producer organisms such as *Lactobacillus*, producing lactic acid (Stern et al., 2018) and *Corynebacterium glutamicum*, producing glutamic acid and other amino acids (Anusree, Wendisch, & Nampoothiri, 2016).

2.2 Design—Choosing enzymes

The next step is to choose the set of enzymes to be expressed. Here we will focus on the degradation of cellulose, the most appealing target component of biomass, due to its homogeneous composition and release of glucose as a product. Hemicellulose is more soluble and degradable, but due to its heterogeneous structure, requires a large battery of different enzymes and releases a mixture of sugars and other products, many of which are not used by common industrial microorganisms (though it appears to be considerably easier to engineer this capacity than degradation of cellulose). From the discussion above, it would seem that a minimal set of enzymes required for cellulose degradation, following suitable pre-treatment to expose the polysaccharides, would consist of at least one endoglucanase, one or two exoglucanases (perhaps one reducing-end-directed and one non-reducing-end-directed), and either a secreted β-glucosidase or an uptake system for cellobiose (if not already present in the host). A wide variety of enzyme combinations have been tested, and those used in some recent reports are summarized in Table 1. Considering the numerous families of enzymes involved in natural biomass degradation, as seen in the CAZy database, it is clear that there is a large enzyme space still to explore, especially when the synergistic effects of different enzyme blends are considered, and here engineering biology concepts such as large-scale combinatorial assembly (discussed below under "Build") may play an important role. It is also interesting to note the modular nature of many of these enzymes, which lends itself to domain swapping and fusion experiments (Duedu & French, 2016).

2.3 Design—Secretion and surface display

For degradation of insoluble extracellular substrates, it is normally necessary that the enzymes should be secreted, either freely into the medium, or attached to the surface of the cell. *E. coli* has been used extensively for recombinant protein production, but this bacterium is not generally considered a good protein secretor because it has an outer membrane and most strains lack the Type II Secretion System (also known as the Main Terminal Branch of the General Secretory Pathway). However, there are plenty of examples demonstrating good production of secreted recombinant proteins (Kleiner-Grote, Risse, & Friehs, 2018) where the peptide of interest is secreted due to the host cellular secretion machinery recognising the secretion signal peptide (SP). In some cases, the SP leads to the periplasmic space; in other cases, secretion to the extracellular medium occurs through known or unknown mechanisms. The signal peptide can be native to the protein of interest, or it can be taken from a different gene and fused to the protein cargo to be secreted. In some cases, the protein of interest is fused to a whole protein known to be secreted to the extracellular medium by *E. coli.*

Different known mechanisms translocate proteins through the inner or outer membrane of bacteria (Burdette, Leach, Wong, & Tullman-Ercek, 2018; Green & Mecsas, 2016; Kleiner-Grote et al., 2018). The most successfully used for secretion of recombinant proteins are the Tat and Sec routes, found in all domains of life that, in *E. coli,* transport proteins into the periplasmic space. Both these pathways use N-terminal secretion signal peptides (SP) to recognise which proteins to secrete. The Tat pathway translocates proteins after folding in the cytoplasm, whereas the Sec route translocates proteins before folding with the SecYEG channel. The Sec route is often sub-divided depending on whether the protein is translocated post-translationally (SecA/SecB route, such as with *E. coli* proteins DsbA, TorT, TolB) or co-translationally (Signal Recognition Particle or SRP route, such as with MalE or maltose-binding protein) (Lee, Lee, Lee, Yim, & Jeong, 2016). The name of the Tat route comes from the twin-arginine motif in the signal peptide in proteins transported through this route (twin-arginine translocation). Sec signal peptides are generally short (~20 residues) and contain a similar structure: an N-terminal region with positively charged residues, a central hydrophobic region recognised by either SRP or SecA, and a mainly polar C-terminal region that contains the signal cleavage motif.

In *S. cerevisiae*, recombinant protein secretion is normally achieved by use of the mating peptide secretion N-terminal signals. Post-translation, the alpha mating factor 1 leader peptide (MFα1pp) directs the protein to the endoplasmic reticulum. This peptide has two functional regions, the pre and pro regions. During the process of extrusion into the ER, the proprotein is formed via the cleavage of the pre region by the signal peptidase. Three asparagine residues within the pro sequence are then glycosylated. It has been shown that removal of these glycosylation sites significantly slows secretion but does not abolish it (Caplan, Green, Rocco, & Kurjan, 1991). Export from the ER to the Golgi is complex and dictated by many factors including glycosylation and protein stability and has been summarised by Ellgaard and Helenius (2003). Within the Golgi, the Kex2 endoprotease cleaves the pro-region resulting in the mature protein ready to be secreted. A mutant version of MFα1pp generated via directed evolution showed reduced Kex2 cleavage efficiency, slowing down maturation but, counter-intuitively, improving overall secretion 16-fold compared to the wild type (Rakestraw, Sazinsky, Piatesi, Antipov, & Wittrup, 2009).

Alternative signal peptides originating from other organisms can also be used to target proteins for secretion in *S. cerevisiae*. Examples include the leader sequence from *Candida albicans*'s glucoamylase (Livi et al., 1991) or that of inulinase from *Kluyveromyces marxianus* (Kang, Nam, Kwon, Chung, & Yu, 1996). Synthetic leaders have been engineered to further improve secretion levels, however which leader to use appears to be context dependent. Overall, secretion is complex. Optimisation, as summarised by Hou, Tyo, Liu, Petranovic, and Nielsen (2012) can be at the DNA, RNA or protein or secretion system level and depends on the protein of choice.

Instead of secretion, proteins can also be displayed on the surface of *S. cerevisiae*. This can improve enzyme stability, potentially reduce costs for purification of proteins, and generate benefits from co-localisation of substrate and enzymes. Several methods exist that allow proteins to be anchored into the cell wall. While different anchor proteins can be fused to the target protein, most contain a putative glycosyl-phosphatidylinositol (GPI) anchor motif. Designing the fusion of target and anchor protein should be carefully considered. There are several native *S. cerevisiae* anchor proteins that can be used including α-agglutinin, Cwp1p, Cwp2p, Tip1p, Flo1p (Van der Vaart et al., 1997), Aga1p and Dan4p (Yang et al., 2019). Which anchor is used can impact total amount of protein displayed. Which end of the protein of interest it is fused to can have implications on enzyme activity, e.g., if the active site is close to the C-terminal, then a C-terminal fusion can block activity. Use of linkers can help improve enzyme activity and can also be a target for optimisation (Tanaka & Kondo, 2015).

The Aga1p/Aga2p anchoring system from α- agglutinin is a popular mechanism of anchoring proteins and consists of two components each with their native secretion signals. Hereby Aga1p is expressed and integrates into the cell wall via the GPI anchor. Aga2p is expressed as an in-frame fusion to the protein of choice and is secreted. Aga2p then binds to Aga1p via two disulphide bonds (Boder & Wittrup, 1997). Alternatively, a one component system can be used. Here the alpha factor secretion signal sequence is expressed followed by the 3' half sequence of agglutinin (AGα1) fused to the protein of choice (Wei, Zhang, Guo, & Ma, 2016). Often the protein of choice is further fused with a c-myc tag to confirm surface display via immunofluorescent staining.

Artificial cellulosomes are a yeast surface display strategy which allows the co-localisation of multiple enzymes via a scaffold. This consolidates bioprocessing into a mulitenzymatic complex to improve pathway *efficiency* especially popular for the expression of cellulases. In principle, a cellulosome is constructed by using an anchor protein such as Aga1p that allows surface display of a scaffold. This scaffold consists of a series of proteins such as tAga1 (Aga1p without the GPI anchor) (Tang et al., 2018) or cohesins (Tsai, DaSilva, & Chen, 2013). The respective binding partners are then fused to the proteins of interest and can therefore bind to the scaffold, co-localising multiple enzymes. As shown in Fig. 3, Aga2p binds to tAga1 and dockerins to the cohesins. Working with a variety of cohesin/dockerin pairs from different organisms such as *Acetovibrio cellulolyticus* and *Clostridium thermocellum* allows for complex, enzymatically denser, cellulosomes to be constructed. For example multiple scaffolds can be stacked via cohesin/dockerin pairs, each with binding sites for enzymes (Tsai et al., 2013).

3 Build

3.1 Build—Assembling genetic constructs for expressing combinations of enzymes

Having chosen a set of enzymes, it is then necessary to assemble the required DNA cassettes, including coding sequences for enzymes, signal peptides and fusion partners along with suitable promoters, ribosome binding sites (for bacteria), and other necessary features. "Building" methods to generate new DNA molecules are used to introduce the genetic changes to the host. Despite the tremendous reduction in the cost of synthesising DNA molecules with custom sequences, researchers still depend on various methodologies to generate complex multi-gene constructs (Casini, Storch, Baldwin, & Ellis, 2015). The multi-gene DNA constructs are built and then introduced to the host. These constructs or devices are often referred to as pathways or circuits because often they contain metabolic pathways or genetic circuits (gene regulatory networks). In synthetic biology, approaching living systems from an engineering perspective, simple DNA elements (or parts) are characterised and rebuilt to generate new, more complex circuits (Cameron, Bashor, & Collins, 2014). The simple parts include, for example, sequences coding for promoters, ribosome binding sites (RBS), coding sequences, and transcription terminators. By incorporating modularity and standardisation in synthetic biology, the DNA assembly methods improve sharability of the parts, reliability of the assembly, and characterisation of the parts (Beal et al., 2020). Building (and designing) the candidates benefit from other advances in synthetic biology, such as new tools for computer-assisted design (Du Lac et al., 2020; Nielsen et al., 2016), and automation (Chao et al., 2017); however, achieving the desired goal remains challenging due to the complexity of the manipulations required.

While there are now many methods available for assembly of large and complex DNA constructs (Casini et al., 2015), two particularly widely used families of methods are those based on *Gibson assembly* and *Golden Gate* (MoClo)-like methods. Gibson assembly (Gibson et al., 2009) requires parts to be provided in the form of linear DNA (such as PCR products), with each part sharing DNA homology of around 40 base pairs with parts which are to be adjacent in the assembly. Assembly is then accomplished by mixing sets of parts together with a blend of enzymes which usually includes T5 exonuclease, a thermostable DNA polymerase, and Taq DNA ligase, in a suitable reaction buffer which provides dNTPs, and incubating at 50 °C. T5 exonuclease will chew back the exposed 5' ends leaving single stranded regions, before being inactivated by the heat. This allows adjacent parts to anneal. The thermostable polymerase will then fill in gaps and the thermostable ligase will join the ends. This allows rapid and reliable assembly of multiple parts in a single reaction, and is highly suitable when the exact nature of the construct is clear. However, it may lack flexibility since it requires terminal homology between adjacent parts, so parts must be remade (for example, by PCR) to add new homology ends for each different assembly. A variety of other DNA assembly methods, such as SLiCE (Seamless Ligation Cloning Extract) and CPEC (Circular Polymerase Extension Cloning), are similar to Gibson assembly in assembling parts based on end homology, but work by different mechanisms.

Golden Gate-like methods, by contrast, are based on the activity of Type IIS restriction endonucleases such as BsaI and BsmBI (Engler, Gruetzner, Kandzia, & Marillonnet, 2009). These cleave adjacent to their asymmetrical recognition sequence, leaving a sticky end of 3 (EarI, SapI) or 4 (BsaI, BsmBI) bases with sequence defined by the DNA adjacent to the recognition site. Thus, a single enzyme can generate many different sticky ends depending on the sequence adjacent to the recognition site. A typical Golden Gate assembly will include a number of parts of different types, each in a donor plasmid which has the part flanked by Type IIS restriction sites which generate a different set of overhangs for each part type. For example, simple kits may include the part types Promoter, Ribosome Binding Site, Coding Sequence, and Terminator. The donor plasmids for a given assembly are all mixed together with an acceptor plasmid, which has a different antibiotic selection marker to the donor plasmids, and has a screening cassette such as GFP also flanked by type IIS restriction sites facing in the opposite orientation. The mixture is then cleaved with the relevant Type IIS enzyme, which excises all of the parts and generates their characteristic overhangs. These can now anneal and assemble in the correct order in the acceptor plasmid, replacing the screening cassette. Ligation occurs, and the ligation mixture is then re-cleaved with the same enzyme. In correctly ligated products, the ligated parts are no longer adjacent to restriction recognition sites, and thus are not cleaved, but where parts have re-ligated to their original vector, they are once again adjacent to restriction recognition sites, and are therefore re-cleaved. After multiple cycles of cleavage and ligation, usually accomplished by temperature cycling, highly efficient assembly of multiple parts is achieved. Generally, around 4 to 10 parts are assembled per reaction, though some reports describe assembly of up to 52 parts in a single reaction (Pryor, Potapov, Bilotti, Pokhrel, & Lohman, 2022).

In most Golden Gate assembly systems, multiple levels of assembly can be used to generate larger and more complex constructs. Typically, individual parts such as promoters and coding sequences are designated Level 0. Multiple Level 0 parts (usually 4–8) are combined in a Level 1 assembly, which is often a Transcription Unit. Multiple Level 1 assemblies can then be combined in a Level 2 assembly, for expression of multiple proteins, and multiple level 2 assembles can then be combined in a level 3 assembly for larger pathways or complexes.

Golden Gate assembly is ideally suited to *combinatorial assembly*, in which many different combinations of parts are generated in a single reaction (Engler et al., 2009). This is easily done by including multiple different parts of each type in the assembly reaction. All possible combinations of the different parts will then be generated in a pool or library, and can be screened as described below under "Test". This may be particularly relevant in complex situations such as CBP, in which unpredictable interactions between the different components are known to be important.

Many different Golden Gate-based kits are now available, aimed at different host organisms including *E. coli* (Moore et al., 2016), *B. subtilis* (Radeck, Meyer, Lautenschläger, & Mascher, 2017), a wide variety of different bacteria (Valenzuela-Ortega & French, 2021), and yeasts such as *S. cerevisiae* (Garcia-Ruiz, Auxillos, Li, Dai, & Cai, 2018; Guo et al., 2015; Lee, DeLoache, Cervantes, & Dueber, 2015), *Pichia pastoris* (Prielhofer et al., 2017) and *Yarrowia lipolytica* (Egermeier, Sauer, & Marx, 2019; Larroude et al., 2019). These kits generally include a set of promoters, ribosome binding sites, N- and C-terminal fusion tags, and transcription termination sequences, plus useful ORFs such as fluorescent proteins. While many of the earlier kits are not compatible with each other, due to use of different overhangs and enzymes at each construction level, there has recently been an increasing tendency to follow a "common syntax" initially proposed for plants (Patron et al., 2015), but later adapted for bacteria, yeast and other hosts (Pollak et al., 2020). Level 0 parts from any kit adhering to this common syntax, such as JUMP (Valenzuela-Ortega & French, 2021) and uLOOP (Pollak et al., 2020), can be used interchangeably, though the rules for assembly at Level 2 and above may be different. This increasing level of standardization and interoperability will be increasingly important in engineering biology.

3.2 Build—Introducing and maintaining the DNA in the chosen host

It is then necessary to choose a method for introduction and maintenance of the genes. This may be accomplished using plasmids or integration into chromosomes. In *E. coli* and related bacteria, plasmids are usually used initially, and a wide variety of plasmids with different replication origins (and thus different copy numbers) are available as vectors; for example the *SEVA* (Standard European Vector Architecture) (Martínez-García et al., 2020; Martínez-García, Aparicio, Goñi-Moreno, Fraile, & de Lorenzo, 2015; Silva-Rocha et al., 2013) collection includes a set of replication origins and antibiotic resistance cassettes which can be combined to generate plasmids with different characteristics, functioning in a wide variety of bacterial hosts. It is also possible to insert DNA onto the chromosome via recombineering and similar methods, or by homologous recombination following CRISPR cleavage of a target site. In *B. subtilis*, a number of plasmid vector systems are available, but it is also very easy to introduce large cassettes onto the chromosome, since *B. subtilis* becomes naturally competent at certain points of its life cycle, and will then spontaneously take up large pieces of DNA from the surrounding environment, degrading one strand and incorporating the other into its chromosome by homologous recombination. Other bacterial hosts which have been used in CBP experiments all have their own idiosyncrasies, and various host-specific or broad host-range vectors are available. For example, the JUMP DNA assembly kit (Valenzuela-Ortega & French, 2021) includes acceptor plasmids with replication origins from the SEVA collection, including broad host range origins which are reported to allow replication in a wide variety of Gram negative bacteria, including *Pseudomonas*, and even in Gram positive bacteria.

In yeast, plasmids are of two types: those containing autonomous replication sequences and centromeres, and those with the replication origin of the 2-μm plasmid. It is also straightforward to insert genes onto the chromosomes at a suitable locus, since yeast will readily perform homologous recombination. This may be a suitable option for large pathways and to improve stability. Popular insertion sites are located between genes essential for growth to help prevent loop out and are located on chromosomes X, XI and XII (Mikkelsen et al., 2012). These sites have further been curated and modified into the plasmid set "EasyClone" for simpler integration (Jensen et al., 2014). A drawback of this strategy is that expression may be limited due to genes being single copy. To overcome this, the pathway can instead be integrated into the Ty retrotransposon delta sites of the yeast genome. To improve efficiency of multi-copy integration via homology-directed repair, double stranded DNA breaks are introduced using a CRISPR-cas9 system. This system was used to successfully introduce 18 copies of a 24 kb pathway into *S. cerevisiae* (Shi, Liang, Zhang, Ang, & Zhao, 2016). In general, CRISPR-Cas9 tools can be used to greatly increase the efficiency of multiplexed, markerless insertions and knock-outs (Zhang et al., 2019). The use of CRISPR-Cas9 and similar RNA-guided endonuclease tools has greatly facilitated genetic engineering of *S. cerevisiae* by generating double-strand breaks at specific loci to encourage homology-directed repair, and automatically selecting against cells which do not incorporate the desired DNA, since in this case the target will be repeatedly cleaved preventing growth.

4 Test

Screening of synthetic biology candidates is based on linking a genotype that has the desired function with a measurable or selectable phenotype. Isolation of clones allows them to be analysed and characterized individually. Screening is sometimes the limiting step in the research cycle (Jeschek, Gerngross, & Panke, 2017). This is especially true for developing Ideal CBP host candidates because the selection of candidates for their cellulolytic capability is limited by the microcrystalline nature of cellulose (French, 2009). Since it is not soluble in culture media, cellulose is often introduced in the form of suspensions which cause optical turbidity and heterogeneity. Turbidity blocks most optical measurements and analytical techniques, which is often circumvented using soluble cellulose analogues (Ostafe, Prodanovic, Commandeur, & Fischer, 2013) or applying separation steps (Mühlmann et al., 2017), such as centrifugation to partition the insoluble cellulose from the sample supernatant (SN) containing soluble molecules. Unfortunately, every required step reduces the technique's throughput capacity.

Assessment of CBP candidates is done by quantifying elements of the CBP process (Fig. 4), but experimental conditions can be far from the real process. Similar to the screening of microbes engineered with multiple genetic modifications, the measured trait can provide information on the expression and activity of individual gene modifications or it can be a trait that depends on several of the gene modifications, thus giving a global picture of the system's alteration.

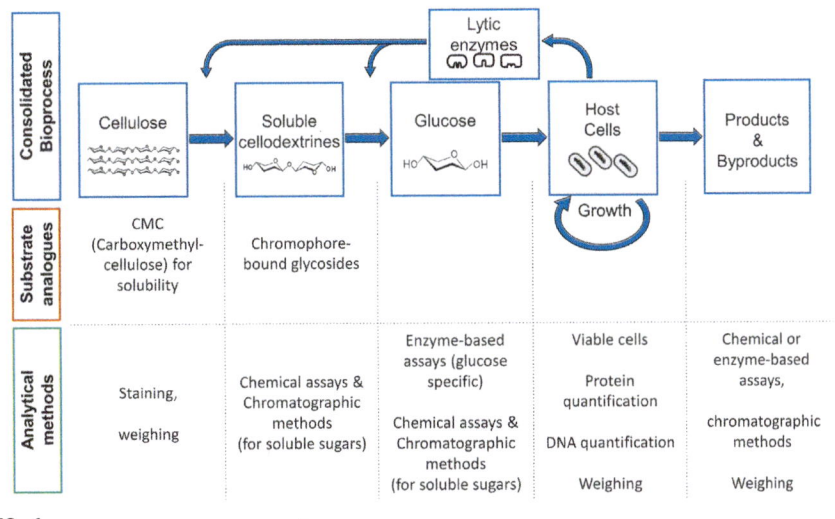

FIG. 4

Existing cellulase characterisation strategies applied at different steps of a consolidated bioprocess.

4.1 Substrate analogues and assaying individual enzyme activities

Functional characterisation of individual cellulases often uses substrate analogues with higher solubility than cellulose or linked to chromophores to directly quantify enzymatic cleavage (Table 2). For example, carboxymethyl-cellulose (CMC) has carboxymethyl groups ($-CH_2-COOH$) bound to some hydroxyl groups of glucose with various degrees of substitution, which impede chain-to-chain interactions thus avoiding crystallinity. Chromogenic substrates are cleaved by specific enzymes releasing a measurable chromophore, such as 4-nitrophenyl-glycosides, which are hydrolysed by different enzymes releasing 4-nitrophenol, which exhibits absorbance at 405 nm.

Table 2 Substrate analogues recognised by cellulases.

Substrate	Enzymatic activity detected	Reference
CMC	EG	Fan et al. (2016)
CMC	CBH	Tang et al. (2018)
Azo-CMC	EG (requires substrate separation)	Bokinsky et al. (2011) Lakhundi et al. (2017)
Azo-xylan	Endoxylanase (requires substrate separation)	Bokinsky et al. (2011)
4-Methylumbelliferyl-β-D-glucopyranoside	CBH (on agar plates or in liquid)	Lakhundi et al. (2017)
4-Methylumbelliferyl-β-D-glucopyranoside	Total cellulase activity	Chang et al. (2018)
4-Methylumberiferyl β-D-lactopyranoside	CBH	Davison, den Haan, and van Zyl (2016)
4-Nitrophenyl-β-D-cellobioside	CBH	Tozakidis et al. (2016)
4-Nitrophenyl-β-D-glucopyranoside	BG	Anusree et al. (2016); Tozakidis et al. (2016)
Azurine-cross-linked hydroxyethylcellulose (AZCL-HE-Cellulose)	EG	Sasaki, Mitsui, Yamada, and Ogino (2019)
4,6-O-(3-Ketobutylidene)-4-nitrophenyl-β-D-cellopentaoside	EG and BG	Wightman, Kroukamp, Pretorius, Paulsen, and Nevalainen (2020)
4,6-O-(3-Ketobutylidene)-4-nitrophenyl-β-D-cellopentaoside	EG (supplied with additional BG)	Claes, Deparis, Foulquié-Moreno, and Thevelein (2020)

CMC, carboxymethylcellulose; CBH, cellobiohydrolase; EG, endoglucanase; BG, β-glucosidase.

4.1.1 Endoglucanase

Secreted endoglucanase activity can be conveniently assayed in a semi-quantitative way using agar containing around 2 g/L CMC. Colonies are grown to allow expression and secretion of the endoglucanase. The plate is then flooded with 1 g/L aqueous solution of Congo Red, left for several minutes, then the Congo Red is poured off into a suitable waste container for safe disposal (in addition to being contaminated with recombinant cells, it is also chemically toxic) and the plate is flooded with 1 M NaCl. Congo Red has an affinity to bind CMC, which causes development of a red colour wherever CMC is present, with zones of clearing surrounding colonies which have secreted active endoglucanase. The size of the clear halo is related to the amount of activity secreted.

For a quantitative assay of endoglucanase activity in supernatants, dye-derivatized cellulose may be used (such as Azo-CMC, AZCL-HE-Cellulose, see Table 2). Degradation of the cellulose releases the dye following an incubation, and then the remaining insoluble cellulose substrate is removed by centrifugation and the cleaved dye's absorbance is read. Similar substrates are commercially available for endo-hemicellulases such as endoxylanase (Table 2).

4.1.2 Exoglucanase and β-glucosidase

Exoglucanase (cellobiohydrolase) activity is conveniently assayed using either chromogenic or fluorogenic substrates, which are commercially available. 3NP-cellobioside or 4NP-cellobioside are cleaved by exoglucanases to release 3-nitrophenol or 4-nitrophenol, both of which absorb strongly at 405–420 nm. Alternatively, MU-cellobioside is cleaved to release methylumbelliferone, which has strong blue fluorescence on ultraviolet illumination (excitation peak and emission peak around 360 and 460 nm, respectively). The activity of β-glucosidases is measured in the same way using MUG or 2NPC (methylumbelliferyl-β-D-glucopyranoside or 2-nitrophenyl-β-D-glucopyranoside).

Similar substrates are also available for exo-acting hemicellulases; for example, exoxylanase activity can be assayed using MUX (methylumbelliferyl β-D-xylofuranoside) or 2NPX (2-nitrophenyl-β-D-xylofuranoside). A wide range of such substrates are commercially available.

4.1.3 LPMO and expansins

Assay of LPMO is more challenging, and these enzymes are generally less well studied than hydrolytic cellulases, having been discovered much more recently. One assay to characterise this monooxygenase activity measures the hydrogen peroxide generated in the presence or absence of different substrates (Guo, Duquesne, Bozonnet, Nicaud, et al., 2017).

Expansins and similar amorphogens are not easily assayed for their own activity. Activity of CBM2 was inferred from electron microscopy of treated cellulose samples, for example (Jäger et al., 2011). Perhaps the most practical solution at present is to conduct a sugar release assay using a standard endoglucanase solution and an insoluble substrate, and measure whether the presence of the putative expansin-containing material increases sugar release. Such synergistic assays, which can also be used to determine the activity of LPMOs, are exemplified by Guo, Duquesne, Bozonnet, Nicaud, et al. (2017).

4.2 High-throughput screening assays with analogue substrates

High-throughput screening (HTS) assays might overcome the screening bottleneck for engineering microorganisms for different applications, but this is hindered for the purpose of biomass degradation due to the crystalline nature of cellulose. Because it must be extracellularly digested, lytic enzymes must be secreted, which impedes directed evolution experiments where clones are not separated physically. Moreover, microdroplet-based screening has not been applied in the presence of cellulose due to its non-homogeneous nature in cultures, and it is necessary to isolate clones in individual cultures.

Using model substrates has allowed some HTS to screen cellulases (Table 3). Nonetheless, activity towards analogue substrates might not correlate with activity towards real substrates and may also overestimate activity compared with more realistic conditions where the substrate will be less soluble and accessible.

4.3 Assaying total polysaccharide degradation capability

The activity of one or multiple cellulases towards realistic substrates can be screened by quantifying either the substrate or any of the products resulting from the substrate hydrolysis, as shown in Fig. 4. This includes quantification of undigested substrate, released soluble sugars, microbial biomass or the final product (Table 4). While such indirect techniques do not inform of individual enzyme activity, they allow a combined assessment of many relevant characteristics of the candidate, including: enzyme secretion, substrate digestion, and (for some techniques), use of the released sugars.

Soluble sugars can be measured with chromatographic methods or colorimetric techniques which generate an optical measure proportional to the concentration of the analyte, such as 3,5-dinitrosalicylic acid (DNS) which reacts with all soluble reducing sugars generating a signal at 450 nm. However, these methods depend on the analyte not being consumed, hence they cannot be applied in process-like conditions where the microbial host should be using the sugars.

Other approaches assess the global cellulolytic activity by indirectly quantifying the sugars consumed by the microbial candidate, such as measuring product generated (e.g. ethanol) with methods that often have a limited throughput due to the need for separation steps (Table 4). Another way of quantifying use of sugars released from cellulose is by measuring microbial growth (Table 4). The archetypal way of measuring microbe concentration through optical density (OD) caused by light scattering (Stevenson, McVey, Clark, Swain, & Pilizota, 2016) is completely impeded by cellulose turbidity. Researchers have turned to counting colony forming units (Lakhundi et al., 2017) or quantifying specific biomolecules. Total protein

Table 3 High-throughput screening methods applied to cellulases and cellulolytic organisms.

Enzyme or candidates screened	Screening scale	Analytical or selection principle	Reference
Intracellular cellulase (different classes)	Survival and selection	Yeast three-hybrid system adapted to detect a cello-oligosaccharides with ligands bound at each end	Peralta-Yahya, Carter, Lin, Tao, and Cornish (2008)
Secretion of cellulases	Microdroplets and fluorescence-assisted sorting	Detection of CMC hydrolysis to glucose using hexose oxidase, which generates H_2O_2 which acts on a reagent generating fluorophore	Ostafe et al. (2013)
Bioprospecting of soil bacteria with CBH activity	Microdroplets and fluorescence-assisted sorting	Fluorogenic substrate (β-D-cellobioside-6,8-difluoro-7-hydroxycoumarin-4-methanesulfonate)	Najah et al. (2014)
Biomining enzyme screening from metagenomics-based libraries of recombinant clones	Robotics-assisted microplate assays	Recombinant expression assay of cell lysates with Azo-CMC, CMC (sugars detected with PAHBAH), and MUC	Mühlmann et al. (2017)
Selection of EG library generated by random mutagenesis	Agar and microplate assays	CMC and Congo red staining (agar), Azo-CMC (microplate cultures)	Cecchini, Pepe, Pennacchio, Fagnano, and Faraco (2018)

Table 4 Analytical methods used to detect elements in the CBP process.

Analyte	Assay	Reference
Avicel	Congo red staining in agar plates	Wei et al. (2014)
CMC	Congo red staining in agar plates	Gao, Luan, et al. (2015)
CMC	Viscosity measurement of liquid	Fan et al. (2016)
Cellobiose	Iodine staining in agar plates	Anusree et al. (2016)
Reducing sugars	Dinitrosalicylic acid (DNS) assay of supernatants	Anusree et al. (2016); Gao, Luan, et al. (2015); Tozakidis et al. (2016)
Reducing sugars	4-Hydroxybenzhydrazide (PAHBAH) assay of supernatants	Mühlmann et al. (2017)
Hydrolysed soluble sugars (xylose, xylobiose, xylotriose, glucose, cellobiose, and cellotriose)	High-performance anion exchange chromatography of supernatants	Stern et al. (2018)
Glucose	High-performance Liquid Chromatography of supernatants	Lee et al. (2017)
Cellobiose	High-performance Liquid Chromatography of supernatants	Hetzler et al. (2013)
Glucose	Enzymatic assay commercial kit (glucose oxidase) of supernatants	Liang et al. (2014)
Cell mass and solids (Avicel)	Weighing centrifuged pellet without supernatant (the cellulose fraction was calculated with posterior acid hydrolysis and HPLC analysis)	Guo, Duquesne, Bozonnet, Cioci, et al. (2017); Guo, Duquesne, Bozonnet, Nicaud, et al. (2017)
Microbial cells	Haemocytometre counting	Liang et al. (2014)
Viable cells	Counting colony forming units after plating in agar	Bokinsky et al. (2011)
Cell density	OD (with CMC as substrate)	Kickenweiz et al. (2018)
Total protein	Bradford assay	Salinas (2017)
DNA	Fluorescent staining with SYBR I Green nucleic acid gel stain and propidium iodide	Duedu and French (2017)
Fatty acids	GC/MS of supernatants	Bokinsky et al. (2011)
Pinene, butanol	GC of supernatants	Bokinsky et al. (2011)
Ethanol	HPLC of supernatants	Tang et al. (2018)
Ethanol	GC of supernatants	Shin et al. (2015)
Ethanol	Enzymatic assay (ADH) of supernatants	Lewicka et al. (2014)
CO_2	Weighting whole culture (CO_2 release reduces weight)	Claes et al. (2020)
Cellulase detection after polyacrylamide gel electrophoresis	Zymogram (gel is submerged in 4-methylumbelliferyl-β-D-cellobioside solution to detect cellulase activity, or laid over Agar-CMC and stained with Congo red)	Chang et al. (2018)

concentration has been quantified with colorimetric assay and DNA can be precisely quantified in the presence of cellulose using fluorescent dyes (Duedu & French, 2017). This last method was especially interesting as it showed that fluorescence could be measured despite turbidity.

5 Learn and general considerations

5.1 Maximizing information from multi-factorial experiments: Sequential vs non-sequential optimisation

Generating the microbe with the desired functions can be framed as a problem with multiple variables or dimensions to be optimised qualitatively and quantitatively (Jeschek et al., 2017). Consequently, Design of Experiments (DoE) principles can be applied (Jeschek et al., 2017). In the case of engineering a microorganism to digest biomass, qualitative (or identity) variables include the types of enzymes to be tested and the homologous genes chosen to be tested, or the secretion signal peptides.

Quantitative variables include the factors controlling the stages in the expression of those genes: gene dosage, transcription, and translation. The different variables must be tuned to achieve maximum product formation while reducing metabolic burden, which will reduce the host growth and the product yield (Jeschek et al., 2017). A sequential optimisation approach (of one gene at a time) might only reach a local optimum, and it requires an in-depth knowledge of the interactions in the system and ways to measure them. The sequential approach also requires more iterations of the cycle, making it costly and time-consuming. Recently, different strategies have been developed to optimise multiple variables of a system at the same time, such as combinatorial optimisation. This strategy is based on building and testing multiple combinations of the DNA elements controlling each variable.

In terms of designing and building, quantitative variation of gene expression is easily reached using different promoters controlling transcription (Moore et al., 2016) and ribosome binding sites (RBS) controlling translation (Salis, Mirsky, & Voigt, 2009). DNA assembly from modular DNA elements (Naseri & Koffas, 2020) can easily generate combinatorial libraries that contain all possible combinations of elements. The size of combinatorial libraries (the different individuals that can exist) increases exponentially with the number of pathway components being co-optimised, leading to a "combinatorial explosion" (Jeschek et al., 2017). On the other side, qualitative variation is reached through variation of coding DNA sequences and is more complicated to perform (Jeschek et al., 2017). DNA mutagenesis generates large libraries that require high-throughput screening methods, and testing several homologues for coding sequence is complicated because preparing the DNA parts is costly and very laborious.

Libraries can only be as diverse as the "test" method allows screening of the candidates generated, often making this the bottleneck of the research cycle. Moreover, the screening method must reflect the activity of the multiple elements or genes being co-optimised, such as those described in Section 4.3.

5.2 Case studies

Previous efforts at engineering microorganisms have followed different approaches, and representative CBP conditions have been summarised in Table 5. In most cases, a set of genes encoding different enzyme types are introduced in the host and tested individually following a sequential optimisation. Then, the best genes of each type are put together in one or a few combinations and tested. Individual enzymes, which are displayed in the cell envelope or secreted to the culture supernatant (SN), are often tested with substrate analogues to allow a higher throughput, which has the limitations described above. The exception is BG enzymes, which can easily be screened since their substrate, cellobiose, is soluble, allowing growth measurements from the resulting glucose. Even using substrate analogues, few reports have tested more than a dozen candidate individual genes. An influential early report by Bokinsky et al. (2011) set an example of systematic screening of multiple variables. The authors tested 16 BG or xylobiosidase genes for the candidate's growth, and 10 EG fused to one secretion partner were tested for activity towards an analogue. The identified best sequences were tested with 4 to 5 different promoters, but only one combination EG+BG was built for cellulose degradation.

In the reviewed studies, just a few combinations of cellulase genes in one host were tested. Consequently, emerging interactions due to context-dependency were ignored, and this sequential optimisation only found local activity maxima when a much-improved capacity is required. The best example of non-sequential optimisation is the publication of Wightman et al. (2020). One hundred and sixty clones of a strain library with different copy numbers of an EG gene and BG gene were screened using a substrate analogue (4,6-O-(3-ketobutylidene)-4-nitrophenyl-β-D-cellopentaoside) that must be cleaved by both enzymes to generate the signal. However, it is unlikely that the BG/EG ratio that best cleaves a soluble 5-glucose substrate is the optimal ratio required for digesting more realistic cellulosic substrates, and this method can not be applied to combinations including other enzymes.

Few combinations of various secretion partners with enzyme-coding sequences have been examined. The report with most enzyme-secretion partner combinations explored was presented by Lee et al. (2017), where 11 enzyme sequences for different enzymes were screened with 24 secretion partners. The authors avoided constructing a strain expressing multiple genes by combining single enzyme-secreting strains as consortia, which they tried to optimise quantitatively using different ratios of the strains in the inoculum, but also noting that this ratio varied during fermentation.

Table 5 Literature summary of systematic construction and screening of IBPM candidates.

Host (reference)	Single enzymes screened	Enzyme combinations screened
E. coli (Bokinsky et al., 2011)	10 EG combined with 1 secretion partner, tested for activity in SN. The best one was tested with 5 promoters 4 BG tested for allowing growth from cellobiose. The best 2 were tested with 4 promoters 12 xylobiosidases tested for growth from xylan. The best one was tested with 4 promoters	1 EG + BG combination tested for growth and product formation from cellulose 1 xylobiosidase+ 1 (previously characterised) EX combination tested for growth and product formation from hemicellulose
B. subtilis (Chang et al., 2018)	(No gene tested individually)	Two constructs built (with 8 genes of *Clostridium thermocellum* cellulosome) with different gene orders. Tested for protein expression and secretion, individual activity, and for sugar release from Napier grass
Geobacillus thermoglucosidasius (Daas et al., 2018)	2 possible EG genes from the host characterised for activity (recombinant in *E. coli*) 1 of the two EG above, plus 3 EG, 3 CBH tested for CMC hydrolysis (from cell lysate)	(No combinations tested)
Lactobacillus plantarum (Stern et al., 2018)	2 EX and 2 EG combined with 2 secretion partners, tested for activity and for binding to cohesin	4 consortia tested for hydrolysis of wheat straw (using concentrated SN)
Corynebacterium glutamicum (Anusree et al., 2016)	(No gene tested individually) 2 secreted EG cloned in one plasmid, 1 BG (secreted free or displayed in the cell wall) cloned in another compatible plasmid	4 (2x2) combinations of plasmids built, expressing 1 EG and 1 BG respectively. Tested for individual enzymatic activity in the SN, tested for hydrolysis of cellulose (rice straw), and fermentation of cellobiose and CMC (measuring OD and lysine produced)
S. cerevisiae (Lee et al., 2017)	7 CBH type I, 3 CBH type II, 1 EG, and 1 BG were fused with 24 secretion partners; they were tested for secretion (SDS-PAGE) and activity	A 4-strain consortium with the best strain expressing each of the four enzyme types was cultured and characterised for cellulosic substrates' hydrolysis Different-ratios consortia were tested for fermentation of Avicel (in the presence of added cellulase)
S. cerevisiae (Sasaki et al., 2019)	Three libraries of 1 EG gene (varying integration, promoter or both) were tested by: screening 400 clones by halo formation in agar plates; 45-64 clones tested for activity in culture SN (96-well plate); best 8 clones tested for activity in culture SN (shake flask, triplicate). The same analogue, AZCL-hydroxyethyl-cellulose, used in three screening steps	(No combinations tested)

Continued

Table 5 Literature summary of systematic construction and screening of IBPM candidates.—cont'd

Host (reference)	Single enzymes screened	Enzyme combinations screened
S. cerevisiae (Fan et al., 2016)	3 BG tested for growth from cellodextrins (presence of a transporter gene) 8 scaffoldin variants tested for GFP display 8 EG and 8 CBH displayed in cellulosome individually tested for hydrolysis of CMC	3 EG combined with 5 CBH and tested for ethanol fermentation from CMC, Avicel and PASC (in the presence of galactose)
S. cerevisiae (Tang et al., 2018)	1 BG and 3 CBH tested with and without dockerin domain (activity) 1 BG, 1 EG, 1 CBH with two different displays tested for activity 1BG, 1 EG, 1 CBH combined with 3 secretion partners, and tested for activity and immunofluorescence To improve secretion, 3 secretion-related genes were co-expressed with BG or CBH genes, and activity was tested	3 strain consortia (before improving secretion and after, and with improved ratios) tested for ethanol fermentation from PASC (in rich medium)
S. cerevisiae (Davison et al., 2016)	30 strains tested for the secretion of 1 BG gene (for activity) 7 best strains additionally tested for secreted activity of 1 CBH or 1 EG (separately) Also tested the effect of various stresses on growth rate (such as growth inhibitors)	(No combinations tested)
S. cerevisiae (Claes et al., 2020)	In a xylose-using strain, 1 BX and 1 BG were tested for SN activity and co-fermentation of glucose, xylose, cellobiose and xylan 3EX, 4 bifunctional acetyl xylan esterases/EX, 6 CBH type I, 6 CBH type II, and 6 EG were tested for SN activity	The best gene of each type was integrated sequentially and confirmed for activity Strains were tested for: co-fermentation of cellobiose, glucose, xylose and xylan; and for hydrolytic activity (of culture SN) of PASC, FP, CMC and xylan
S. cerevisiae (Wightman et al., 2020)	(No gene tested individually)	Library of clones with different gene copy numbers of 1 EG and 1 BG were generated and tested for individual activity and synergistic activity (using substrate analogs)
P. pastoris (Kickenweiz et al., 2018)	2 BG tested for growth from cellobiose	3 strains with different combinations of BG, CBH and EG genes were tested for growth on CMC, and SN tested for hydrolysis of Avicel

Reference		
Y. lipolytica (Wei et al., 2014)	1 EG, 1 CBH type II and 4 CBH type I tested for secretion (SDS-PAGE and western blot), then for growth from CMC 1 of the CBH I genes was also purified and tested for hydrolysis of Avicel (in the presence of added BG and EG)	1 consortium (1 EG + CBH II) tested for clearing of Avicel in rich-medium agar plates, and SN tested for PASC hydrolysis 1 consortium (1 EG + 1 CBH I + 1 CBH II) compared with the previous consortium for Avicel fermentation (measuring growth and fatty acid produced)
Y. lipolytica (Guo, Duquesne, Bozonnet, Cioci, et al., 2017)	2 EG, 2 CBH type I, and 1 CBH type II tested for secretion (SDS-PAGE and western blot) and purified enzyme activity (using a his-tag) 1 CBH I and 1 CBH II genes with a different promoter tested for secretion (SDS-PAGE)	6 strains (with different combinations of the 2 EG and 4 CBH genes, and always 2 BG) were tested for hydrolytic activity in the SN towards Avicel, PASC, and CMC, and for growth from Avicel Work continued with the strain from the previous report (see above)
Y. lipolytica (Guo, Duquesne, Bozonnet, Nicaud, et al., 2017)	1 LPMO purified and tested for activity alone and with added cellulases 1 swollenin purified and tested with added cellulases (it was found that pre-incubation with swollenin was more effective than expression during saccharification)	Authors changed the promoter of 1 EG; then they introduced a swollenin, an EX, an LPMO, or the LPMO and the EX The five new strains were compared to the parental one in SN hydrolytic activity towards CMC, PASC, Avicel and industrial cellulose pulp, and compared in growth from Avicel

Activity testing of individual enzymes means characterisation using substrate analogues as those listed in Table 2.

6 Conclusions

An ideal research cycle strategy should build and test multiple combinations of cellulolytic genes in CBP-like conditions: an insoluble cellulose substrate is digested by enzymes secreted *in situ,* and the resulting sugars are used by the microorganism for growth and/or product formation. Multiple variables of each gene are relevant and must be included in the optimisation process: coding sequence of the mature enzyme, secretion partner, and regulatory elements such as promoter and ribosome binding site. Unfortunately, testing all these variables for the different genes becomes an intractable problem due to a combinatorial explosion. Therefore, an initial screen of individual genes is necessary to select the promising genes among all the possible combinations. In contrast to previous studies, this initial selection should also be made in CBP-relevant conditions; individual genes should be tested with natural substrates which require the presence of additional cellulases to be digested.

The methods of engineering biology, such as Golden Gate-based DNA assembly systems, are ideally suited to generating many different combinations of enzymes with different expression levels, secretion tags or fusion partners, and other variations. This generates a large library of candidates to be screened, either individually or as a pool.

An ideal screening method should be applicable in experimental conditions that are relevant and informative of the behaviour of microbial candidates in the final production conditions. In the case of engineering CBP candidates, those conditions should include *in situ* secretion of enzymes, cellulolysis and use of the resulting sugars. Such methodology should allow combined assessment of additional relevant characteristics of the candidate in addition to cellulase activity (such as secretion capacity, growth, and genetic stability), that is to say, a "global" measurement. Additionally, the screening approach should be of high-throughput capacity to allow screening for non-sequential optimization approaches (such as combinatorial assembly). Unfortunately, none of the existing methods fulfil all these requirements for an ideal screening system.

In our opinion, the cost-effective conversion of non-food biomass to useful products is one of the most critical issues in moving towards a circular bio-based economy. This problem has not yet been solved, but new tools and approaches offer promising routes to explore.

References

Anusree, M., Wendisch, V. F., & Nampoothiri, K. M. (2016). Co-expression of endoglucanase and β-glucosidase in Corynebacterium glutamicum DM1729 towards direct lysine fermentation from cellulose. *Bioresource Technology, 213*, 239–244. https://doi.org/10.1016/j.biortech.2016.03.019.

Atiwesh, G., Parrish, C. C., Banoub, J., & Le, T. T. (2022). Lignin degradation by microorganisms: A review. *Biotechnology Progress, 38*(2), e3226. https://doi.org/10.1002/btpr.3226.

Banner, A., Toogood, H. S., & Scrutton, N. S. (2021). Consolidated bioprocessing: Synthetic biology routes to fuels and fine chemicals. *Microorganisms, 9*(5). https://doi.org/10.3390/microorganisms9051079.

Barth, A., Hendrix, J., Fried, D., Barak, Y., Bayer, E. A., & Lamb, D. C. (2018). Dynamic interactions of type I cohesin modules fine-tune the structure of the cellulosome of Clostridium thermocellum. *Proceedings of the National Academy of Sciences of the United States of America, 115*(48), E11274–e11283. https://doi.org/10.1073/pnas.1809283115.

Bäumgen, M., Dutschei, T., & Bornscheuer, U. T. (2021). Marine polysaccharides: Occurrence, enzymatic degradation and utilization. *Chembiochem, 22*(13), 2247–2256. https://doi.org/10.1002/cbic.202100078.

Beal, J., Goñi-Moreno, A., Myers, C., Hecht, A., de Vicente, M. D. C., Parco, M., et al. (2020). The long journey towards standards for engineering biosystems: Are the molecular biology and the biotech communities ready to standardise? *EMBO Reports, 21*(5), e50521. https://doi.org/10.15252/embr.202050521.

Bernardes, A., Pellegrini, V. O. A., Curtolo, F., Camilo, C. M., Mello, B. L., Johns, M. A., et al. (2019). Carbohydrate binding modules enhance cellulose enzymatic hydrolysis by increasing access of cellulases to the substrate. *Carbohydrate Polymers, 211*, 57–68. https://doi.org/10.1016/j.carbpol.2019.01.108.

Bischof, R. H., Ramoni, J., & Seiboth, B. (2016). Cellulases and beyond: The first 70 years of the enzyme producer Trichoderma reesei. *Microbial Cell Factories, 15*(1), 106. https://doi.org/10.1186/s12934-016-0507-6.

Bissaro, B., Várnai, A., Røhr, Å. K., & Eijsink, V. G. H. (2018). Oxidoreductases and reactive oxygen species in conversion of lignocellulosic biomass. *Microbiology and Molecular Biology Reviews, 82*(4). https://doi.org/10.1128/mmbr.00029-18.

Boder, E. T., & Wittrup, K. D. (1997). Yeast surface display for screening combinatorial polypeptide libraries. *Nature Biotechnology, 15*(6), 553–557. https://doi.org/10.1038/nbt0697-553.

Bokinsky, G., Peralta-Yahya, P. P., George, A., Holmes, B. M., Steen, E. J., Dietrich, J., et al. (2011). Synthesis of three advanced biofuels from ionic liquid-pretreated switchgrass using engineered Escherichia coli. *Proceedings of the National Academy of Sciences of the United States of America, 108*(50), 19949–19954. https://doi.org/10.1073/pnas.1106958108.

Boraston, A. B., Bolam, D. N., Gilbert, H. J., & Davies, G. J. (2004). Carbohydrate-binding modules: Fine-tuning polysaccharide recognition. *Biochemical Journal, 382*(Pt. 3), 769–781. https://doi.org/10.1042/bj20040892.

Brenelli, L., Squina, F. M., Felby, C., & Cannella, D. (2018). Laccase-derived lignin compounds boost cellulose oxidative enzymes AA9. *Biotechnology for Biofuels and Bioproducts, 11*, 10. https://doi.org/10.1186/s13068-017-0985-8.

Burdette, L. A., Leach, S. A., Wong, H. T., & Tullman-Ercek, D. (2018). Developing Gram-negative bacteria for the secretion of heterologous proteins. *Microbial Cell Factories, 17*(1), 196. https://doi.org/10.1186/s12934-018-1041-5.

Cameron, D. E., Bashor, C. J., & Collins, J. J. (2014). A brief history of synthetic biology. *Nature Reviews Microbiology, 12*(5), 381–390. https://doi.org/10.1038/nrmicro3239.

Cantarel, B. L., Coutinho, P. M., Rancurel, C., Bernard, T., Lombard, V., & Henrissat, B. (2009). The carbohydrate-active enzymes database (CAZy): An expert resource for glycogenomics. *Nucleic Acids Research, 37*(Database issue), D233–D238. https://doi.org/10.1093/nar/gkn663.

Caplan, S., Green, R., Rocco, J., & Kurjan, J. (1991). Glycosylation and structure of the yeast MF alpha 1 alpha-factor precursor is important for efficient transport through the secretory pathway. *Journal of Bacteriology*, *173*(2), 627–635. https://doi.org/10.1128/jb.173.2.627-635.1991.

Casini, A., Storch, M., Baldwin, G. S., & Ellis, T. (2015). Bricks and blueprints: Methods and standards for DNA assembly. *Nature Reviews Molecular Cell Biology*, *16*(9), 568–576. https://doi.org/10.1038/nrm4014.

Cecchini, D. A., Pepe, O., Pennacchio, A., Fagnano, M., & Faraco, V. (2018). Directed evolution of the bacterial endo-β-1,4-glucanase from Streptomyces sp. G12 towards improved catalysts for lignocellulose conversion. *AMB Express*, *8*(1), 74. https://doi.org/10.1186/s13568-018-0602-7.

Chang, J. J., Anandharaj, M., Ho, C. Y., Tsuge, K., Tsai, T. Y., Ke, H. M., et al. (2018). Biomimetic strategy for constructing Clostridium thermocellum cellulosomal operons in Bacillus subtilis. *Biotechnology for Biofuels and Bioproducts*, *11*, 157. https://doi.org/10.1186/s13068-018-1151-7.

Chao, R., Mishra, S., Si, T., & Zhao, H. (2017). Engineering biological systems using automated biofoundries. *Metabolic Engineering*, *42*, 98–108. https://doi.org/10.1016/j.ymben.2017.06.003.

Claes, A., Deparis, Q., Foulquié-Moreno, M. R., & Thevelein, J. M. (2020). Simultaneous secretion of seven lignocellulolytic enzymes by an industrial second-generation yeast strain enables efficient ethanol production from multiple polymeric substrates. *Metabolic Engineering*, *59*, 131–141. https://doi.org/10.1016/j.ymben.2020.02.004.

Cosgrove, D. J. (2017). Microbial expansins. *Annual Review of Microbiology*, *71*, 479–497. https://doi.org/10.1146/annurev-micro-090816-093315.

Daas, M. J. A., Nijsse, B., van de Weijer, A. H. P., Groenendaal, B., Janssen, F., van der Oost, J., et al. (2018). Engineering Geobacillus thermodenitrificans to introduce cellulolytic activity; expression of native and heterologous cellulase genes. *BMC Biotechnology*, *18*(1), 42. https://doi.org/10.1186/s12896-018-0453-y.

Das, M., Patra, P., & Ghosh, A. (2020). Metabolic engineering for enhancing microbial biosynthesis of advanced biofuels. *Renewable and Sustainable Energy Reviews*, *119*, 109562.

Davison, S. A., den Haan, R., & van Zyl, W. H. (2016). Heterologous expression of cellulase genes in natural Saccharomyces cerevisiae strains. *Applied Microbiology and Biotechnology*, *100*(18), 8241–8254. https://doi.org/10.1007/s00253-016-7735-x.

Del Vecchio, D. (2015). Modularity, context-dependence, and insulation in engineered biological circuits. *Trends in Biotechnology*, *33*(2), 111–119. https://doi.org/10.1016/j.tibtech.2014.11.009.

Du Lac, M., Duigou, T., Herisson, J., Carbonell, P., Swainston, N., Zulkower, V., et al. (2020). Galaxy-SynBioCAD: Synthetic biology design automation tools in galaxy workflows. *BioRxiv*.

Duedu, K. O., & French, C. E. (2016). Characterization of a *Cellulomonas fimi* exoglucanase/xylanase-endoglucanase gene fusion which improves microbial degradation of cellulosic biomass. *Enzyme and Microbial Technology*, *93-94*, 113–121. https://doi.org/10.1016/j.enzmictec.2016.08.005.

Duedu, K. O., & French, C. E. (2017). Two-colour fluorescence fluorimetric analysis for direct quantification of bacteria and its application in monitoring bacterial growth in cellulose degradation systems. *Journal of Microbiological Methods*, *135*, 85–92. https://doi.org/10.1016/j.mimet.2017.02.006.

Egermeier, M., Sauer, M., & Marx, H. (2019). Golden Gate-based metabolic engineering strategy for wild-type strains of Yarrowia lipolytica. *FEMS Microbiology Letters, 366*(4). https://doi.org/10.1093/femsle/fnz022.

Ellgaard, L., & Helenius, A. (2003). Quality control in the endoplasmic reticulum. *Nature Reviews Molecular Cell Biology, 4*(3), 181–191. https://doi.org/10.1038/nrm1052.

Endalur Gopinarayanan, V., & Nair, N. U. (2019). Pentose metabolism in Saccharomyces cerevisiae: The need to engineer global regulatory systems. *Biotechnology Journal, 14*(1), e1800364. https://doi.org/10.1002/biot.201800364.

Engler, C., Gruetzner, R., Kandzia, R., & Marillonnet, S. (2009). Golden gate shuffling: A one-pot DNA shuffling method based on type IIs restriction enzymes. *PLoS One, 4*(5), e5553. https://doi.org/10.1371/journal.pone.0005553.

Fan, L. H., Zhang, Z. J., Mei, S., Lu, Y. Y., Li, M., Wang, Z. Y., et al. (2016). Engineering yeast with bifunctional minicellulosome and cellodextrin pathway for co-utilization of cellulose-mixed sugars. *Biotechnology Biofuels, 9*, 137. https://doi.org/10.1186/s13068-016-0554-6.

Frederix, M., Mingardon, F., Hu, M., Sun, N., Pray, T., Singh, S., et al. (2016). Development of an: E. coli strain for one-pot biofuel production from ionic liquid pretreated cellulose and switchgrass. *Green Chemistry, 18*(15), 4189–4197. https://doi.org/10.1039/c6gc00642f.

French, C. E. (2009). Synthetic biology and biomass conversion: A match made in heaven? *Journal of the Royal Society, Interface, 6*(Suppl. 4), S547–S558. https://doi.org/10.1098/rsif.2008.0527.focus.

Gao, D., Luan, Y., Wang, Q., Liang, Q., & Qi, Q. (2015). Construction of cellulose-utilizing Escherichia coli based on a secretable cellulase. *Microbial Cell Factories, 14*, 159. https://doi.org/10.1186/s12934-015-0349-7.

Garcia-Ruiz, E., Auxillos, J., Li, T., Dai, J., & Cai, Y. (2018). YeastFab: High-throughput genetic parts construction, measurement, and pathway engineering in yeast. *Methods in Enzymology, 608*, 277–306. https://doi.org/10.1016/bs.mie.2018.05.003.

Gibson, D. G., Young, L., Chuang, R. Y., Venter, J. C., Hutchison, C. A., 3rd, & Smith, H. O. (2009). Enzymatic assembly of DNA molecules up to several hundred kilobases. *Nature Methods, 6*(5), 343–345. https://doi.org/10.1038/nmeth.1318.

Gilkes, N. R., Kwan, E., Kilburn, D. G., Miller, R. C., Antony, R., & Warren, J. (1997). Attack of carboxymethylcellulose at opposite ends by two cellobiohydrolases from Cellulomonas fimi. *Journal of Biotechnology, 57*(1-3), 83–90.

Green, E. R., & Mecsas, J. (2016). Bacterial secretion systems: An overview. *Microbiology Spectrum, 4*(1). https://doi.org/10.1128/microbiolspec.VMBF-0012-2015.

Guo, Y., Dong, J., Zhou, T., Auxillos, J., Li, T., Zhang, W., et al. (2015). YeastFab: the design and construction of standard biological parts for metabolic engineering in Saccharomyces cerevisiae. *Nucleic Acids Research, 43*(13), e88. https://doi.org/10.1093/nar/gkv464.

Guo, Z. P., Duquesne, S., Bozonnet, S., Cioci, G., Nicaud, J. M., Marty, A., et al. (2017). Conferring cellulose-degrading ability to Yarrowia lipolytica to facilitate a consolidated bioprocessing approach. *Biotechnology Biofuels, 10*, 132. https://doi.org/10.1186/s13068-017-0819-8.

Guo, Z. P., Duquesne, S., Bozonnet, S., Nicaud, J. M., Marty, A., & O'Donohue, M. J. (2017). Expressing accessory proteins in cellulolytic Yarrowia lipolytica to improve the conversion yield of recalcitrant cellulose. *Biotechnology Biofuels, 10*, 298. https://doi.org/10.1186/s13068-017-0990-y.

Gupta, V. K., Steindorff, A. S., de Paula, R. G., Silva-Rocha, R., Mach-Aigner, A. R., Mach, R. L., et al. (2016). The post-genomic era of Trichoderma reesei: What's next? *Trends in Biotechnology*, *34*(12), 970–982. https://doi.org/10.1016/j.tibtech.2016.06.003.

Gyorgy, A., Jiménez, J. I., Yazbek, J., Huang, H. H., Chung, H., Weiss, R., et al. (2015). Iso-cost lines describe the cellular economy of genetic circuits. *Biophysical Journal*, *109*(3), 639–646. https://doi.org/10.1016/j.bpj.2015.06.034.

Haigler, C. H., & Roberts, A. W. (2019). Structure/function relationships in the rosette cellulose synthesis complex illuminated by an evolutionary perspective. *Cellulose*, *26*(1), 227–247.

Hassan, S. S., Williams, G. A., & Jaiswal, A. K. (2018). Emerging technologies for the pre-treatment of lignocellulosic biomass. *Bioresource Technology*, *262*, 310–318. https://doi.org/10.1016/j.biortech.2018.04.099.

Herburger, K., Franková, L., Pičmanová, M., Loh, J. W., Valenzuela-Ortega, M., Meulewa-eter, F., et al. (2020). Hetero-trans-β-glucanase produces cellulose-xyloglucan covalent bonds in the cell walls of structural plant tissues and is stimulated by expansin. *Molecular Plant*, *13*(7), 1047–1062. https://doi.org/10.1016/j.molp.2020.04.011.

Hetzler, S., Bröker, D., & Steinbüchel, A. (2013). Saccharification of cellulose by recombinant Rhodococcus opacus PD630 strains. *Applied Environmental Microbiology*, *79*(17), 5159–5166. https://doi.org/10.1128/aem.01214-13.

Hou, J., Tyo, K. E., Liu, Z., Petranovic, D., & Nielsen, J. (2012). Metabolic engineering of recombinant protein secretion by Saccharomyces cerevisiae. *FEMS Yeast Research*, *12*(5), 491–510. https://doi.org/10.1111/j.1567-1364.2012.00810.x.

Jäger, G., Girfoglio, M., Dollo, F., Rinaldi, R., Bongard, H., Commandeur, U., et al. (2011). How recombinant swollenin from Kluyveromyces lactis affects cellulosic substrates and accelerates their hydrolysis. *Biotechnology Biofuels*, *4*(1), 33. https://doi.org/10.1186/1754-6834-4-33.

Jayani, R. S., Saxena, S., & Gupta, R. (2005). Microbial pectinolytic enzymes: A review. *Process Biochemistry*, *40*(9), 2931–2944.

Jensen, M. K., & Keasling, J. D. (2015). Recent applications of synthetic biology tools for yeast metabolic engineering. *FEMS Yeast Research*, *15*(1), 1–10. https://doi.org/10.1111/1567-1364.12185.

Jensen, N. B., Strucko, T., Kildegaard, K. R., David, F., Maury, J., Mortensen, U. H., et al. (2014). EasyClone: Method for iterative chromosomal integration of multiple genes in Saccharomyces cerevisiae. *FEMS Yeast Research*, *14*(2), 238–248. https://doi.org/10.1111/1567-1364.12118.

Jeschek, M., Gerngross, D., & Panke, S. (2017). Combinatorial pathway optimization for streamlined metabolic engineering. *Current Opinion in Biotechnology*, *47*, 142–151. https://doi.org/10.1016/j.copbio.2017.06.014.

Kachroo, A. H., Kancherla, A. K., Singh, N. S., Varshney, U., & Mahadevan, S. (2007). Mu-tations that alter the regulation of the chb operon of Escherichia coli allow utilization of cellobiose. *Molecular Microbiology*, *66*(6), 1382–1395. https://doi.org/10.1111/j.1365-2958.2007.05999.x.

Kang, H. A., Nam, S. W., Kwon, K. S., Chung, B. H., & Yu, M. H. (1996). High-level secretion of human alpha 1-antitrypsin from Saccharomyces cerevisiae using inulinase signal se-quence. *Journal of Biotechnology*, *48*(1-2), 15–24. https://doi.org/10.1016/0168-1656(96)01391-0.

Kickenweiz, T., Glieder, A., & Wu, J. C. (2018). Construction of a cellulose-metabolizing Komagataella phaffii (Pichia pastoris) by co-expressing glucanases and β-glucosidase. *Applied Microbiology and Biotechnology*, *102*(3), 1297–1306. https://doi.org/10.1007/s00253-017-8656-z.

Kleiner-Grote, G. R. M., Risse, J. M., & Friehs, K. (2018). Secretion of recombinant proteins from E. coli. *Engineering in Life Sciences*, *18*(8), 532–550. https://doi.org/10.1002/elsc.201700200.

Kuusk, S., Kont, R., Kuusk, P., Heering, A., Sørlie, M., Bissaro, B., et al. (2019). Kinetic insights into the role of the reductant in H(2)O(2)-driven degradation of chitin by a bacterial lytic polysaccharide monooxygenase. *Journal of Biological Chemistry*, *294*(5), 1516–1528. https://doi.org/10.1074/jbc.RA118.006196.

Lakhundi, S. S., Duedu, K. O., Cain, N., Nagy, R., Krakowiak, J., & French, C. E. (2017). Citrobacter freundii as a test platform for recombinant cellulose degradation systems. *Letters in Applied Microbiology*, *64*(1), 35–42. https://doi.org/10.1111/lam.12668.

Larroude, M., Park, Y. K., Soudier, P., Kubiak, M., Nicaud, J. M., & Rossignol, T. (2019). A modular golden gate toolkit for Yarrowia lipolytica synthetic biology. *Microbial Biotechnology*, *12*(6), 1249–1259. https://doi.org/10.1111/1751-7915.13427.

Lee, M. E., DeLoache, W. C., Cervantes, B., & Dueber, J. E. (2015). A highly characterized yeast toolkit for modular, multipart assembly. *ACS Synthetic Biology*, *4*(9), 975–986. https://doi.org/10.1021/sb500366v.

Lee, Y. J., Lee, R., Lee, S. H., Yim, S. S., & Jeong, K. J. (2016). Enhanced secretion of recombinant proteins via signal recognition particle (SRP)-dependent secretion pathway by deletion of rrsE in Escherichia coli. *Biotechnology and Bioengineering*, *113*(11), 2453–2461. https://doi.org/10.1002/bit.25997.

Lee, C. R., Sung, B. H., Lim, K. M., Kim, M. J., Sohn, M. J., Bae, J. H., et al. (2017). Co-fermentation using recombinant Saccharomyces cerevisiae yeast strains hyper-secreting different cellulases for the production of cellulosic bioethanol. *Science Reports*, *7*(1), 4428. https://doi.org/10.1038/s41598-017-04815-1.

Lewicka, A. J., Lyczakowski, J. J., Blackhurst, G., Pashkuleva, C., Rothschild-Mancinelli, K., Tautvaišas, D., et al. (2014). Fusion of pyruvate decarboxylase and alcohol dehydrogenase increases ethanol production in Escherichia coli. *ACS Synthetic Biology*, *3*(12), 976–978. https://doi.org/10.1021/sb500020g.

Li, J., He, Z., Liang, Y., Peng, T., & Hu, Z. (2022). Insights into algal polysaccharides: A review of their structure, depolymerases, and metabolic pathways. *Journal of Agricultural and Food Chemistry*, *70*(6), 1749–1765. https://doi.org/10.1021/acs.jafc.1c05365.

Liang, Y., Si, T., Ang, E. L., & Zhao, H. (2014). Engineered pentafunctional minicellulosome for simultaneous saccharification and ethanol fermentation in Saccharomyces cerevisiae. *Applied Environmental Microbiology*, *80*(21), 6677–6684. https://doi.org/10.1128/aem.02070-14.

Liu, W., Bevan, D. R., & Zhang, Y. H. (2010). The family 1 glycoside hydrolase from Clostridium cellulolyticum H10 is a cellodextrin glucohydrolase. *Applied Biochemistry and Biotechnology*, *161*(1-8), 264–273. https://doi.org/10.1007/s12010-009-8782-x.

Liu, Z., Inokuma, K., Ho, S. H., Haan, R., Hasunuma, T., van Zyl, W. H., et al. (2015). Combined cell-surface display- and secretion-based strategies for production of cellulosic ethanol with Saccharomyces cerevisiae. *Biotechnology for Biofuels*, *8*, 162. https://doi.org/10.1186/s13068-015-0344-6.

Livi, G. P., Lillquist, J. S., Miles, L. M., Ferrara, A., Sathe, G. M., Simon, P. L., et al. (1991). Secretion of N-glycosylated interleukin-1 beta in Saccharomyces cerevisiae using a leader peptide from Candida albicans. Effect of N-linked glycosylation on biological activity. *Journal of Biological Chemistry*, *266*(23), 15348–15355.

Loose, J. S., Forsberg, Z., Kracher, D., Scheiblbrandner, S., Ludwig, R., Eijsink, V. G., et al. (2016). Activation of bacterial lytic polysaccharide monooxygenases with cellobiose dehydrogenase. *Protein Science*, *25*(12), 2175–2186. https://doi.org/10.1002/pro.3043.

Lynd, L. R., Weimer, P. J., van Zyl, W. H., & Pretorius, I. S. (2002). Microbial cellulose utilization: Fundamentals and biotechnology. *Microbiology and Molecular Biology Reviews*, *66*(3), 506–577. table of contents https://doi.org/10.1128/mmbr.66.3.506-577.2002.

Martinez-Anaya, C. (2016). Understanding the structure and function of bacterial expansins: A prerequisite towards practical applications for the bioenergy and agricultural industries. *Microbial Biotechnology*, *9*(6), 727–736. https://doi.org/10.1111/1751-7915.12377.

Martínez-García, E., Aparicio, T., Goñi-Moreno, A., Fraile, S., & de Lorenzo, V. (2015). SEVA 2.0: An update of the standard European vector architecture for de-/re-construction of bacterial functionalities. *Nucleic Acids Research*, *43*(Database issue), D1183–D1189. https://doi.org/10.1093/nar/gku1114.

Martínez-García, E., Goñi-Moreno, A., Bartley, B., McLaughlin, J., Sánchez-Sampedro, L., Pascual Del Pozo, H., et al. (2020). SEVA 3.0: An update of the Standard European vector architecture for enabling portability of genetic constructs among diverse bacterial hosts. *Nucleic Acids Research*, *48*(D1), D1164–d1170. https://doi.org/10.1093/nar/gkz1024.

Mikkelsen, M. D., Buron, L. D., Salomonsen, B., Olsen, C. E., Hansen, B. G., Mortensen, U. H., et al. (2012). Microbial production of indolylglucosinolate through engineering of a multi-gene pathway in a versatile yeast expression platform. *Metabolic Engineering*, *14*(2), 104–111. https://doi.org/10.1016/j.ymben.2012.01.006.

Moore, S. J., Lai, H. E., Kelwick, R. J., Chee, S. M., Bell, D. J., Polizzi, K. M., et al. (2016). EcoFlex: A multifunctional MoClo Kit for E. coli synthetic biology. *ACS Synthetic Biology*, *5*(10), 1059–1069. https://doi.org/10.1021/acssynbio.6b00031.

Morais, E. S., Lopes, A., Freire, M. G., Freire, C. S. R., Coutinho, J. A. P., & Silvestre, A. J. D. (2020). Use of ionic liquids and deep eutectic solvents in polysaccharides dissolution and extraction processes towards sustainable biomass valorization. *Molecules*, *25*(16). https://doi.org/10.3390/molecules25163652.

Mühlmann, M., Kunze, M., Ribeiro, J., Geinitz, B., Lehmann, C., Schwaneberg, U., et al. (2017). Cellulolytic robolector—Towards an automated high-throughput screening platform for recombinant cellulase expression. *Journal of Biological Engineering*, *11*, 1. https://doi.org/10.1186/s13036-016-0043-2.

Müller, G., Chylenski, P., Bissaro, B., Eijsink, V. G. H., & Horn, S. J. (2018). The impact of hydrogen peroxide supply on LPMO activity and overall saccharification efficiency of a commercial cellulase cocktail. *Biotechnology for Biofuels*, *11*, 209. https://doi.org/10.1186/s13068-018-1199-4.

Najah, M., Calbrix, R., Mahendra-Wijaya, I. P., Beneyton, T., Griffiths, A. D., & Drevelle, A. (2014). Droplet-based microfluidics platform for ultra-high-throughput bioprospecting of cellulolytic microorganisms. *Chemistry & Biology*, *21*(12), 1722–1732. https://doi.org/10.1016/j.chembiol.2014.10.020.

Nakatani, Y., Yamada, R., Ogino, C., & Kondo, A. (2013). Synergetic effect of yeast cell-surface expression of cellulase and expansin-like protein on direct ethanol production from cellulose. *Microbial Cell Factories*, *12*, 66. https://doi.org/10.1186/1475-2859-12-66.

Naseri, G., & Koffas, M. A. G. (2020). Application of combinatorial optimization strategies in synthetic biology. *Nature Communications*, *11*(1), 2446. https://doi.org/10.1038/s41467-020-16175-y.

Nielsen, A. A., Der, B. S., Shin, J., Vaidyanathan, P., Paralanov, V., Strychalski, E. A., et al. (2016). Genetic circuit design automation. *Science*, *352*(6281), aac7341. https://doi.org/10.1126/science.aac7341.

Ostafe, R., Prodanovic, R., Commandeur, U., & Fischer, R. (2013). Flow cytometry-based ultra-high-throughput screening assay for cellulase activity. *Analytical Biochemistry*, *435*(1), 93–98. https://doi.org/10.1016/j.ab.2012.10.043.

Parisutham, V., Chandran, S. P., Mukhopadhyay, A., Lee, S. K., & Keasling, J. D. (2017). Intracellular cellobiose metabolism and its applications in lignocellulose-based biorefineries. *Bioresource Technology*, *239*, 496–506. https://doi.org/10.1016/j.biortech.2017.05.001.

Patron, N. J., Orzaez, D., Marillonnet, S., Warzecha, H., Matthewman, C., Youles, M., et al. (2015). Standards for plant synthetic biology: A common syntax for exchange of DNA parts. *The New Phytologist*, *208*(1), 13–19. https://doi.org/10.1111/nph.13532.

Peralta-Yahya, P., Carter, B. T., Lin, H., Tao, H., & Cornish, V. W. (2008). High-throughput selection for cellulase catalysts using chemical complementation. *Journal of American Chemical Society*, *130*(51), 17446–17452. https://doi.org/10.1021/ja8055744.

Peretó, J. (2020). Transmetabolism: The non-conformist approach to biotechnology. *Microbial Biotechnology*. https://doi.org/10.1111/1751-7915.13691.

Pollak, B., Matute, T., Nuñez, I., Cerda, A., Lopez, C., Vargas, V., et al. (2020). Universal loop assembly: open, efficient and cross-kingdom DNA fabrication. *Synthetic Biology (Oxford, England)*, *5*(1), ysaa001. https://doi.org/10.1093/synbio/ysaa001.

Pretorius, I. S., & Boeke, J. D. (2018). Yeast 2.0-connecting the dots in the construction of the world's first functional synthetic eukaryotic genome. *FEMS Yeast Research*, *18*(4). https://doi.org/10.1093/femsyr/foy032.

Prielhofer, R., Barrero, J. J., Steuer, S., Gassler, T., Zahrl, R., Baumann, K., et al. (2017). GoldenPiCS: A golden gate-derived modular cloning system for applied synthetic biology in the yeast Pichia pastoris. *BMC Systems Biology*, *11*(1), 123. https://doi.org/10.1186/s12918-017-0492-3.

Pryor, J. M., Potapov, V., Bilotti, K., Pokhrel, N., & Lohman, G. J. S. (2022). Rapid 40 kb genome construction from 52 parts through data-optimized assembly design. *ACS Synthetic Biology*, *11*(6), 2036–2042. https://doi.org/10.1021/acssynbio.1c00525.

Radeck, J., Meyer, D., Lautenschläger, N., & Mascher, T. (2017). Bacillus SEVA siblings: A golden gate-based toolbox to create personalized integrative vectors for Bacillus subtilis. *Science Reports*, *7*(1), 14134. https://doi.org/10.1038/s41598-017-14329-5.

Ragauskas, A. J., Beckham, G. T., Biddy, M. J., Chandra, R., Chen, F., Davis, M. F., et al. (2014). Lignin valorization: Improving lignin processing in the biorefinery. *Science*, *344*(6185), 1246843. https://doi.org/10.1126/science.1246843.

Rakestraw, J. A., Sazinsky, S. L., Piatesi, A., Antipov, E., & Wittrup, K. D. (2009). Directed evolution of a secretory leader for the improved expression of heterologous proteins and full-length antibodies in Saccharomyces cerevisiae. *Biotechnology and Bioengineering*, *103*(6), 1192–1201. https://doi.org/10.1002/bit.22338.

Sadler, J. C. (2020). The bipartisan future of synthetic chemistry and synthetic biology. *Chembiochem*, *21*(24), 3489–3491. https://doi.org/10.1002/cbic.202000418.

Salehi Jouzani, G., & Taherzadeh, M. J. (2015). Advances in consolidated bioprocessing systems for bioethanol and butanol production from biomass: A comprehensive review. *Biofuel Research Journal*, *2*(1), 152–195.

Salinas, A. (2017). *A synthetic biology approach for green macroalgal biomass depolymerization. (Doctor of Philosophy)*. The University of Edinburgh.

Salis, H. M., Mirsky, E. A., & Voigt, C. A. (2009). Automated design of synthetic ribosome binding sites to control protein expression. *Nature Biotechnology*, *27*(10), 946–950. https://doi.org/10.1038/nbt.1568.

Sasaki, Y., Mitsui, R., Yamada, R., & Ogino, H. (2019). Secretory overexpression of the endoglucanase by Saccharomyces cerevisiae via CRISPR-δ-integration and multiple promoter shuffling. *Enzyme and Microbial Technology*, *121*, 17–22. https://doi.org/10.1016/j.enzmictec.2018.10.014.

Satari, B., Karimi, K., & Kumar, R. (2019). Cellulose solvent-based pretreatment for enhanced second-generation biofuel production: A review. *Sustainable Energy & Fuels, 3*(1), 11–62.

Scheller, H. V., & Ulvskov, P. (2010). Hemicelluloses. *Annual Review of Plant Biology, 61,* 263–289. https://doi.org/10.1146/annurev-arplant-042809-112315.

Shi, S., Liang, Y., Zhang, M. M., Ang, E. L., & Zhao, H. (2016). A highly efficient single-step, markerless strategy for multi-copy chromosomal integration of large biochemical pathways in Saccharomyces cerevisiae. *Metabolic Engineering, 33,* 19–27. https://doi.org/10.1016/j.ymben.2015.10.011.

Shin, S. K., Hyeon, J. E., Kim, Y. I., Kang, D. H., Kim, S. W., Park, C., et al. (2015). Enhanced hydrolysis of lignocellulosic biomass: Bi-functional enzyme complexes expressed in Pichia pastoris improve bioethanol production from Miscanthus sinensis. *Biotechnology Journal, 10*(12), 1912–1919. https://doi.org/10.1002/biot.201500081.

Silva, J. P., Ticona, A. R. P., Hamann, P. R. V., Quirino, B. F., & Noronha, E. F. (2021). Deconstruction of lignin: From enzymes to microorganisms. *Molecules, 26*(8). https://doi.org/10.3390/molecules26082299.

Silva-Rocha, R., Martínez-García, E., Calles, B., Chavarría, M., Arce-Rodríguez, A., de Las Heras, A., et al. (2013). The standard European vector architecture (SEVA): A coherent platform for the analysis and deployment of complex prokaryotic phenotypes. *Nucleic Acids Research, 41*(Database issue), D666–D675. https://doi.org/10.1093/nar/gks1119.

Stern, J., Moraïs, S., Ben-David, Y., Salama, R., Shamshoum, M., Lamed, R., et al. (2018). Assembly of synthetic functional cellulosomal structures onto the cell surface of Lactobacillus plantarum, a potent member of the gut microbiome. *Applied and Environmental Microbiology, 84*(8). https://doi.org/10.1128/aem.00282-18.

Stevenson, K., McVey, A. F., Clark, I. B. N., Swain, P. S., & Pilizota, T. (2016). General calibration of microbial growth in microplate readers. *Scientific Reports, 6,* 38828. https://doi.org/10.1038/srep38828.

Tanaka, T., & Kondo, A. (2015). Cell-surface display of enzymes by the yeast Saccharomyces cerevisiae for synthetic biology. *FEMS Yeast Research, 15*(1), 1–9. https://doi.org/10.1111/1567-1364.12212.

Tang, H., Wang, J., Wang, S., Shen, Y., Petranovic, D., Hou, J., et al. (2018). Efficient yeast surface-display of novel complex synthetic cellulosomes. *Microbial Cell Factories, 17*(1), 122. https://doi.org/10.1186/s12934-018-0971-2.

Tozakidis, I. E., Brossette, T., Lenz, F., Maas, R. M., & Jose, J. (2016). Proof of concept for the simplified breakdown of cellulose by combining Pseudomonas putida strains with surface displayed thermophilic endocellulase, exocellulase and β-glucosidase. *Microbial Cell Factories, 15*(1), 103. https://doi.org/10.1186/s12934-016-0505-8.

Tsai, S. L., DaSilva, N. A., & Chen, W. (2013). Functional display of complex cellulosomes on the yeast surface via adaptive assembly. *ACS Synthetic Biology, 2*(1), 14–21. https://doi.org/10.1021/sb300047u.

Valenzuela-Ortega, M., & French, C. E. (2019). Engineering of industrially important microorganisms for assimilation of cellulosic biomass: Towards consolidated bioprocessing. *Biochemical Society Transactions, 47*(6), 1781–1794. https://doi.org/10.1042/bst20190293.

Valenzuela-Ortega, M., & French, C. (2021). Joint universal modular plasmids (JUMP): A flexible vector platform for synthetic biology. *Synthetic Biology (Oxford, England), 6*(1), ysab003. https://doi.org/10.1093/synbio/ysab003.

Van der Vaart, J. M., te Biesebeke, R., Chapman, J. W., Toschka, H. Y., Klis, F. M., & Verrips, C. T. (1997). Comparison of cell wall proteins of Saccharomyces cerevisiae as anchors for cell surface expression of heterologous proteins. *Applied Environmental Microbiology, 63*(2), 615–620. https://doi.org/10.1128/aem.63.2.615-620.1997.

Walls, L. E., & Rios-Solis, L. (2020). Sustainable production of microbial isoprenoid derived advanced biojet fuels using different generation feedstocks: A review. *Frontiers in Bioengineering and Biotechnology, 8,* 599560. https://doi.org/10.3389/fbioe.2020.599560.

Wei, H., Wang, W., Alahuhta, M., Vander Wall, T., Baker, J. O., Taylor, L. E., 2nd, et al. (2014). Engineering towards a complete heterologous cellulase secretome in Yarrowia lipolytica reveals its potential for consolidated bioprocessing. *Biotechnology for Biofuels, 7*(1), 148. https://doi.org/10.1186/s13068-014-0148-0.

Wei, Q., Zhang, H., Guo, D., & Ma, S. (2016). Cell surface display of four types of Solanum nigrum metallothionein on Saccharomyces cerevisiae for biosorption of cadmium. *Journal of Microbiology and Biotechnology, 26*(5), 846–853. https://doi.org/10.4014/jmb.1512.12041.

Weng, C., Peng, X., & Han, Y. (2021). Depolymerization and conversion of lignin to value-added bioproducts by microbial and enzymatic catalysis. *Biotechnology for Biofuels, 14*(1), 84. https://doi.org/10.1186/s13068-021-01934-w.

Wightman, E. L. I., Kroukamp, H., Pretorius, I. S., Paulsen, I. T., & Nevalainen, H. K. M. (2020). Rapid optimisation of cellulolytic enzymes ratios in Saccharomyces cerevisiae using in vitro SCRaMbLE. *Biotechnology for Biofuels, 13*(1), 182. https://doi.org/10.1186/s13068-020-01823-8.

Yang, X., Tang, H., Song, M., Shen, Y., Hou, J., & Bao, X. (2019). Development of novel surface display platforms for anchoring heterologous proteins in Saccharomyces cerevisiae. *Microbial Cell Factories, 18*(1), 85. https://doi.org/10.1186/s12934-019-1133-x.

Yong, W. T. L., Thien, V. Y., Rupert, R., & Rodrigues, K. F. (2022). Seaweed: A potential climate change solution. *Renewable and Sustainable Energy Reviews, 159,* 112222.

Zaky, A. S., Carter, C. E., Meng, F., & French, C. E. (2021). A preliminary life cycle analysis of bioethanol production using seawater in a coastal biorefinery setting. *Processes, 9*(8), 1399.

Zhang, Y., Wang, J., Wang, Z., Zhang, Y., Shi, S., Nielsen, J., et al. (2019). A gRNA-tRNA array for CRISPR-Cas9 based rapid multiplexed genome editing in Saccharomyces cerevisiae. *Nature Communications, 10*(1), 1053. https://doi.org/10.1038/s41467-019-09005-3.

Zhao, C., Zhang, Y., & Li, Y. (2019). Production of fuels and chemicals from renewable resources using engineered Escherichia coli. *Biotechnology Advances, 37*(7), 107402. https://doi.org/10.1016/j.biotechadv.2019.06.001.

Zhou, S., & Ingram, L. (2001). Simultaneous saccharification and fermentation of amorphous cellulose to ethanol by recombinant Klebsiella oxytoca SZ21 without supplemental cellulase. *Biotechnology Letters, 23*(18), 1455–1462.

Zhu, Y., & McBride, M. J. (2017). The unusual cellulose utilization system of the aerobic soil bacterium Cytophaga hutchinsonii. *Applied Microbiology and Biotechnology, 101*(19), 7113–7127. https://doi.org/10.1007/s00253-017-8467-2.

Recombineering

Asheemita Bagchi, Shreyoshi Karmakar, Virendra Swarup Bisaria, and Preeti Srivastava*

Department of Biochemical Engineering and Biotechnology, Indian Institute of Technology Delhi, Delhi, India

Corresponding author: e-mail address: preeti@dbeb.iitd.ac.in

Abbreviations

ALFIRE	alternative and enhanced method for large fragment recombineering
BAC	bacterial artificial chromosome
BRED	Bacteriophage Recombineering of Electroporated DNA
CRISPR-Cas	clustered regularly interspaced short palindromic repeats and associated proteins
DADA-PCR	deletion amplification assay PCR
DSB	double-stranded break
flp	Flippase gene
FRT	Flippase recognition target
FRUIT	flexible recombineering using integration of *thyA*
galK	galactokinase gene
GFP	green fluorescent protein
IPTG	isopropylthio-β-galactoside
kanR	kanamycin resistance gene
MAMA-PCR	Mismatch Amplification Mutation Assay PCR
MCS	multiple cloning site
mTERT	mouse telomerase reverse transcriptase
ORBIT	oligonucleotide mediated recombineering followed by Bxb1 integrase targeting
PAC	P1 Artificial Chromosome
PCR	polymerase chain reaction
SDS	sodium dodecyl sulfate

Methods in Microbiology, Volume 52, ISSN 0580-9517, https://doi.org/10.1016/bs.mim.2023.01.005

1 Introduction

Recombineering, as the name itself suggests, is engineering by recombination. It is a very modern *in vivo* method in the genome editing field which does not depend on restriction sites and ligase-mediated cloning strategies. In contrast, it depends on homologous recombination, which uses substrates like single-stranded DNA, double-stranded DNA and PCR-amplified substrates (Warming et al., 2005; Zhang, Buchholz, Muyrers, & Stewart, 1998). Homologous recombination occurs between two fragments of DNA having identical sequences. It is a very adept pathway of repair mechanism in higher and lower organisms in case of any mutation or breakage in DNA. It leads to precise exchange of genetic material between two molecules with seamless joining of the two (Court, Sawitzke, & Thomason, 2002). Recombineering trumps *in vitro* cloning methods of creating a recombinant DNA molecule in a way that saves time and the intensive labour demanding nature that is accompanied with *in vitro* cloning. It also does not need a plasmid or even a phage DNA material to be made; instead, it can even make use of synthetically synthesised oligonucleotides which bear homology to the target sequence. Genetic engineering which is performed with phage-encoded recombination factors that utilise short homologies has been named 'recombineering'. Recombineering has a plethora of uses (Sharan, Thomason, Kuznetsov, & Court, 2009). Some of them are listed below:

- insertion of a selectable marker
- insertion of a non-selectable marker
- securing and retrieval of a particular fragment in the genomic DNA of an organism
- changing the genetic makeup subtly, for instance, by introducing very precise point mutations, insertions or even deletions in a genetic material, chromosomal or episomal.

2 History and development of recombineering

Though recombineering was developed in bacterial hosts, much of the work was done using *Saccharomyces cerevisiae* as early as 1983 (Orr-Weaver, Szostak, & Rothstein, 1983). Herein, the authors designed a linear plasmid where a double-stranded break can be introduced by a restriction endonuclease to initiate the integration of the plasmid into a particular locus of the chromosomal DNA of yeast. Soon after this, in 1988, Moerschell, Tsunasawa, and Sherman (1988) successfully transformed yeast DNA with synthetically synthesised oligonucleotides as short as only 20 base pairs. The reason for choosing yeast as the primary host for carrying out recombineering is that it houses a very efficient pathway for double-stranded break and repair recombination, as noted by Yu et al. (2000). But as they also remarked, the manipulation of the integrated DNA after recombination takes place is difficult to perform, as opposed to working out the same process with a simple prokaryote, like

Escherichia coli. So, gradually over time, scientists working on recombineering started to accommodate many prokaryotes in the arena. As will be discussed later in the chapter, recombineering has been achieved on many Gram-negative strains, namely *E. coli*, *Klebsiella*, *Yersinia*, *Salmonella*, *Shigella*, etc.

3 Molecular tools for recombineering

Before getting into the details of the tools needed for recombineering, one needs to have a thorough knowledge of the various recombination repair systems present in *E. coli*. It primarily has two categories of recombination systems, RecBCD and RecF (Court et al., 2002). Both of these systems require the usage of a protein called RecA, a single stranded binding protein, which binds to the ssDNA ends and forms a complex with the DNA which is capable of searching homologies across the length of the DNA molecule.

3.1 The RecBCD system

The RecBCD system is credited with participating in 99% of recombination events at the points of double-stranded breaks (DSBs) in Wild Type *E. coli*. The enzyme has many activities, like single stranded *exo*- and *endo*-nucleases, DNA-dependent ATPase and helicase activity (Kowalczykowski, 2000). It is characterised by unwinding double-stranded DNA (dsDNA), while also adeptly cleaving the single stranded DNA, and stabilising it by helping single stranded binding proteins (SSBs) to bind to the ssDNA. Kowalczykowski (2000), also schematically depicts how the RecBCD enzyme loads the RecA protein on to the single stranded DNA (ssDNA), which further helps RecA to initiate strand exchange and consequent recombination (Court et al., 2002).

3.2 The RecF pathway

The RecF pathway partakes in the same dsDNA end repair but at a much reduced frequency, and it involves proteins which more or less have similar functions and nature as the RecBCD pathway.

   Alongside this, *E. coli* also gains recombination systems by a phage integrated in its genome.

3.3 Lambda phage Red functions

This includes *exo* (also known as *redα*) and beta (*redβ*), assisted by *gam* shown in Fig. 1. These three genes are located right next to each other in the *pL* operon of the Lambda phage.

   As the name suggests, *exo* is a 5′ to 3′ dsDNA dependent exonuclease, which degrades the DNA from the 5′ end of the dsDNA break site, and *beta* binds to the

| attL | int | xis | hin | **exo** | **bet** | **gam** | kil | T | N | pL |

FIG. 1

The pL operon of Lambda prophage: The *att* sites in the phage genome are where the recombination takes place. It is on these *att* sites are sites for the Int protein (gene product of *int*, next to *attL* in the diagram) to bind, facilitating the recombination event. Xis, the gene product of *xis*, is needed for the excision event of the phage genome, to get integrated inside the bacterial genome (Hsu, Ross, & Landy, 1980). The hin gene imparts 'host inhibitory' effect, altering the host cell's physiology in a number of ways (Court, Gottesman, & Gallo, 1980). The gene product of the next gene, *exo*, is a key player in the recombination event. Being a 5′ to 3′ exonuclease, it degrades the 5′ end of the linear dsDNA, leaving 3′ single stranded ends. Following that is the gene *bet*, coding for the Beta product, whose function is to bind the protruding ssDNA ends and help in single stranded annealing between two complementary DNA molecules. Gam, the gene product of *gam*, provides protection to the same 3′ single stranded ends of the DNA from being degraded by the host cell's RecBCD system (Sharan et al., 2009). Kil, the gene product of the gene *kil*, right next to *gam*, essentially kills the host bacterial cell. Haeusser et al., 2014 reports that the protein is capable of stopping *E. coli* cell division, making the cells assume long filamentous shape and the eventually die. Protein N stimulates the lytic phase of the Lambda phage (Conant, Goodarzi, Weitzel, & von Hippel, 2008). It allows extension of transcription from the promoter *pL* (Court et al., 1980).

Figure adapted from Court, Donald L., James A. Sawitzke, and Lynn C. Thomason. (2002). "Genetic engineering using homologous recombination." Annual Review of Genetics 36: 361–88.

ssDNA tails which are not degraded by *exo*. It prepares the ssDNA for the imminent homologous strand invasion (Fig. 2) (Yu et al., 2000). Beta has the ability to promote pairing and annealing of two complementary single strands of DNA. It is interesting to note that these two proteins interact with each other, as has been confirmed by co-purification (Court et al., 2002). Another protein encoded by the gene, *gam*, protects the cellular linear DNA from random attacks by the RecBCD complex. By virtue of this ability, gam serves to protect the integrity of the transformed genetic material (sometimes a linear DNA) in the cellular environment (Sharan et al., 2009).

Phage mediated recombination can be dependent on RecA as well as completely independent of it. This is because Lambda phage encodes its own recombination functions, which enable homologous recombination to take place even in an *E. coli* strain which bears a mutation in *rec*A, rendering it non-functional.

3.4 Rac prophage encoded RecE and RecT

These genes are encoded by a cryptic Lambda like phage in the *E. coli* genome, called the Rac prophage. Their function is analogous to the Lambda encoded *exo* and *beta* genes (Court et al., 2002); in fact, they act cooperatively as *exo* and *bet* genes in Lambda phage.

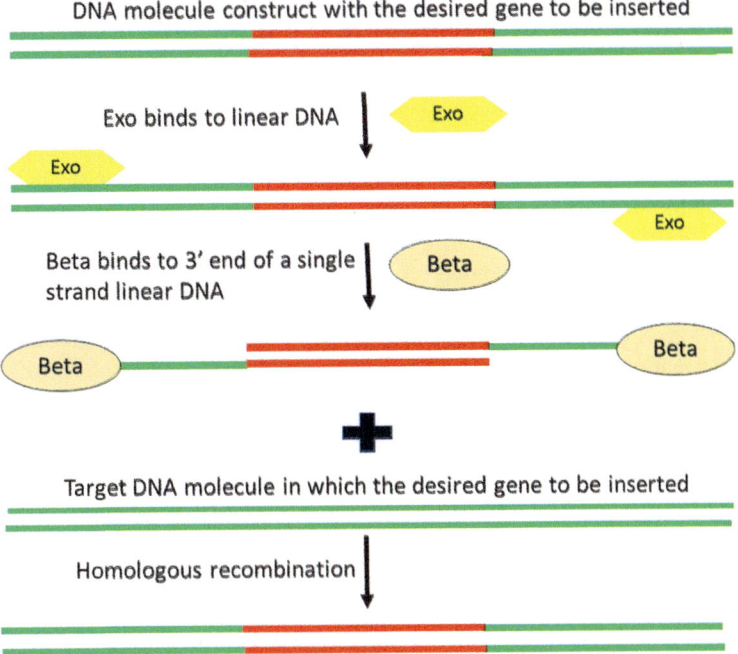

DNA molecule construct with the desired gene to be inserted

Exo binds to linear DNA

Beta binds to 3′ end of a single strand linear DNA

Target DNA molecule in which the desired gene to be inserted

Homologous recombination

FIG. 2

Homologous recombination mediated recombineering: In step 1. As the protein Exo binds to the ends of the dsDNA molecule, by virtue of its 5′ to 3′ exonuclease activity it leaves 3′ single stranded overhangs (green lines in the figure denotes these single stranded ends), which are then bound by the Beta protein, thus promoting single stranded annealing between two complementary DNA molecules. The Gam protein (not shown in the figure) essentially helps in providing protection to the 3′ ends of the linear DNA molecules from getting degraded by the host's exonuclease complex, RecBCD.

Lambda-mediated recombineering was first adapted in *E. coli* and 50 bases of homologous sequence was used (Copeland, Jenkins, & Court, 2001; Ellis, Daiguan, DiTizio, & Court, 2001; Yu et al., 2000; Zhang et al., 1998; Zhang, Muyrers, Testa, & Stewart, 2000). After that, this system was developed for other Gram-negative bacteria like *Salmonella* and *Shigella* (Marinelli, Hatfull, & Piuri, 2012). Later on, this system was also adapted for BAC (Bacterial artificial chromosome) which was developed from a low-copy number F1 plasmid in *E. coli* for cloning larger DNA fragments (up to 300 kb) (Shizuya et al., 1992). Similarly, it was also adapted in P1 Artificial Chromosome (PAC) which was originally developed by Sternberg (1992). These recombineered BAC and PAC were also thoroughly used to produce conditional and targeted genome wide knock-out and knock-in in embryonic stem cells to produce transgenic mice (Chan et al., 2007; Liu, Jenkins, & Copeland, 2003; Sharan et al., 2009).

4 Steps involved in a typical recombineering experiment
4.1 Substrate DNA and its meticulous design

The DNA substrates vary depending upon the type of application. They can be linear ssDNA or dsDNA. For example, for inserting a selectable marker into the host, the targeting substrate DNA needs to be a linear dsDNA with the gene for the selectable marker flanked by regions bearing homology to the host's genetic material. In the case of generating precise subtle mutations (say, point mutations), we need to design single stranded oligonucleotides or denatured PCR products as the substrate. For retrieving a genomic fragment from the chromosomal DNA of the organism, the substrate will be linear dsDNA having a selectable marker (like an antibiotic marker) flanked by homology arms and a plasmid origin of replication (Sharan et al., 2009).

4.2 Provision for the Lambda Red recombination genes

There are several ways to provide the prophage, the recombination proteins in the *E. coli* or other host cells. The preference depends on the target DNA that needs to be electroporated. They are described below:

4.2.1 For bacterial chromosomal DNA

For this, replication defective Lambda phage (λ-TetR), Mini-λ or mobile recombineering systems like pSIM plasmids can be used. Mini-λ and pSIM can be used to manipulate the chromosomal DNA and BAC. These two elements provide Red recombination factors and endogenous phage controlling elements.

(a) Mini-λ is a defective, non-replicating prophage, having *att* sites where attachment occurs, *int* and *xis* sites through which the phage integrates into the genome and excises from the genome, respectively. It also has a majority of the lytic phase cycle genes deleted which further ensures stable maintenance of the phage in the lysogenized state (Sharan et al., 2009). They are capable of recombineering due to the presence of *exo, bet* and *gam* genes. It also confers the ability for selection of the transformants because it carries an antibiotic resistance gene.

(b) pSIM vectors also have all the elements needed for recombineering and are available with a number of origins of replication from different plasmids but unlike the Mini-λ, it needs drug selection for stable maintenance (Datta, Costantino, & Court, 2006).

(c) λTetR is a Lambda prophage carrying a tetracycline resistance gene. As soon as it gets introduced into a BAC containing strain, the prophage remains stable without the need for constant drug selection.

4.2.2 *For high and low copy number plasmids*

A bacterial strain, named DY380 with a defective Lambda prophage stably integrated and maintained in the chromosome may be used for this. These strains also have a mutated *rec*A gene. These strains can also be used for recombineering with BACs (Chan et al., 2007; Lee et al., 2001).

4.3 Inducing the Red genes

Induction essentially means the way in which phage production is triggered in a lysogen inside the host bacterium. This process helps to switch the phage from a lysogenic cycle to the lytic cycle. So, through induction, we can bring about phage-mediated lysing of the host cells to obtain intact phages in the cell lysate having the desired recombineered gene or mutant gene (Kowalczykowski, 2000; Rokney et al., 2008; Serra-Moreno et al., 2006; Svenningsen, Costantino, Court, & Adhya, 2005). Usage of a temperature- sensitive mutation in the cI_{857} repressor gene to do away with the inconveniences of plasmid instability in recombinants was stated by Chen, Lin, and Lim (1995). The idea was to use the switch between lysogenic state and lytic stage for target protein production by increasing the temperature from 28 degrees to 42 degrees. The gene, cI_{857} is susceptible to denaturation at 42 degrees, which is why it is the optimum temperature to initiate the switch from lysogenic to lytic phase. Thus, induction of the Red recombination genes is brought about by incubating the culture (harboring either the pSIM series of plasmids, Mini-λ or defective prophage) in its mid-log phase at 42 °C for 15 min. It is absolutely necessary to prevent the growth of the culture beyond its logarithmic phase, as recombination may prove to be difficult in stationary phase. After the heat induction, it is important to chill the cells in ice-cold water.

4.4 Electroporation of the construct in the desired host

Electroporation of the target substrate DNA should be done to the chilled cells after washing them with ice cold sterile water to remove the salts which may impede the path of electric pulse given during electroporation. Immediately after the pulse, sterile Luria broth is added to the cells to initiate the recovery.

4.5 Growing and maintaining the electroporated cells

A minimum of 30-min recovery period for *E. coli* strains is needed for the electroporated cells before they are plated. This recovery period varies according to the doubling time of the host bacterium. Often, an outgrowth period is required to allow the expression of antibiotic resistance gene cassettes.

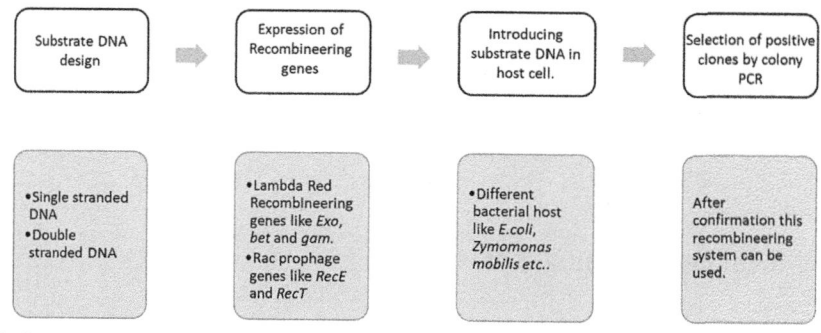

FIG. 3

General steps of a recombination experiment: The first step of a typical recombineering experiment is to design the substrate DNA depending upon the purpose, that is single stranded (a denatured PCR product, or synthetically synthesised oligonucleotide) or double-stranded DNA molecule (might contain a gene for a selectable marker bound by homology arms). The next step is the induction or expression of the Lambda recombineering genes. This ensures that the genes of the phage required for the recombination event to be expressed inside the host bacterial cell. The third step of the protocol dictates that the substrate DNA be introduced into the host bacterial cell, usually by means of electroporating the construct. The positive colonies are then used for downstream purposes.

4.6 Selection and recombination of the clones

Drug resistant recombinants can be identified in the presence of a suitable antibiotic and then further confirmed by performing a PCR with the gene specific primers. Restriction digestion can also be carried out to confirm the same. For large inserts like BACs, Southern Blotting may be carried out. All of these can then be followed by DNA sequencing (Fig. 3).

5 Uses of recombineering

5.1 Recombineering methods for inserting a selectable marker into the bacterial chromosome

Recombineering can be used for inserting a selectable marker into the bacterial chromosome. The desired selectable marker (e.g. antibiotic resistance) is constructed by PCR amplification. The linear DNA coding for the antibiotic resistance gene cassette should be flanked by 50 base pairs of sequence homology which can recombine with the target sequence. These primers are about 70 bp and chimeric in nature, meaning they have two different types of sequence specifically constructed to bind to two different regions of the template. As is expected, one part will bear homology to the target region of the DNA where the recombination is about to take place, and the other part will contribute to the amplification of the selectable marker gene cassette (say, antibiotic resistance, Neomycin or kanamycin resistance). This strategy can also be used to disrupt a gene in the host chromosomal or episomal DNA (Sharan et al., 2009).

The PCR-generated construct is electroporated in a bacterial cell where recombineering genes are induced. After the transformation of an electrocompetent cell, recombinant clones are first selected by plating them on an antibiotic plate, followed by confirmation of the clone by colony PCR. However, Lee et al. (2001), noted the disadvantage of using such a selectable marker as it sometimes may hamper the functionality of the targeted loci. This problem can be solved by using a different set of recombinase proteins, Flp and Cre, which help in removing the antibiotic resistance gene cassette, as will be explained later.

5.2 Recombineering can be used for inserting non-selectable DNA fragments (Sharan et al., 2009)

A non-selectable DNA fragment is usually a reporter gene (*lacZ*), and a transgene (*flp*, or *GFP* or *His*-tag). The difference between a selectable marker gene and a non-selectable gene is that if the host houses the selectable marker gene, it will be allowed to grow in a specific media. For example, if the host bacterium is a recombinant having a kanamycin resistance gene, it will be able to grow on a kanamycin plate, while the non-recombinants will fail to do so. However, a non-selectable gene will not exactly inhibit the growth of the non-recombinant, but will definitely impart some properties (biochemical, physiological or even morphological) to the recombinant which will help in distinguishing it from the non-recombinant. In the above example, GFP (Green Fluorescent Protein) is a non-selectable reporter gene which imparts the property of fluorescence to the recombinant bacterium in contrast to the non-recombinant. In the case of *lacZ*, the recombinant will be able to utilise lactose as a carbon source. Two types of methods are used for inserting this into the bacterial chromosome:

5.2.1 Seamless method

Seamless method makes use of a marker which can code for both positive as well as negative selection. As shown in Fig.4, the marker can be a gene coding for the enzyme, galactokinase (*galK*), which confers the ability to process lactose sugar. They allow selection for the absence of a marker instead of its presence. The process consists of two steps:

(1) Insertion of the counter-selectable (which will inhibit the growth of the host cell in case of its absence) gene cassette into the target site.

(2) A second recombination event which will place another gene to destroy the property of the marker inserted in the first step.

There are several counter-selectable markers that are used by researchers. For example, *sacB* gene (from *B. subtilis* encoding *sacB* which produces levansucrase to convert sucrose to levans) which is linked to a drug marker such as *kan* (conferring

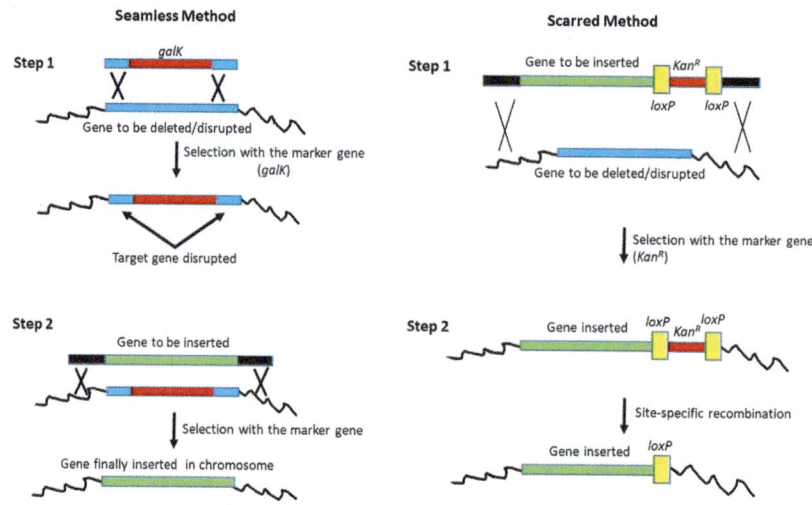

FIG. 4

Recombineering strategies: Seamless method and Scarred method: In the first step of the seamless method, the red box denotes a counter-selectable gene, or *galK*, which confers positive selection to the host colonies in which it is present. It is flanked by blue arms which bear homology to the target gene that needs to be deleted/disrupted at the end of Step 1 (read: Section 5.2.1). So, step 1 ends with the counter-selectable marker, or *galK* inserted in between the target gene (blue). Now, in step 2, a green box denotes the gene to be ultimately inserted in the final recombinant DNA molecule, flanked by black arms bearing homology to the final DNA molecule obtained from Step 1 (black curved lines). As a result, the final recombinant DNA molecule, lacks the gene cassette for the counter-selectable marker. In the scarred method, the red box denoted a counter selectable marker, *Kan^R*, flanked by *loxP* (denoted by yellow rectangle in the figure) sites, or *FRT* sites, which are the sites of proteins like Cre recombinase and Flippase proteins, respectively. The green box denotes the gene to be inserted, and along with the non-selectable (yellow) and selectable (red) markers, it is flanked with DNA sequences homologous to the black regions flanking the gene to be deleted/disrupted (denoted in blue). After the first step, the fragment having gene to be inserted as well as the marker genes get inserted. Cre protein then acts upon the *loxP* sites flanking the Kanamycin resistance gene, starting recombination. Leading to the final product without the Kanamycin resistance gene with one *loxP* site remaining after recombination between the two homology bearing black curved lines in the two DNA molecules (as is shown in the figure). This sole *loxP* site leaves a 'scar' in the DNA molecule.

kanamycin resistance), *galK* (galactokinase gene which confers the ability to process lactose) (Warming et al., 2005), *rpsL* (encoding the ribosomal subunit protein (S12) target of streptomycin) (Rivero-Müller, Lajić, & Huhtaniemi, 2007), *thyA* (encoding thymidylate synthetase, which confers sensitivity to trimethoprim) (Wong et al., 2005) and *tolC* [encoding 1.5 kb monomer of a homotrimer pore providing resistance to SDS and conferring sensitivity to bactericidal colicin E1 (colE1)]. In this method, the property of the counter-selectable marker is used at every step, first the recombinant is selected against the presence of the counter-selectable or selectable marker gene (in Fig. 4, it is denoted as *galK*, in red). After the second recombination step, the final recombinant ultimately does not have the marker gene (or *galK*). Since the marker gene or any of its fragment isn't found in the final sequence (in the form of a "scar" in the genetic material of the host), this method is deemed 'seamless'.

5.2.2 Scarred method

This method is based on *loxP* or *FRT* sites which utilises excision by expression of the *Cre* and *Flp* proteins (Yu et al., 2000). *Cre-Lox* recombination involves targeting a specific region of DNA employing the Cre-recombinase enzyme. *Cre* is a site-specific DNA recombinase which catalyses recombination in a specific DNA site (Gronostajski & Sadowski, 1985; Schnütgen, Doerflinger, Calléja, et al., 2003). These sites are known as *loxP* sequence. *Flp-FRT* recombination is another kind of site-specific recombination. *Flp* (Flippase) is a site-specific DNA recombinase which catalyses site-specific recombination. These sites are known as FRT sequences (Gronostajski & Sadowski, 1985).

In the 'scarred' method, a linear DNA with a non-selectable marker (yellow box in Fig. 4), linked to a selectable marker (red box denoting the gene for kanamycin resistance in Fig. 4) is electroporated inside the host (Fig. 4). As noted by Lee et al. (2001), the removal of the selectable marker gene may be required which can be performed by using the flanking non-selectable sequences, like the *loxP* sites or FRT sites which will be acted upon by the Cre recombinase proteins(Sharan et al., 2009). These sites will then be acted upon by the Cre recombinase or Flippase proteins and upon their action, the selectable marker gene cassette will be removed, leaving, however, a single *loxP/FRT* site termed as a 'scar' in the final recombinant DNA molecule, unlike the case of the 'seamless' method.

6 Regulation and expression of recombineering gene

The recombineering system uses phage-derived recombination proteins like *Exo, Beta*, and *Gam* for Lambda Red System and *RecE* and *RecT* for Rac Prophage (Zhang et al., 1998). As explained before (Section 3.3, Fig. 2), Exo is an exonuclease that creates a single-stranded end from double-stranded DNA through its $5'$ to $3'$ dsDNA exonuclease activity. *Beta* is a gene product of the *Bet* gene. It is a ssDNA-binding protein that helps to anneal two complementary DNA molecules. Gam (gene product of the *gam* gene) prevents RecBCD, which has nuclease activity for degrading linear DNA fragments. Gam gene helps to prevent this RecBCD nuclease (Karu, Sakaki, Echols, & Linn, 1975).

It is observed that the expression of the Lambda recombineering genes is to be regulated strictly to reduce the toxic effect of Gam protein and also to ensure that random and undesirable genomic rearrangements do not take place due to a prolonged

Table 1 The most commonly used modes of regulating the expression of recombineering genes.

Serial No.	Mode of regulation	Advantages	Disadvantages	References
1	IPTG-inducible lac promoter	The longer exposure time of IPTG can increase the recombination frequency multifold	The *lacI^q* repressor gene, which in *cis* is very tightly regulated	Murphy (1998) and Murphy, Campellone, and Poteete (2000)
2	Arabinose-inducible pBAD promoter	Not a tightly regulated promoter which gives ease of induction and repression	It still fails to provide a fool-proof tight regulation of the expression of the recombineering genes	Datsenko and Wanner (2000)
3	Lambda's own endogenous regulation, *pL* promoter	No leaky expression, very tightly regulated		Sharan et al. (2009)

recombination event (Sergueev, Yu, Austin, & Court, 2001; Sharan et al., 2009). Thus, it is recommended generally to supply a high-level induction of the Lambda recombination genes but only for a brief period of time (Sergueev et al., 2001). Following are the ways the regulation of these genes can be modulated (also tabulated in Table 1).

6.1 *Lac* promoter

Most recombineering systems involve the recombineering genes located in multicopy plasmids. Gene replacement using Lambda phage recombination takes place within the *lacZ* gene in Wild Type *E. coli* strains with a rate which is 15–130 times higher than that of the strains which are usually used for recombination, i.e., *recBC* negative (Murphy, 1998). The recombineering genes are under the control of an IPTG-inducible *lac* promoter, the *lacI^q* repressor gene, which in *cis* is very tightly regulated (Murphy, 1998; Murphy et al., 2000).

6.2 Arabinose promoter

An arabinose-inducible P_{BAD} promoter might prove to be more suitable for usage as it is not tightly regulated (Datsenko & Wanner, 2000). It must be noted that when pBAD plasmids are used for recombineering, arabinose is added generally for making electrocompetent cells (Datsenko & Wanner, 2000; Guzman, Belin, Carson, & Beckwith, 1995). The P_{BAD} promoter gives a less tight repression system than using the *lac* promoter (Guzman et al., 1995; Wang et al., 2006). Even then, with all its ease of induction and decent repression, it still fails to provide a tight control on the recombineering genes. Therefore, scientists started utilising the phage's endogenous regulatory system, which is described in the next point.

6.3 Lambda phage's own promoter-repressor system

The lambda repressor binds cooperatively at the three operator sites in p_L and p_R promoters in the phage DNA sequence. These repressor proteins interact with each other to give rise to a complex of 12 repressor proteins "handcuff"-ed with each other. To replicate this strict regulation, these sites are provided in every prophage construct designed for recombineering, be it in the host chromosome, or a plasmid (Sharan et al., 2009). The repressor is active at low temperatures of 30 °C to 34 °C, and as a result, the recombination genes are not expressed. But at 42 °C, repressors become inactive by denaturation, and recombination genes are expressed at a high level from the *p*L promoter. Again if the temperature is decreased below 42 °C, the repressor is renatured and tight repression is restored (Murphy & Campellone, 2003; Sergueev et al., 2001; Sharan et al., 2009). The advantage of using the endogenous repression system is its excellent control over the recombination genes and circumventing any chances of leaky expression that is meted out by heterologous promoters.

7 Recombineering in various systems
7.1 In BAC (bacterial artificial chromosome)

Recombineering is now used as a technique to perform subcloning. This technique allows precise insertion of a large DNA fragment in a BAC system. For large genomic DNA insertion, BAC (Bacterial Artificial Chromosome) and PAC (P1 Artificial Chromosome) systems are used. In the PAC system, the insert can be 50–100 kb whereas for BAC, the insert size can be >100 kb. (Sharan et al., 2009). Furthermore, BACs offer the ease of working with simple DNA purification tools, the consequence of this is that they are ideal for cloning not just the ORF of a gene, but its promoters and regulatory elements, like terminators, enhancers, etc. This is useful for geneticists when developing transgenic mice, by enabling the study of expression patterns and functional studies of the gene cloned within its endogenous expression pattern (Heaney, Rettew, & Bronson, 2004). For such advantages BAC systems have been used in Embryonic Stem (ES) cell technology and to knock-in and knock-out mice.

BAC recombineering bypasses the limitations of PCR amplification and the difficulties of purifying a larger DNA construct. It uses rare restriction enzyme sites and positive selection markers to exchange very large genetic fragments between a donor BAC molecule to an acceptor BAC molecule. The previously established BAC recombination system made use of negative selection markers and was perfectly capable of excising and integrating sequences up to 2 kb in length. However, with this technology developed by Zhao, Wang, and Zhu (2011) the group has worked with short sequences of homologous regions (called Homology Arms or 'HA'), and a temperature sensitive prophage, homing endonucleases and have developed a technology to insert large genomic fragments of the human gene telomerase (in nature, a reverse transcriptase, termed as *TERT*) into the corresponding genetic loci of mouse. They used two selection markers, kanamycin resistance gene, and *galK*.

BAC recombineering system was developed with different strategies like three-step, four-step, and ALFIRE recombineering. All of these help in larger DNA fragment manipulation by using positive and negative selection markers into the ends of DNA fragments (Sharan et al., 2009).

7.1.1 Three step strategy

This technique was reported by Zhu and co-workers (Zhao et al., 2011). Here, the 5′ intergenic region of the gene, *mTERT* in mice was replaced by the corresponding human gene. This was achieved in three steps. The crucial part for the success of the method is to design primers which is described here. In the first step, (as is shown in Fig. 5), a selectable marker (for example, *galK* gene) is inserted into one end of the donor sequence, which is the human gene. This intermediate product is then amplified using PCR primers. The primers are designed in such a way so as to introduce a restriction site, few sequences homologous to the human gene, and a longer stretch homologous to the mouse gene. Here, the *galK* cassette has been amplified from a plasmid pGalK (Warming et al., 2005). The resultant fragment was then electroporated into electrocompetent cells and selected on galactose plates. Following this, the *galk* gene was deleted by recombination between the adjoining human and mouse gene sequences; thus the final selection, or the third step, is done against galactose (a detailed explanation of this method can be seen in Fig. 5) (Zhao et al., 2011).

7.1.2 Four step strategy

This strategy is used to insert 14.6 kb of the mouse intergenic region in place of the 6.6 kb human counterpart. Instead of only one selectable marker, here two markers are used, *galK* and Kanamycin resistance cassette (amplified from pGalK and pREP4, Warming et al., 2005).

This approach gives us a strategy by positive/negative selection markers without leaving any unwanted sequences (Zhao et al., 2011) (Fig. 6).

7.1.3 ALFIRE (assisted large fragment insertion with red/ET recombination)

Although three steps, four-steps, and multi-steps recombineering were used for larger DNA fragments, they were still not very successful. So, to overcome this challenge ALFIRE was established (Rivero-Müller et al., 2007). ALFIRE stands for Assisted Large Fragment Insertion with Red/ET-recombination. In this recombineering technique, we can modify longer DNA fragments by introducing I-*Sce*I recognition sites into BACs. I-*Sce*I is a homing endonuclease capable of recognising and cutting within an 18base pair target region, which tolerates few base substitutions. Homing endonucleases are dsDNA recognising DNases which have a large,

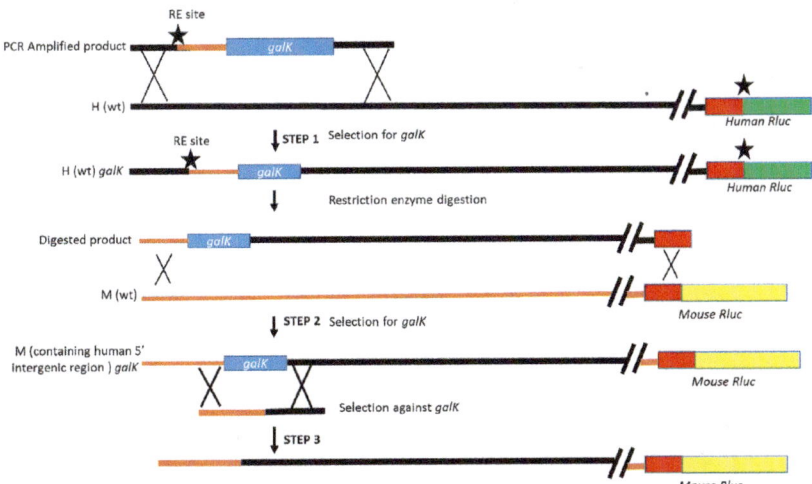

FIG. 5

The 3-step strategy: In this figure, the black bars denote human (wt) sequences and the orange bars indicate that of mouse (wt). At first a dsDNA PCR product is made which contains *galK* gene in the middle (blue box), sequences homologous to the intergenic region of the mouse *TERT* gene (*mTERT*) which has been shown in orange, and flanking these on both sides are sequences homologous to the intergenic region of the human *TERT* gene (*hTERT*). Just in the junction of the mTERT and hTERT sequences, a site for a restriction endonuclease has been inserted (denoted by a star). Once electroporated in human cell lines (1st step), they undergo homologous recombination, thus having the *galK* gene along with the *mTERT* region successfully inserted in the human chromosome. Now, in the first exon of both the mouse and the human gene is a cassette denoted as *Rluc*, which is a reporter gene. This region will be referred to as the HA, or Homology Arm). Both the green and the red parts indicate the entirety of this same cassette (the sequences in the cassette which are different in human and mouse are shown in green and yellow, respectively). The star again, depicts the site for the same restriction endonuclease as was inserted in the PCR product. After digestion with the restriction endonuclease, we have a molecule with the *galK* gene (blue), *mTERT* region (orange bar), *hTERT* region (black), followed by the 5′ end sequence of the *Rluc* HA (depicted in red). This product is then capable of recombination with the M(wt) chromosome (denoted in orange with the *Rluc* cassette here denoted in red and yellow). So, in step 2, this entire fragment is then inserted in the mouse genome, leading to the insertion of the human intergenic sequence (black bar), as well as a *galK* marker. Following this, an oligonucleotide is then designed bearing homology to the human and the mouse sequences (black and orange parts). In the 3rd step, the human intergenic sequence has been inserted with the removal of *galK* marker, after homologous recombination with the oligonucleotide.

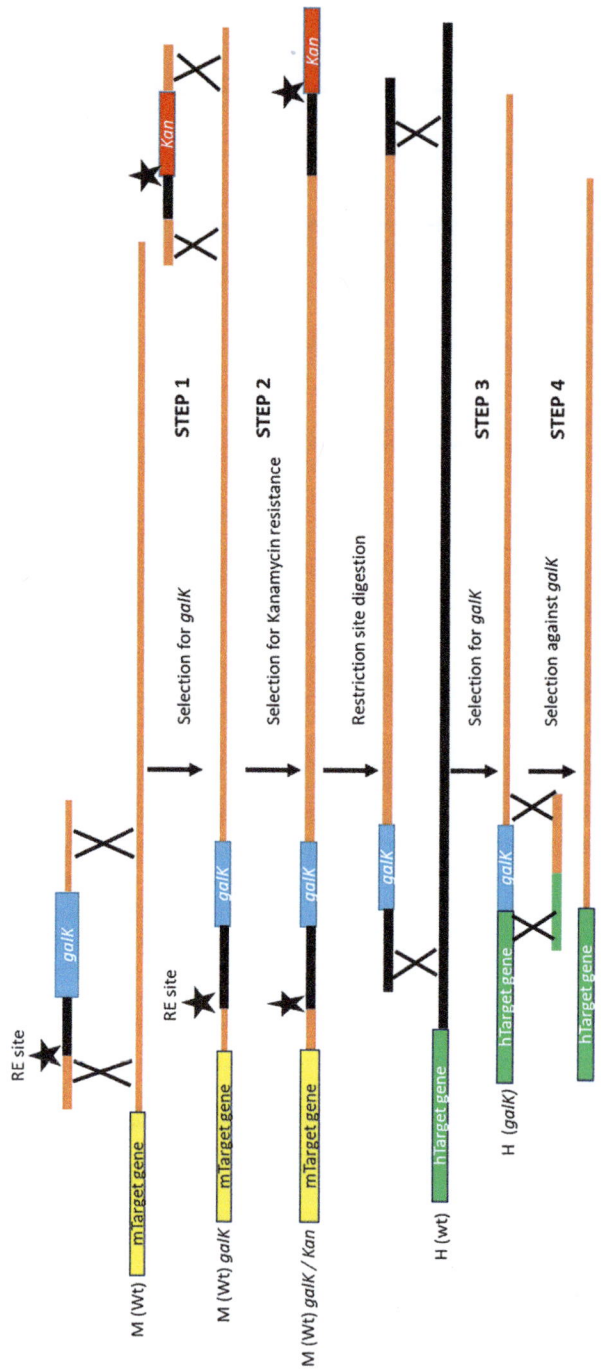

FIG. 6

The 4-step scheme: This strategy was used to replace a 6.6kb human intergenic fragment (black bars following a gene, coloured green) with a 14.6kb counterpart from the mouse (orange bars, following a gene shown in yellow). However, unlike in the 3 step strategy, one more selectable marker, a kanamycin resistance cassette is also inserted (shown in red). While constructing the kanamycin resistance cassette, primers are designed in such a way so as to incorporate sequences homologous to the mouse as well as human sequences (orange and black bars, respectively) as well as a restriction enzyme site. After both the markers have been inserted into the mouse chromosome in a way described in 3 step strategy, a restriction digestion is performed (the restriction sites are denoted by star). This gives rise to a product having a *galK* (blue) gene followed by a region having mouse sequences (orange), flanked by human sequences (in black). However, the kanamycin cassette gets removed after the restriction digestion. This molecule then undergoes homologous recombination with the human chromosome (black bars), resulting in incorporation of the *galK* cassette (blue) and the mouse sequences (orange) replacing the corresponding part in the human genome. In step 4, an oligonucleotide having homology to the human gene (green) and the mouse sequence (orange) is made to undergo recombination, thus removing the *galK* (blue) gene from the human chromosome.

asymmetrical target sequence. This creates a double-stranded break and helps increase homologous recombination efficiency, increasing the insert size to as large as 55 kb (Zhao et al., 2011).

7.2 Recombineering in *E. coli* phages

This section briefly describes a method for engineering lambda phage and construction and screening of functional phage mutants from the lysate of infected *E. coli* cells. Two model systems can be used for recombineering such as Lambda phage and mycobacteriophage (Marinelli et al., 2012).

In the lambda phage mediated system, *E. coli* cells are infected with phage to express the recombineering proteins. Substrate like ssDNA or dsDNA, housing a deletion or a point mutation in its sequence, is introduced in electrocompetent cells of the infected *E. coli*. After electroporation, the cultures are allowed to assemble phage. Following this, the lysate (carrying the phages) is screened for the presence of the desired mutant. Screening can be done with the recombinant lysate utilising the Double-Layer Agar (DLA) technique. The DLA technique has been used for years to identify, quantitate and isolate mutant phages. It is based on the concept of lysing of host bacterial cells by the phage producing clear zones or "plaques" on agar (Santos et al., 2009). As the name suggests, double layer agar means two layers of agar (of different composition). Bacteriophage mixture is mixed with the host bacterial cells in soft agar. Together, they are now overlaid on another layer of agar (hard agar) (Panec & Katz, 2006). French-Canadian microbiologist, Felix d'Herelle, is credited with the development of this method (d'Herelle, 1917). The usage of this DLA method to compare the plating pattern of the recombinant lysate to the control lysate produced without addition of substrate DNA was reported (Marinelli et al., 2012). Following this, a plaque hybridization assay can be carried out to detect the presence of the desired mutant phage wherein a probe designed against the mutant is expected to bind to only the mutant and not to the Wild Type phage (Marinelli et al., 2012).

7.3 Construction of Mycobacteriophage mutants by recombineering

Studying the molecular biology of the human pathogen *Mycobacterium tuberculosis* was vastly aided by using phages which infect *Mycobacterium*. Despite its uses, generating mutations in this bacteriophage is problematic. Moreover, screening with an antibiotic selectable marker is not really feasible because the cells will be lysed anyway due to the lytic nature of the phages. Adapting a strategy similar to the ones used for manipulating the common *E. coli* phages was not recommended because mycobacteriophages are not typically lysogen forming in nature.

For years, a strategy that was employed to manipulate these phages for recombineering was to use phage-plasmid hybrids, or shuttle vectors. These were then packaged into Lambda particles and *E. coli* were then infected with these. A group used the strategy to introduce the mutation in the gene in *E. coli* by

Lambda Red recombination but recovering the mutant phage in *M. smegmatis* (Marinelli et al., 2012). Unfortunately, there still are certain disadvantages like construction of hybrid phage and plasmid is very time-consuming with demanding, copious and laborious steps. Coupled with that, packing capacity is also less, as the whole mycobacterial genome is too large to prepare a properly functional plasmid. To overcome this problem, BRED was discovered (Marinelli et al., 2012) which has been elaborated below:

7.3.1 Bacteriophage Recombineering of Electroporated DNA (BRED)

This technique was developed to overcome the low transduction efficiency or DNA uptake efficiency of Mycobacteria, even when electrocompetent cells of the genus are used. BRED is like a ssDNA recombineering strategy where a selected plasmid is co-transformed with a ssDNA substrate. In these methods transformed cells can be screened by using MAMA-PCR (Mismatch Amplification Mutation Assay PCR) (Marinelli et al., 2012), a technique which can selectively amplify the mutant allele. BRED works well for detecting single point mutations (Aggarwal et al., 2012).

BRED depends on a co-selection procedure. In the first step, electrocompetent *Mycobacterium smegmatis* cells are prepared. These cells are then co-transformed by both the genomic DNA of the phage and a PCR amplified substrate. In order to identify recombinant lysate an infectious centre assay is performed. In this approach, only those cells that have taken up the phage DNA will give a plaque in a bacterial lawn (Fields & Joklik, 1969; Hirst & Pons, 1973). This plaque contains phages and a few of them contain the mutant allele that can be screened by PCR in the next step. Mutant phages can be isolated by serial dilution plating of the mixed plaques which can be further screened by PCR (Marinelli et al., 2012). This method has many applications, for example:

- Performing insertion and deletion. Insertion and deletion both can be identified by PCR. In insertion, a primer is used that can anneal with the inserted region and another one can anneal to the downstream of the phage chromosome. To detect deletions primers that flank the deleted region are used.
- Creating a point mutation and gene replacement: Targeted gene replacement mutants are performed by using linear dsDNA fragments containing regions of homology upstream and downstream to the target gene flanking an antibiotic resistance marker gene, which are electroporated and the recombinants are then subsequently selected against the antibiotic.

However, in some cases for defective mutant phages, the flanking primers are unable to identify the deletion. So in that case DADA-PCR is used (Marinelli et al., 2012).

7.3.2 DADA-PCR: Deletion amplification assay PCR

This is a PCR technique that can be used to identify deletion mutant phages for defective phage (Fig. 7). DADA-PCR primers are used that can anneal to the new junction after deletion at 3′ end and another primer can anneal to the flanking region

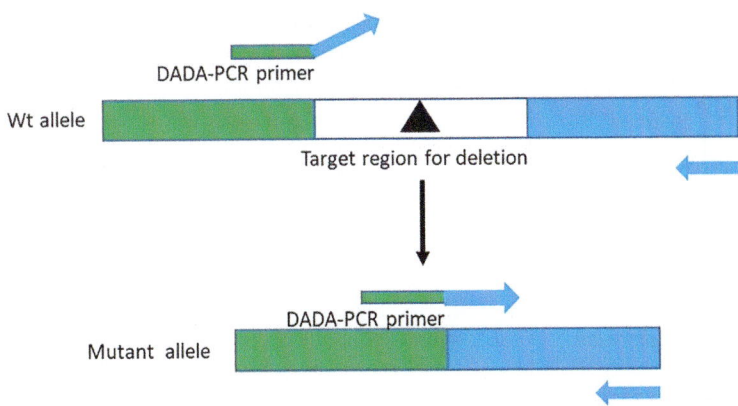

FIG. 7

Schematic diagram of DADA PCR: This was developed to detect the small population of mutant alleles which carry the deletion of the desired gene. Here, the green and the blue regions denote the upstream and the downstream regions, respectively, of the gene whose deletion we are trying to detect. One primer is designed so that its 3′ end anneals to the junction formed if the gene deletion has taken place, i.e., it will anneal to the 5′ end of the downstream region (blue region). So, in the WT allele, the primer is not expected to anneal, thus failing to amplify the WT allelic fragment under stringent PCR conditions, whereas the mutant allele is expected to get amplified.

(Marinelli et al., 2012). Owing to the sequence of the primers and consequently the region to which they are designed to anneal to, PCR amplification will fail to take place if the Wild type allele remains and will only amplify if a successful deletion event has occurred.

7.3.3 BRED for point mutation

BRED is also used for generating point mutations. A synthetically synthesised ssDNA substrate is required. The oligonucleotide should be about 71 base pairs in length. MAMA-PCR is used here for the screening and detection. Also, as antibiotic resistance cassettes are not useful in lytic phages, fluorescent reporter phages are used (Marinelli et al., 2012).

7.4 Recombineering in other strains

A total of nine genes have been discovered from Gram-positive and Gram-negative bacteria and their phages that are similar to Lambda and Rac prophage. There are many prokaryotic hosts in which recombineering has been established (Table 2).

Table 2 The different prokaryotic hosts in which recombineering has been established.

Serial. No.	System	Application of recombineering	References
1	*Escherichia coli*	For insertion, deletion and point mutation of different genes in *E. coli*	Marinelli et al. (2012)
3	*Salmonella enterica*	To produce scar-free point mutations, deletions, epitope tags and promoters in *S. enterica*	Stringer et al. (2012)
4	*Klebsiella pneumoniae*	To delete the dhak1 gene in *K. pneumoniae*	Wei, Wang, Shi, and Hao (2012)
5	*Zymomonas mobilis*	To delete pyruvate decarboxylase gene in *Z. mobilis*	Khandelwal et al. (2018)
6	*Pseudomonas aeruginosa*	To create large chromosomal deletions or transferring mutated genes in PA14 transposon library	Lesic and Rahme (2008)
7	*Vibrio cholerae*	To create a null mutation in regulatory genes, i.e., *toxT* in *V. cholerae*	Yamamoto et al. (2009)

7.5 Gram negative bacteria

7.5.1 Recombineering in Shewanella

Shewanella oneidensis prophage uses an endogenous phage recombination system by using W3 Beta recombinase which has 55% identity with Lambda Red Beta. It uses ssDNA oligo mediated recombination. This protein was functional in *E. coli* with efficiencies comparable to Lambda Red Beta in *E. coli* (Corts, Thomason, Gill, & Gralnick, 2019; Fels, Gevaert, & Van Damme, 2020).

7.5.2 Recombineering in Vibrio natriegens

A marine bacterium *Vibrio natriegens* encodes a recombinase protein in the SXT mobile genetic element. SXT-Beta and SXT-Exo have 46.5% and only 22.9% identity with the Lambda Beta and Lambda Exo protein. SXT-Beta catalysed recombination with 90 nucleotide ssDNA but it still needed SXT-Exo and the Lambda Gam expression gene (Chan et al., 2007).

7.5.3 Recombineering in Vibrio cholerae

Homologous recombination-mediated gene deletion was introduced in *V. cholerae* (Yamamoto et al., 2009).

Donor DNA is prepared through one-step PCR or two-step PCR (Figs 8 and 9). Antibiotic selectable markers with FRT sequence are amplified with the primer of short homologies of the target genes (i.e. *ctxB* to *toxT*). In two-step PCR, upstream and downstream of the target gene and FRT flanked antibiotic selectable markers are

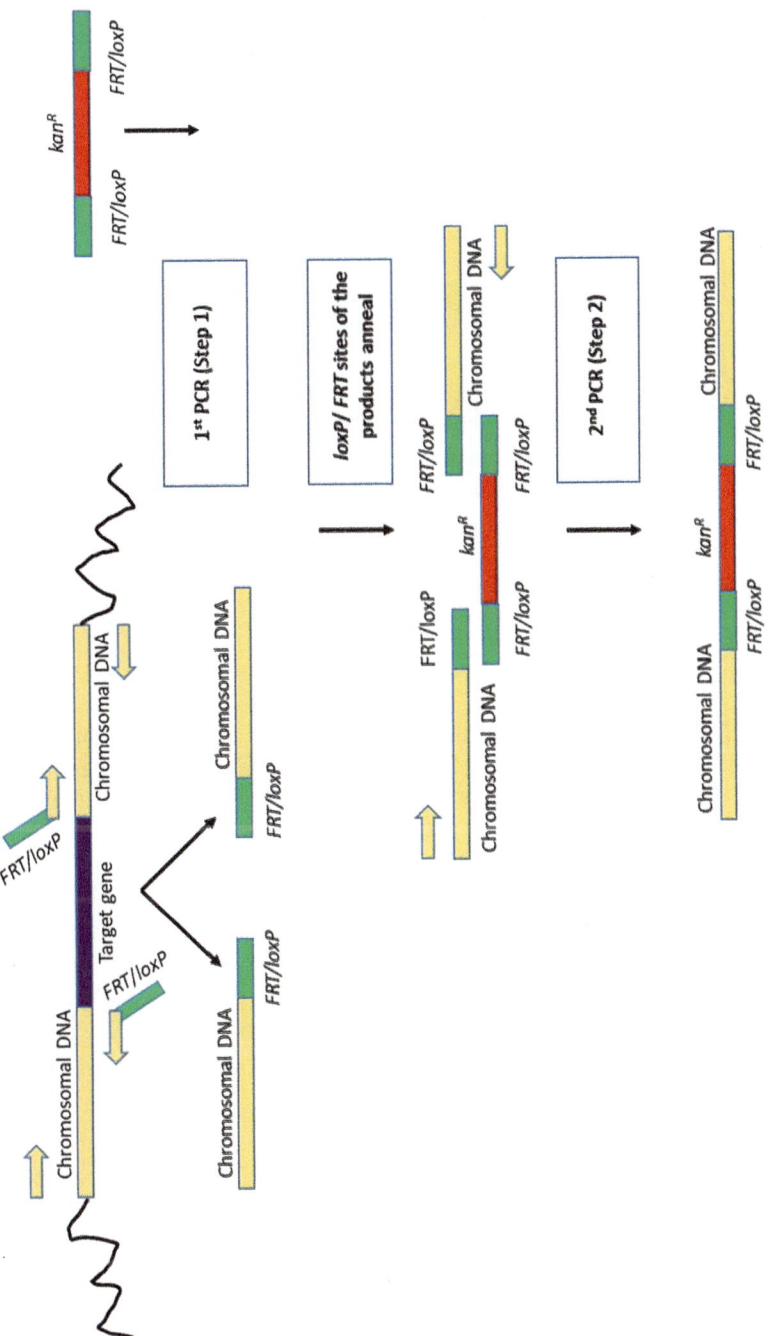

FIG. 8

Schematic diagram of Two Step PCR to generate donor DNA. In step 1, two independent PCR reactions are set up. In one, the upstream and the downstream region of the double-stranded DNA sequence of the target gene (denoted in purple) are amplified by designing primers (yellow arrows) with an *FRT* or a *loxP* site in it (denoted in green). The black squiggly lines denote the rest of the chromosomal DNA. This leads to two fragments, the amplified upstream fragment with an *FRT/loxP* site at its 3′ end, and a downstream region with an *FRT* or a *loxP* site at its 5′ end. In the other reaction in the first step itself, an antibiotic cassette (as an example, *kan^R* has been shown here in red) is amplified, with both forward and reverse primers having an *FRT* or a *loxP* site (shown in green as above). When these three PCR products are mixed together at equimolar concentrations, the *FRT* or *loxP* sites anneal, and the second PCR reaction is then performed which leads to the construction of the final molecule, that is an antibiotic cassette (red) having two homologous regions upstream and downstream (both in yellow) to it.

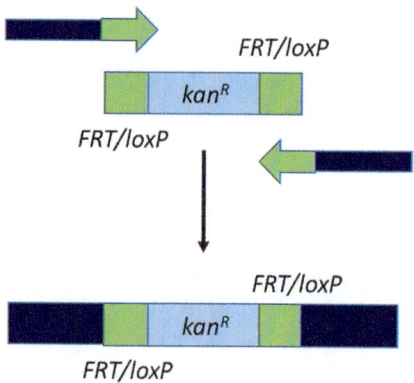

FIG. 9

Schematic diagram of Single Step PCR to generate donor DNA: In One step PCR method, both the forward and the reverse primers are designed to have homologous sequences to the sequence of the gene that needs to be deleted (denoted in dark blue) at their 5′ ends, and few sequences homologous to the *FRT* or *loxP* site (green). The resultant amplicon after the PCR will be an antibiotic cassette (here, kanamycin resistance cassette has been denoted in light blue), flanked by *FRT* or *loxP* sites (shown in green), further flanked by sequences homologous to the gene that needs to be deleted.

separately amplified through PCR. After that, these three PCR products are mixed, and a second PCR is run using primers of the upstream and downstream regions. Finally, an antibiotic-selectable marker flanked by upstream and downstream homologous regions is generated. The upstream and downstream regions can be 50–1000 base pairs in length (Yamamoto et al., 2009).

In the next step, this PCR amplified donor DNA is electroporated in a *Vibrio cholerae* strain where Lambda Red protein gene is encoded in another vector. After electroporation of the PCR product, the target gene is deleted by homologous recombination using Lambda Red Recombineering. In the next step, the antibiotic selectable marker is removed through a helper plasmid which can express FLP recombinase (Yamamoto et al., 2009).

7.5.4 Recombineering in Photorhabdus luminescens

In *Photorhabdus luminescens* Plu2934, Plu2935, Plu2936 phage proteins are shown to be effective, which are encoded by a Lambda Red-like operon present in the organism. Plu2936, 5′–3′ exonuclease is reported to be equivalent to Red-alpha protein of the Lambda phage. Plu2935 is a single strand annealing protein equivalent to Red-beta, and Plu2934 is thought to be functionally equivalent to Red-Gamma protein, which is capable of inhibiting the *E. coli* exonuclease RecBCD.

7.5.5 *Recombineering in* Pseudomonas

Lambda *red* operon and *Plu* operon were inapplicable in *Pseudomonas* strains, but the BAS operon, a Lambda Red like operon which was identified from *Pseudomonas aeruginosa* phage Ab31, was applicable in a broad range of *Pseudomonas* strains. Plu-Gam and Lambda-Gam improved the recombination efficiency of RecTEPsy which is a RecET-like operon from *Pseudomonas syringae* and Lambda Red like BAS operon from *Pseudomonas aeruginosa* phage Ab31. These are applicable in *Pseudomonas aeruginosa* and *Pseudomonas fluorescens*. But interestingly this BAS operon is not efficient for *Pseudomonas putida* and *Pseudomonas syringae* (Fels et al., 2020; Swingle, Bao, Markel, Chambers, & Cartinhour, 2010; Yin et al., 2019).

Recombineering has been successfully applied to delete a single gene. Single gene deletion through recombineering in *Pseudomonas aeruginosa* has been done with antibiotic cassettes that have been obtained by different strategies like single step PCR and three step PCR (see Figs 10 and 11 for single step and 3 step PCR methods, respectively). It is already reported that three-step PCR needs 400–500 nucleotides homologous to the target sequence whereas single step PCR needs only 100 nucleotides for successful recombineering.

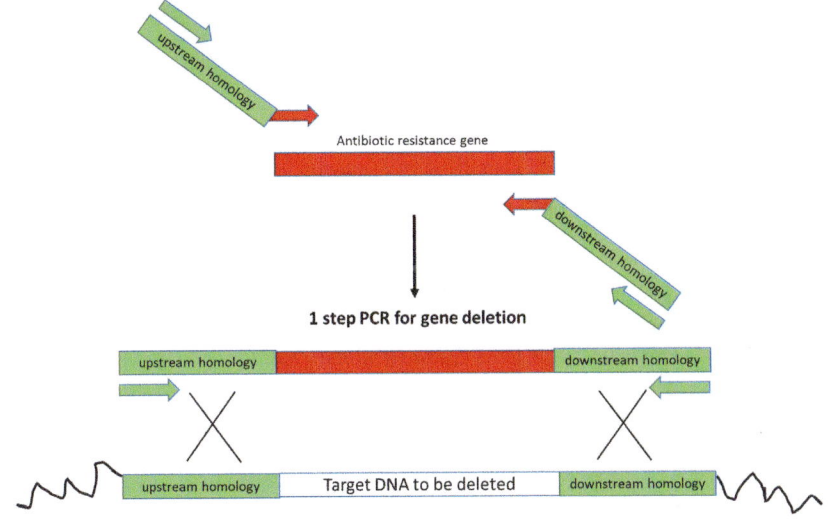

FIG. 10

Schematic diagram of Single Step PCR to generate donor DNA: Here an antibiotic resistance gene cassette (shown in red) is amplified with primers having sequence complementarity to the sequences found upstream and downstream of the target gene in the host chromosome (green) These sequences need not be more than 100 nucleotides. The PCR produces an amplicon with the antibiotic cassette (red) flanked by regions homologous to the upstream and downstream regions (green) of the target gene that is to be deleted. Once this construct is electroporated, it undergoes recombination and the target gene is replaced with the antibiotic cassette. The black squiggly lines denote the adjacent genomic DNA.

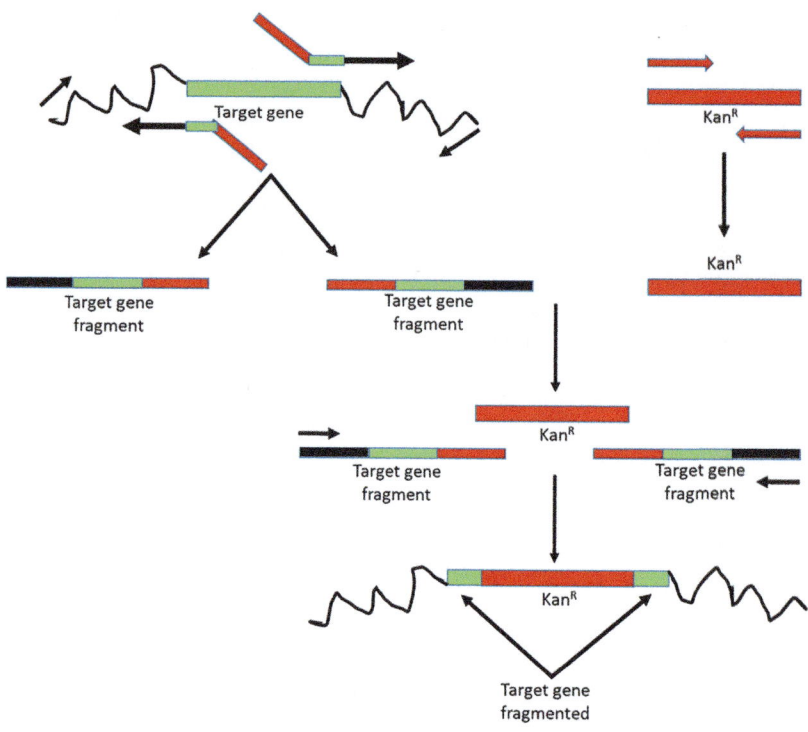

FIG. 11

Schematic diagram of 3-Step PCR to generate donor DNA: In the first step, two independent PCR reactions take place. The target gene is demonstrated in the colour green, and the adjacent genomic DNA in black squiggly lines. Two sets of primers are used to amplify two different amplicons from this genomic sequence. In each set of primer, one of the primers have some sequences complementary to the kanamycin gene cassette (denoted in green), some complementary to the target gene (shown in green), and the rest complementary to the adjacent genomic DNA (about 400–600 nucleotides). The other primers, in both sets are simply complementary to the genomic DNA (see the black arrows binding to the black genomic DNA). In the first step itself, another PCR reaction is performed, wherein a kanamycin resistance gene (red) is amplified with primers complementary to its end sequences. Now, in the 2nd step, all the three different amplicons that were obtained from step 1, are mixed together and are amplified with primers (see the black arrows) against the black regions (signifying genomic DNA). In the 3rd step, this amplicon is then amplified and the final construct is formed which may then be electroporated. A 2.2 kb PCR fragment was electroporated by Lesic & Rahme, 2008.

In the single step PCR technique 2 sets of primers are used—forward and reverse primer of antibiotic resistance contain 5′ 100 nucleotides and 3′22–24 nucleotide homology to the antibiotic cassette. Another set of primers are homologous to the 5′ extremities of another set (Lesic & Rahme, 2008) (Fig. 10).

In the 3-step PCR technique, independently the upstream and downstream regions of the targeted gene of *P. aeruginosa* and the antibiotic cassette are amplified through PCR. These three PCR products are then mixed together to get a product with upstream region of the now fragmented gene, followed by an antibiotic cassette then the downstream region of the same disrupted gene, from 5′ to 3′ (Lesic & Rahme, 2008) (Fig. 11). This recombineering mediated construction of *P. aeruginosa* mutants is very time efficient as it takes less than a week to perform 3-step PCR (Lesic & Rahme, 2008).

For deletion of a large chromosomal region of *P. aeruginosa* 3-step PCR is used with two transposon mutants. These two transposon mutants are used as templates to generate the upstream and downstream regions. Then these two fragments are mixed to get the desired product which is further amplified by forward-upstream and reversed-downstream primers (Lesic & Rahme, 2008).

7.5.6 Recombineering in Salmonella enterica

Recombineering systems in Enterobacteria like *Escherichia coli* and *Salmonella enterica* have some limitations like inaccurate excision of selectable markers because a 50–100 bp 'scar' is left that is not desirable and secondly the efficiency of the process is compromised. So, another technique named FRUIT (Flexible Recombineering Using Integration of *thyA*) was developed where *thyA* gene was used both as a selectable and counter-selectable marker (Stringer et al., 2012). The *thyA* gene is a well-known bacterial gene used for production of thymine. When *thyA* is used for screening in recombineering, it needs trimethoprim. This is an enzyme that is treated as toxic for the cells as it is a dihydrofolate reductase. In the FRUIT method, a *thyA*-containing PCR product is amplified and after recombination with chromosomal DNA it is screened for *thyA*+ cells by growing them on thymine lacking medium. Then Lambda Red recombination or homologous recombination is used to remove the *thyA* marker (Fig. 12). This approach is used for introducing point mutations, deletions, epitope tagging and introducing artificial promoters for *E. coli* K-12 and *S. enterica* (Stringer et al., 2012).

7.5.7 Recombineering in Klebsiella pneumoniae

Recombineering in *Klebsiella pneumoniae* is a kind of markerless gene replacement technique.

Previously the Red recombination system was not successful in *Klebsiella pneumoniae*. Two plasmids pIJ790 (Gust, Challis, Fowler, Kieser, & Chater, 2003), and pCP20 (Cherepanov & Wackernagel, 1995) were used for expression of the Red recombinase system in *E. coli* but they were unable to express homologous recombination in *K. pneumoniae*. No recombinase activity was seen as pCP20 was not able to replicate in *K. pneumoniae*. In order to get a recombineering system in *Klebsiella pneumoniae* Wei et al. (2012) discovered a plasmid system which is able to express the Red recombination gene. This system was able to give a new strategy for recombineering in *Klebsiella pneumoniae*. This plasmid pDK6 (Kleiner, Paul, & Merrick, 1988), was constructed to produce both Red and FLP recombinase in this study (Wei et al., 2012), named pDK6-red and pDK6-flp.

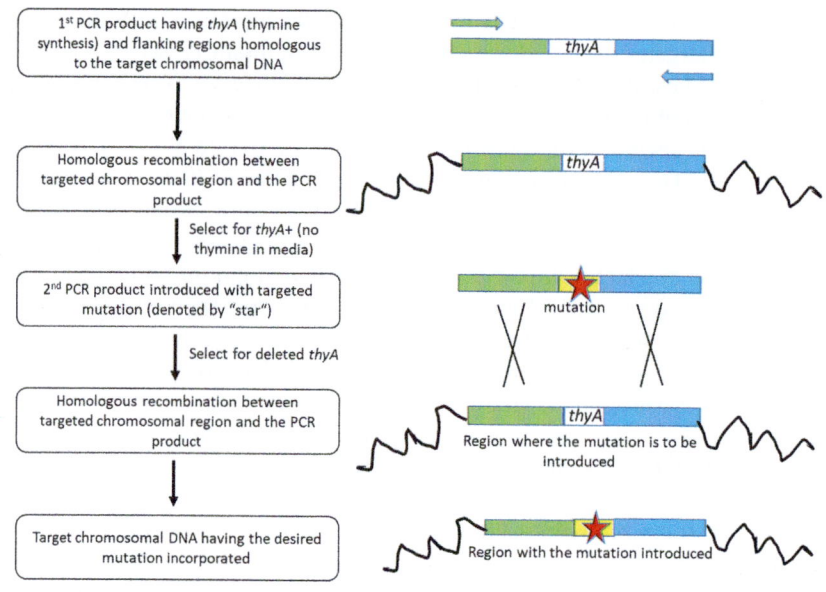

FIG. 12

Schematic diagram for FRUIT for introducing mutation. The first PCR reaction produces an amplicon having the *thyA* gene flanked by regions homologous to the upstream and downstream regions to the gene where the mutation needs to be introduced in the chromosomal DNA (denoted in green and blue, respectively). For convenience, the primers in step 1 are also denoted in green and blue, binding to the green and blue regions. In the next step, this product recombines with the homologous regions (green and blue), leading to the *thyA* gene now incorporated in the genome (the black squiggly lines denote the adjacent genomic DNA). The recombinant colonies are then selected for their ability to synthesise thymine by virtue of the gene, *thyA* (having not provided thymine in the media). Now, another PCR reaction is designed to produce a molecule which has the target mutation (yellow region with the red star) incorporated in it, with the same homologous sequences flanking it (green and blue). Once electroporated, this construct undergoes homologous recombination with the chromosomal DNA and as a consequence, the chromosomal DNA now has the same mutation incorporated in it (black squiggly lines, once again denote the adjacent chromosomal regions).

The Red recombineering gene in pDK6 plasmid is tightly regulated by the lac repressor. Transformation efficiency in *K. pneumoniae* was highly increased through pDK6-red and pDK6-flp mediated recombineering (about 1000-fold as stated by Wei et al. (2012). This system was used for deletion of *dhakl* gene, coding for a subunit of dihydroxyacetone kinase II in *K. pneumoniae*. Plasmid pDK6 *red* and *flp* mediated recombineering was tested under different concentrations of IPTG and different lengths of homologous extensions (Wei et al., 2012).

7.5.8 *Recombineering in* Yersinia pestis

The recombineering technique is a combination of *Lambda* Red recombination and counter selective screening for *Yersinia pestis*. It is used to construct live attenuated vaccines. Previously in this strain suicide vectors were used for gene knockout but transformation efficiency for these suicide vectors is very low in *Yersinia pestis*. Suicide vectors are plasmids which have a particular gene encompassing its Multiple cloning site (MCS) that has the ability to kill the host cell, or is unable to replicate in the host. Once, the desired gene is inserted into the MCS of such vectors, the suicide gene is disrupted and thus the host cell does not die after transformation. This is a very efficient method of determining true recombinants (Sun, Wang, & Roy 3rd Curtiss., 2008).

Lambda Red recombineering was used to edit the genome of *Yersinia pseudotuberculosis*. But antibiotic marker-based gene disruption and FLP recombinase-based techniques give scars in bacterial chromosomes which can decrease the efficiency of introducing the desired mutation into target sites. To avoid these issues, Sun et al. (2008) developed a highly efficient technique for scarless gene deletion and marker less gene insertion. In this technique, in the first step, a linear DNA fragment is constructed which carries an antibiotic resistance marker with a selection marker gene that is flanked by 500 bp homologous to the region flanking the deletion site. This DNA fragment is electroporated into *Yersinia pestis*. After electroporation of the DNA fragment in the strain, the targeted gene is replaced through homologous recombination and the integrant is prepared, now having the markers inserted in place of the target gene. In the next step, in order to delete the antibiotic marker-selection marker cassette, a second DNA fragment is constructed where the desired deletion, or a gene fragment having a point mutation, or even insertion, is prepared and electroporated in the strain. Then through homologous recombination again, the antibiotic marker-selection marker cassette is replaced and the desired deletion or insertion is obtained (Sun et al., 2008) (Fig. 13).

7.5.9 *Recombineering in* Zymomonas mobilis

In *Zymomonas mobilis*, the Lambda Red recombineering system was adapted to delete the desired gene by Khandelwal et al. (2018). They used the pSIM series plasmids (pSIM7 and pSIM9), which can express the Lambda Red Protein to produce homologous recombination. The stability of plasmids pSIM7 and pSIM9 was checked up to 40 generations. These plasmids were electroporated in electrocompetent *Z. mobilis* for expression of Lambda Red recombination protein. Donor DNA was prepared through PCR which contained an antibiotic cassette flanked by a 50 base pair sequence homologous to the upstream and downstream regions of the targeted gene. This PCR amplified product was transferred to the *Z. mobilis* strain. The targeted gene was deleted through homologous recombination. The antibiotic resistance gene cassette was flanked by FRT sites which allows its removal using Flp recombinase (Khandelwal et al., 2018).

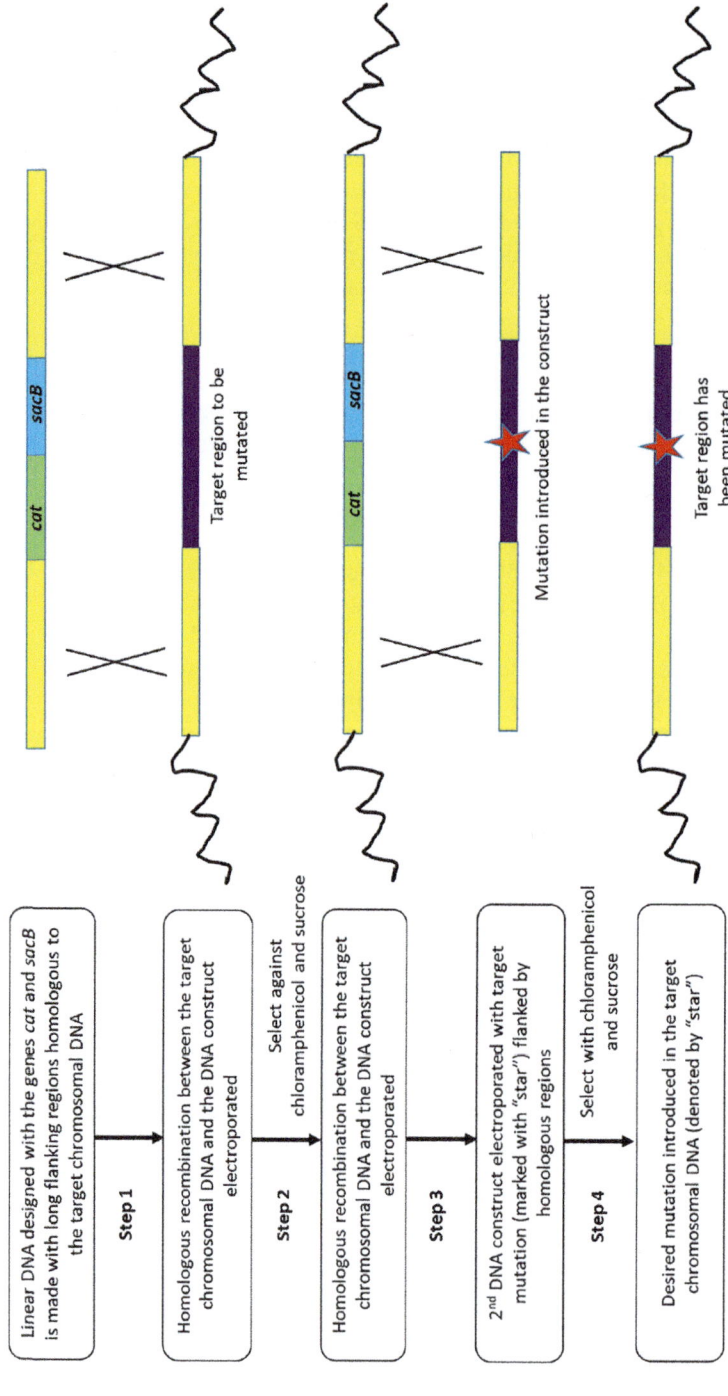

FIG. 13

Recombineering in *Yersinia pestis*. In the 1st step, a DNA fragment is constructed containing *cat-sacB* genes flanked by homology (indicated by the yellow-coloured regions) to the upstream and downstream of the target region (purple). After recombination which takes place once the construct is electroporated into the host cell, the target gene is deleted and the target region is replaced by *cat-sacB* gene cassettes in the chromosomal DNA (indicated by the black curved lines). So, the recombinants are selected against sucrose and chloramphenicol. In the third step, another DNA fragment is introduced which has a point mutation (indicated by the red star) incorporated in the target sequence (purple region). Once again, this construct will have homologous regions (yellow) flanking the target gene. The *cat-sacB* gene cassettes from the chromosome (as is shown in step 3) are deleted through homologous recombination. And finally in the fourth step, the target region with the mutation (purple region with the red star) in incorporated into the host chromosome.

7.6 Gram positive strains

7.6.1 Recombineering in mycobacteria

The lambda Red and Rac prophage system is not applicable for a broad range of bacteria. Mycobacteriophage Che9c encodes gp60 and gp61 genes which have 28% similarity with *recE* gene and 29% similarity with *recT* gene, respectively, of the Rac Prophage. Che9c gp60 has exonuclease activity that creates a single stranded end from double-stranded DNA. Che9c gp61 is a kind of ssDNA annealing protein, which binds short ssDNA that is created by the exonuclease (Fels et al., 2020).

For the Mycobacteriophage Che9c, the genes gp60 and gp61 are applied for deletion of different genes in *Mycobacterium smegmatis* and *Mycobacterium tuberculosis*. Gp61 based recombineering is applicable for introducing an *attP* site in the target site. Phage bxb1 integrase helps in site-specific recombination between *attP* and *attB* sites. This system is known as oligonucleotide mediated recombineering followed by Bxb1 integrase targeting (ORBIT) system. This is helpful in overcoming the barrier of using long flanking homologies (Fels et al., 2020).

8 Future prospects of recombineering

The work of many geneticists and bioinformaticians have yielded a collection of hundreds of bacteriophage genome sequences in the NCBI database. This collective work has proven to be extremely helpful in elucidating all the functions of the phage proteins whose genomes have been sequenced. Their annotation has led scientists to consider various genes and operons as candidates for bringing about recombineering just like the Lambda and Rac bacteriophages. Work is ongoing on to find phage functions which show promise to initiate and carry out species-specific recombination in a plethora of host organisms. This chapter has described various methods in which numerous genetic modifications, including point mutations, gene deletions, insertions, and gene replacements have been carried out using bacteriophage-mediated recombineering. As mentioned by Court and co-workers (2002), recombineering is an excellent and extremely efficient method of inserting modified nucleic acid segments like fluorophore tagged oligonucleotides or even biotinylated DNA fragments directly into the chromosome of the host bacterium for the study of chromosomal events like DNA replication, repair pathways, chromosomal segregation, etc.

Recombineering, owing to its simplicity, is being used not just for prokaryotes but also for higher organisms. Site-specific recombination was carried out previously for gene deletions in crop plants like corn, rice, etc. (Court et al., 2002). Numerous articles have reported the development of transgenic crop plant lines that have been developed through site specific recombination (Mitousis, Thoma, & Musiol-Kroll, 2020). Thus, the application of recombineering in the field of plant genome engineering remains unparalleled.

CRISPR-Cas9 (Clustered regularly interspaced short palindromic repeats and associated proteins) is an adaptive immune system for bacteria to protect themselves by cleaving foreign incoming DNAs like phage or plasmid. Modified CRISPR-Cas9 has been used as a genome editing technique for some time now. Being a targeted nuclease system, it has been used extensively for targeted genome editing with high precision. As shown by Li et al. (2019) who reported an efficient genome editing technique by combining ssDNA-mediated recombineering with CRISPR-Cas9 method, the technique can be used in future for genome editing.

References

Aggarwal, A., et al. (2012). Clinical & immunological erythematosus in systemic lupus Maryam. *Journal of Dental Education, 76,* 1532–1539.

Chan, W., Costantino, N., Li, R., Lee, S. C., Su, Q., Melvin, D., et al. (2007). A recombineering based approach for high-throughput conditional knockout targeting vector construction. *Nucleic Acids Research, 35*(8), e64.

Chen, B. Y., Lin, C. S., & Lim, H. C. (1995). Temperature induction of bacteriophage λ mutants in *Escherichia coli. Journal of Biotechnology, 40*(2), 87–97.

Cherepanov, P. P., & Wackernagel, W. (1995). Gene disruption in *Escherichia coli*: TcR and KmR cassettes with the option of Flp-catalyzed excision of the antibiotic-resistance determinant. *Gene, 158*(1), 9–14.

Conant, C. R., Goodarzi, J. P., Weitzel, S. E., & von Hippel, P. H. (2008). The antitermination activity of bacteriophage λ N protein is controlled by the kinetics of an RNA-looping-facilitated interaction with the transcription complex. *Journal of Molecular Biology, 384*(1), 87–108.

Copeland, N. G., Jenkins, N. A., & Court, D. L. (2001). Recombineering: A powerful new tool for mouse functional genomics. *Nature Reviews. Genetics, 2*(10), 769–779.

Corts, A. D., Thomason, L. C., Gill, R. T., & Gralnick, J. A. (2019). A new recombineering system for precise genome-editing in *Shewanella oneidensis* strain MR-1 using single-stranded oligonucleotides. *Scientific Reports, 9*(1), 1–10.

Court, D., Gottesman, M., & Gallo, M. (1980). Bacteriophage lambda hin function. I. Pleiotropic alteration in host physiology. *Journal of Molecular Biology, 138*(4), 715–729.

Court, D. L., Sawitzke, J. A., & Thomason, L. C. (2002). Genetic engineering using homologous recombination. *Annual Review of Genetics, 36,* 361–388.

d'Herelle, F. (1917). Sur un microbe invisible antagoniste des bacilles dysenterique. *Comptes Rendus de l'Académie des Sciences Paris, 165,* 373–375.

Datsenko, K. A., & Wanner, B. L. (2000). One-step inactivation of chromosomal genes in *Escherichia coli* K-12 using PCR products. *Proceedings of the National Academy of Sciences of the United States of America, 97*(12), 6640–6645.

Datta, S., Costantino, N., & Court, D. L. (2006). A set of recombineering plasmids for gram-negative bacteria. *Gene, 379*(1–2), 109–115.

Ellis, H. M., Daiguan, Y., DiTizio, T., & Court, D. L. (2001). High efficiency mutagenesis, repair, and engineering of chromosomal DNA using single-stranded oligonucleotides. *Proceedings of the National Academy of Sciences of the United States of America, 98*(12), 6742–6746.

Fels, U., Gevaert, K., & Van Damme, P. (2020). Bacterial genetic engineering by means of recombineering for reverse genetics. *Frontiers in Microbiology, 11*(September), 1–19.

Fields, B. N., & Joklik, W. K. (1969). Isolation and preliminary genetic and biochemical characterization of temperature-sensitive mutants of reovirus. *Virology, 37,* 335–342.

Gronostajski, R. M., & Sadowski, P. D. (1985). The FLP recombinase of the *Saccharomyces cerevisiae* 2 microns plasmid attaches covalently to DNA via a phosphotyrosyl linkage. *Molecular and Cellular Biology, 5*(11), 3274–3279.

Gust, B., Challis, G. L., Fowler, K., Kieser, T., & Chater, K. F. (2003). PCR-targeted *Streptomyces* gene replacement identifies a protein domain needed for biosynthesis of the sesquiterpene soil odor geosmin. *Proceedings of the National Academy of Sciences of the United States of America, 100*(4), 1541–1546.

Guzman, L. M., Belin, D., Carson, M. J., & Beckwith, J. (1995). Tight regulation, modulation, and high-level expression by vectors containing the arabinose PBAD promoter. *Journal of Bacteriology, 177*(14), 4121–4130.

Haeusser, D. P., Hoashi, M., Weaver, A., Brown, N., Pan, J., Sawitzke, J. A., et al. (2014). The Kil peptide of bacteriophage λ blocks *Escherichia coli* cytokinesis via ZipA-dependent inhibition of FtsZ assembly. *PLoS Genetics, 10*(3), e1004217.

Heaney, J. D., Rettew, A. N., & Bronson, S. K. (2004). Tissue-specific expression of a BAC transgene targeted to the Hprt locus in mouse embryonic stem cells. *Genomics, 83*(6), 1072–1082.

Hirst, G. K., & Pons, M. W. (1973). Mechanism of influenza recombination. II. Virus aggregation and its effect on plaque formation by so-called noninfective virus. *Virology, 56*(2), 620–631.

Hsu, P. L., Ross, W., & Landy, A. (1980). The λ phage att site: Functional limits and interaction with Int protein. *Nature, 285*(5760), 85–91.

Karu, A. E., Sakaki, Y., Echols, H., & Linn, S. (1975). The gamma protein specified by bacteriophage gamma. Structure and inhibitory activity for the RecBC enzyme of *Escherichia coli. The Journal of Biological Chemistry, 250*(18), 7377–7387.

Khandelwal, R., et al. (2018). Deletion of pyruvate decarboxylase gene in *Zymomonas mobilis* by Recombineering through bacteriophage lambda red genes. *Journal of Microbiological Methods, 151,* 111–117.

Kleiner, D., Paul, W., & Merrick, M. J. (1988). Construction of multicopy expression vectors for regulated over-production of proteins in *Klebsiella pneumoniae* and other enteric bacteria. *Microbiology, 134*(7), 1779–1784.

Kowalczykowski, S. C. (2000). Initiation of genetic recombination and recombination-dependent replication. *Trends in Biochemical Sciences, 25*(4), 156–165.

Lee, E. C., et al. (2001). A highly efficient *Escherichia coli*-based chromosome engineering system adapted for recombinogenic targeting and subcloning of BAC DNA. *Genomics, 73*(1), 56–65.

Lesic, B., & Rahme, L. G. (2008). Use of the lambda red recombinase system to rapidly generate mutants in Pseudomonas aeruginosa. *BMC Molecular Biology, 9*(1), 20. https://doi.org/10.1186/1471-2199-9-20.

Li, J., et al. (2019). Coupling SsDNA recombineering with CRISPR-Cas9 for *Escherichia coli* DnaG mutations. *Applied Microbiology and Biotechnology, 9,* 3559–3570.

Liu, P., Jenkins, N. A., & Copeland, N. G. (2003). A highly efficient recombineering-based method for generating conditional knockout mutations. *Genome Research, 13*(3), 476–484.

Marinelli, L. J., Hatfull, G. F., & Piuri, M. (2012). Recombineering: A powerful tool for modification of bacteriophage genomes. *Bacteriophage, 2*(1), 5–14.

Mitousis, L., Thoma, Y., & Musiol-Kroll, E. M. (2020). An update on molecular tools for genetic engineering of actinomycetes—The source of important antibiotics and other valuable compounds. *Antibiotics, 9*(8), 494.

Moerschell, R. P., Tsunasawa, S., & Sherman, F. (1988). Transformation of yeast with synthetic oligonucleotides. *Proceedings of the National Academy of Sciences of the United States of America, 85*(2), 524–528.

Murphy, K. C. (1998). Use of bacteriophage lambda recombination functions to promote gene replacement in *Escherichia coli. Journal of Bacteriology, 180*(8), 2063–2071.

Murphy, K. C., & Campellone, K. G. (2003). Lambda Red-mediated recombinogenic engineering of enterohemorrhagic and enteropathogenic *E. coli. BMC Molecular Biology, 4,* 11.

Murphy, K. C., Campellone, K. G., & Poteete, A. R. (2000). PCR-mediated gene replacement in *Escherichia coli. Gene, 246*(1–2), 321–330.

Orr-Weaver, T. L., Szostak, J. W., & Rothstein, R. J. (1983). Genetic applications of yeast transformation with linear and gapped plasmids. *Methods in Enzymology, 101*(C), 228–245.

Panec, M., & Katz, D. S. (2006). *Plaque assay protocol.* ASM MicrobeLibrary. Disponible at< Disponible at http://www. microbelibrary. org/component/resource/laboratory-test/3073-plaque-assay-protocols> Access on Jan, 15, 2017.

Rivero-Müller, A., Lajić, S., & Huhtaniemi, I. (2007). Assisted large fragment insertion by Red/ET-recombination (ALFIRE)—an alternative and enhanced method for large fragment recombineering. *Nucleic Acids Research, 35*(10), e78.

Rokney, A., et al. (2008). Host responses influence on the induction of lambda prophage. *Molecular Microbiology, 68*(1), 29–36.

Santos, S. B., et al. (2009). The use of antibiotics to improve phage detection and enumeration by the double-layer agar technique. *BMC Microbiology, 9,* 1–10.

Schnütgen, F., Doerflinger, N., Calléja, C., et al. (2003). A directional strategy for monitoring Cre-mediated recombination at the cellular level in the mouse. *Nature Biotechnology, 21,* 562–565.

Sergueev, K., Yu, D., Austin, S., & Court, D. (2001). Cell toxicity caused by products of the p(L) operon of bacteriophage lambda. *Gene, 272*(1–2), 227–235.

Serra-Moreno, R., et al. (2006). Use of the lambda red recombinase system to produce recombinant prophages carrying antibiotic resistance genes. *BMC Molecular Biology, 7,* 1–12.

Sharan, S. K., Thomason, L. C., Kuznetsov, S. G., & Court, D. L. (2009). Recombineering: A homologous recombination-based method of genetic engineering. *Nature Protocols, 4*(2), 206–223.

Shizuya, H., et al. (1992). Cloning and stable maintenance of 300-Kilobase-pair fragments of human DNA in *Escherichia coli* using an F-factor-based vector. *Proceedings of the National Academy of Sciences of the United States of America, 89*(18), 8794–8797.

Sternberg, N. L. (1992). Cloning high molecular weight DNA fragments by the bacteriophage P1 system. *Trends in Genetics, 8*(1), 11–16.

Stringer, A. M., et al. (2012). FRUIT, a scar-free system for targeted chromosomal mutagenesis, epitope tagging, and promoter replacement in *Escherichia coli* and *Salmonella enterica. PLoS One, 7*(9).

Sun, W., Wang, S., & Roy 3rd Curtiss. (2008). Highly efficient method for introducing successive multiple Scarless gene deletions and Markerless gene insertions into the *Yersinia pestis* chromosome. *Applied and Environmental Microbiology, 74*(13), 4241–4245.

Svenningsen, S. L., Costantino, M., Court, D. L., & Adhya, S. (2005). On the role of Cro in λ prophage induction. *Proceedings of the National Academy of Sciences of the United States of America, 102*(12), 4465–4469.

Swingle, B., Bao, Z., Markel, E., Chambers, A., & Cartinhour, S. (2010). Recombineering using RecTE from *Pseudomonas syringae*. *Applied and Environmental Microbiology, 76*(15), 4960–4968.

Wang, J., et al. (2006). An improved recombineering approach by adding RecA to lambda red recombination. *Molecular Biotechnology, 32*(1), 43–53.

Warming, S., et al. (2005). Simple and highly efficient BAC recombineering using GalK selection. *Nucleic Acids Research, 33*(4), e36.

Wei, D., Wang, M., Shi, J., & Hao, J. (2012). Red recombinase assisted gene replacement in *Klebsiella pneumoniae*. *Journal of Industrial Microbiology & Biotechnology, 39*(8), 1219–1226.

Wong, Q. N. Y., et al. (2005). Efficient and seamless DNA recombineering using a thymidylate synthase a selection system in *Escherichia coli*. *Nucleic Acids Research, 33*(6), e59.

Yamamoto, S., et al. (2009). Application of lambda red recombination system to *Vibrio cholerae* genetics: Simple methods for inactivation and modification of chromosomal genes. *Gene, 438*(1–2), 57–64.

Yin, J., Zheng, W., Gao, Y., Jiang, C., Shi, H., Diao, X., et al. (2019). Single-stranded DNA-binding protein and exogenous RecBCD inhibitors enhance phage-derived homologous recombination in *Pseudomonas*. *Iscience, 14*, 1–14.

Yu, D., et al. (2000). An efficient recombination system for chromosome engineering in *Escherichia coli*. *Proceedings of the National Academy of Sciences of the United States of America, 97*(11), 5978–5983.

Zhang, Y., Buchholz, F., Muyrers, J. P., & Stewart, A. F. (1998). A new logic for DNA engineering using recombination in *Escherichia coli*. *Nature Genetics, 20*(2), 123–128.

Zhang, Y., Muyrers, J. P., Testa, G., & Stewart, A. F. (2000). DNA cloning by homologous recombination in *Escherichia coli*. *Nature Biotechnology, 18*(12), 1314–1317.

Zhao, Y., Wang, S., & Zhu, J. (2011). A multi-step strategy for BAC recombineering of large DNA fragments. *International Journal of Biochemistry and Molecular Biology, 2*(3), 199–206.

Further reading

Ran, F. A., et al. (2013). Genome engineering using the CRISPR-Cas9 system. *Nature Protocols, 8*(11), 2281–2308.

van Kessel, J. C., & Hatfull, G. F. (2007). Recombineering in *Mycobacterium tuberculosis*. *Nature Methods, 4*(2), 147–152.

Wang, Y., Yau, Y. Y., Perkins-Balding, D., & Thomson, J. G. (2011). Recombinase technology: Applications and possibilities. *Plant Cell Reports, 30*(3), 267–285.

CRISPR

Applications of CRISPR/Cas9 in the field of microbiology

5

Iqra Bano[a],* and Adnan Ali[b]

[a]*SBBUVAS, Sakrand Faculty of Bio-sciences, Sakrand, Sindh, Pakistan*
[b]*Faculty of Veterinary & Animal Sciences, Pir Mahar Ali Shah Arid Agriculture University, Rawalpindi, Punjab, Pakistan*
Corresponding author: e-mail address: iqrashafi05@yahoo.com

1 Overview of CRISPR/Cas9 biology

There has been remarkable development in genetics and biology during the past 60 years, particularly in the area of DNA manipulation. The discovery of the DNA double helix by Watson and Crick paved the way for subsequent developments like solid-phase DNA system and genome sequencing technologies that have allowed researchers to identify and investigate genomic organization (Watson & Crick, 1953). Recent developments in genetic engineering technologies have the potential to revolutionize a wide range of disciplines, from finding variability in genomes to creating site-specific modifications and modifying the genome of animals and cells in their native context (Serajian, Ahmadpour, Rodrigues Oliveira, de Lourdes Pereira, & Heidarzadeh, 2021). The discovery of an unusual region of repeated DNA in *Escherichia coli* (*E. coli*) laid the groundwork for subsequent genome editing and significantly advanced our knowledge of bacterial immunology (Donohoue, Barrangou, & May, 2018). Twenty years have passed since the discovery of numerous organisms harbouring these repetitive genomic sequences ranging in length from 20 to 58 base pairs. These components have been named 'clustering regularly interspaced short palindromic repeats' (CRISPR) (Jiang & Doudna, 2017). The CRISPR/Cas9 system is an adaptive immune system based on prokaryotic nucleic acids that enable particular microorganisms to react to and eradicate invading genetic material (Redman, King, Watson, & King, 2016). Microbes that have been subjected to transduction, conjugation, or transformation are prompted to develop defence mechanisms that can detect and eliminate genome-altering foreign DNA. Researchers have confirmed that the integration of foreign DNA sequence fragments into the CRISPR region provides defence (Sashital, Wiedenheft, & Doudna, 2012). Short repeating base sequences (or repeating units) are found in the CRISPR region, and these repeating units are separated by spacers, which are variable regions of

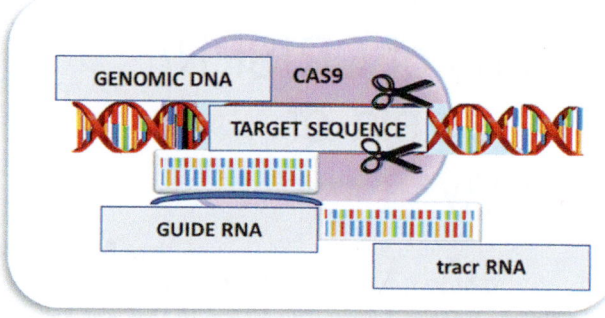

FIG. 1

The representation of CRISPR–CAS9 structure showing target sequence of genomic DNA and guide RNA as well as tracr RNA as a vital components. The GUIDE RNA is attached with the target gene while tracr RNA delivers DNA cutting enzyme from CAS9 to the site.

sequence that have sequence homology with foreign entities like bacteriophages and plasmids (Cui, Xu, Cheng, Liao, & Peng, 2018). The degree of resistance or susceptibility to phages is determined by modifications to the CRISPR locus, such as the insertion or deletion of spacers (Redman et al., 2016). These Cas genes, which code for Cas proteins, are often found on either side of a CRISPR array that has an adenine/thymine (AT) rich leader sequence (Fig. 1). During the adaption phase, the invading DNA is sliced into small segments and inserted into such a CRISPR locus as novel spacers. These new spacers represent the memory track of the infection. Integration of new spacers, which occurs as a consequence of DNA infection, is skewed towards the leading end of the CRISPR locus (Cui et al., 2018).

2 Applications of CRISPR/Cas9

There are different applications of CRISPR/cas9 in the field of drug discovery, animal model tests, epigenetics, human disease management as well as agriculture (Fig. 2). Manipulating genomics is the best hope to date for humanity to overcome monogenic disorders and for finding effective treatments for a wide range of malignancies, infections, and degenerative conditions (Lino, Harper, Carney, & Timlin, 2018). Previously, genetic therapies used technologies like zinc-finger nucleases (ZFNs) and transcription activator-like effectors (TALENs) to edit genes, but with promising technology like the CRISPR/Cas9 system, the capacity to modify genes becomes more specific and adaptable (Gaj, Gersbach, & Barbas, 2013). The CRISPR/Cas9 system also shows potential for use in disease prevention research using animal and human cell cultures as preclinical models. This technology can be applied in a variety of therapeutic contexts, with the optimal strategy depending on the type of editing to be performed, the tissue being edited, and the route of

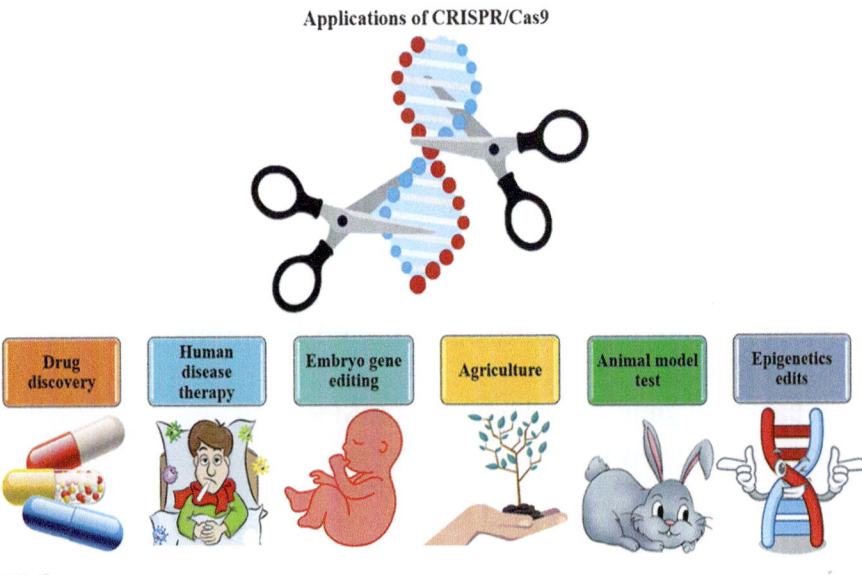

Applications of CRISPR/Cas9

FIG. 2

The representation of different applications of CRISPR/cas9 in the field of drug discovery, animal model tests, epigenetics, human disease management as well as agriculture.

delivery (Zhen et al., 2015). First, cells from the patient's body are extracted, the genes of those cells are manipulated using the CRISPR/Cas9 system, and then the cells are re-implanted into the patient *via* transplantation. Blood or bone marrow can be used to make the necessary changes, depending on the type of tissue that needs to be fixed (Harrison, Jenkins, O'Connor-Giles, & Wildonger, 2014). Haematopoietic stem and progenitor cells, (HSPCs) for instance, are the intended targets of gene editing for the treatment of Wiskott-Aldrich syndrome in autoimmune illnesses like rheumatoid arthritis, in which the regulatory T cells may be the point of attack. The CRISPR/Cas9 system is a promising tool that could hasten these developments (Ferrari, Thrasher, & Aiuti, 2021). The autologous transplantation is another method essential for *in vitro* therapies; however, it would be promising if genome editing were to lead to designed "universal donor" T cells or stem cells. Recently, researchers have revealed that the hemoglobinopathies such as sickle cell disease and thalassemia can be treated with CRISPR/Cas9 since the technology can correct the underlying genetic mutations that underlie these conditions (Safari, Farajnia, Arya, Zarredar, & Nasrolahi, 2018). Multiple therapy options were presented by the research team, including *ex vivo* gene modification of CD34+ haematopoietic cells (Pavel-Dinu et al., 2019). In 2019, CTX001 genomic stem cell treatments were also used to modify the haematopoietic stem cells of a patient called Victoria Gray in an attempt to treat sickle cell disease using CRISPR/Cas9 technology. So far, tests conducted on her at the Sarah Cannon Research Centre in Nashville, Tennessee, have shown that after taking a single dose of CDX001, the red blood cells as well as protein concentration were modified (Mellor, 2022). Furthermore, the CRISPR/Cas9 system has shown promise in treating infectious diseases, liver disorders, and a variety of afflictions affecting the brain and eyes, including Duchenne muscular dystrophy and congenital blindness. Nowadays, the patient's biopsy samples are being analysed using CRISPR/Cas9-based technologies to identify disease-causing mutations (Alves-Bezerra, Furey, Johnson, & Bissig, 2019).

3 Recent uses of CRISPR/Cas9-based technologies in microbiology

3.1 Bacterial gene expression and CRISPR/Cas9

The CRISPR/Cas9 system has many potential uses, one of which is the regulation of gene transcription. Gene transcription can be down- or upregulated with the use of a catalytically dead Cas9 molecule (dCas9) that retains the ability to recognize and bind to a specific DNA sequence (Tian et al., 2017). The use of dCas9 to repress genes is known as CRISPR interference (CRISPRi), while the use of dCas9 to activate genes is known as CRISPR activation (CRISPRa). Instead of cutting the DNA, the dCas9 enzyme is left in place in the target DNA sequence, where it can interfere with the binding of RNA polymerase or transcription factors (Gilbert et al., 2014). CRISPR-Cas technologies can be further developed to target many genes at once or to achieve precise control of gene expression. Understanding the design principles of customized CRISPR/Cas9 signalling pathways in bacteria is crucial because using these tools comes with a number of limitations, including toxicity and off-target consequences. Alternately, certain forms of CRISPR-Cas systems can target RNA and have the potential to be utilized in order to inhibit the expression of genes at the post-transcriptional level (Vigouroux & Bikard, 2020).

3.2 Bacterial resistance and CRISPR/Cas9

The lack of effective antibiotics for drug-resistant illnesses contributes to the growing problem of antibiotic resistance, which is currently one of the greatest public health problems (Serajian et al., 2021). Antimicrobial resistance develops over time without any intervention from humans, however, often genetic modifications help speed up the process. Antimicrobial resistance is typically accelerated by the misuse or incorrect prescription of antibiotics. It can also be triggered by several other factors, including a lack of hygiene, inadequate infection and disease prevention, and related problems (Wan et al., 2021). Researchers claimed that antimicrobial resistance can arise *via* a wide variety of mechanisms, including antibiotic efflux, drug target change, drug inactivation, and drug uptake inhibition. Antibiotic efflux is a bacterial defence mechanism that involves the export of harmful chemicals and the regulation of the cellular microenvironment to lower the intracellular concentration of antibiotics. Bacteria can reduce or eliminate effectiveness of a drug by altering a target mechanism, making the spot where the drug is supposed to attach less effective or nonexistent. Genes can undergo such changes as a result of point mutations (He et al., 2022). Alternating antimicrobials with the use of three primary enzymes, including chloramphenicol acetyl-transferases, β-lactamases, as well as aminoglycoside-modifying enzymes, limits the uptake of a drug; this occurs when bacteria modify their cellular membranes porin channel in a way that minimizes permeability. Therefore, mutations are a contributing factor to the development of resistance to antibiotics (Shukla, Jani, Polra, Kamath, & Patel, 2021). Many potential strategies for combating antibiotic-resistant bacteria have been proposed, including the use of genetically altered synthetic peptides, designed bacteriophages, as well as nano-antibiotics (synthesized virus-like nanoparticles), with eubiotics as growth promoters. However, there is still a long way to go before the effectiveness and consistency of these methodologies can be tested (Ghosh, Sarkar, Issa, & Haldar, 2019). Revolution in gene editing and modification has been achieved by the development of CRISPR/Cas9. Making use of drug-resistant bacteria's innate capacity to build such precise scissions in target genes could shed light on this method. The precision of this method makes it stand out as a viable alternative to the standard methods. Therefore, the system can be programmed with a sgRNA (single-guide RNA) that specifically targets a bacterial gene of interest, allowing for the targeted destruction of only those bacteria that carry that gene (Wan et al., 2021).

3.3 Delivery strategies *via* CRISPR/Cas9

Microinjection, electroporation, adenovirus, and lentivirus delivery, as well as non-viral administration *via* liposomes, nanoparticles, and polyplexes, are only some of the many cargos and delivery systems described for CRISPR/Cas9 (Givens, Naguib, Geary, Devor, & Salem, 2018). Researchers like Lino et al. (2018) have conducted extensive literature reviews of these methods. Hydrodynamic delivery is being studied as an *in vivo* delivery approach because it rapidly pushes a substantial volume (8–10% of body weight) of standard solution gene-editing cargo into the bloodstream (Lino et al., 2018). The kidneys, lungs, muscles, and heart are all severely impacted by this technique. To get passage through impermeable barriers like endothelium and parenchymal cells, this approach briefly raises the pressure in a closed system and is therefore only applicable *in vivo* (Serajian et al., 2021). Success in delivering cargo and obtaining the desired tissues and results using *in vitro* as well as *ex vivo* approaches (e.g., the iTOP method) has been demonstrated, however, this has not yet been translated to clinical situations. For example, viral vectors have been utilized *in vitro, ex vivo*, and *in vivo* delivery methods, but they have unintended consequences such as mutagenesis, low cloning capacity, and immunological reactions (Kholosy et al., 2021). However, optimizing chemical delivery systems to increase their efficacy for *in vivo* gene editing is a lengthy and laborious process. Currently, the most potential delivery agents targeting CRISPR/Cas9 components include nanoparticles made of lipid or gold as well as extracellular vesicle-based methods (such as exosome-based) that cause minimal or no immune response. However, there are problems with the currently available *in vivo* applications, as they can be stressful, produce physiological issues including hepatotoxicity, and have low-efficiency rates in the clinical environment (Aghamiri et al., 2020). The first CRISPR/Cas9 testing was completed in 2016, marking promising futures for nano-delivery methods for clinical genome editing to cure and/or fix genetic disorders. Despite numerous advancements, it is still difficult to develop an *in vivo* delivery method for CRISPR/Cas9 technology that is safe, reliable, and effective (Suzuki et al., 2016).

3.4 Bacterial infections and CRISPR/Cas9

The CRISPR/Cas9 system allows for the distribution of a particular sequence to bacteria through the mediation of phages or numerous other vectors, however, intracellular infections provide a greater delivery challenge. It is necessary for the CRISPR/Cas9 encoding phage to not only enter the host cell but also to preferentially transport cargo to the resident pathogen. This process is complicated, but accuracy is improved with two layers (Verma, Sahu, Singh, & Egbo, 2019). As phage structures get more varied and nanoparticle delivery is phased out, the task of delivery becomes more difficult. Several methods, including liposomal encapsulation, were investigated using the method of Carnes et al. which used evaporation and induced a self-assembly process to encapsulate the bacterial component inside a particle architecture that was based on silica and lipids (Carnes et al., 2010). It is possible for several biological components, such as a protein stabilization or the existence of silica, to assist in the alteration of payloads. For instance, silica can interact with the polymeric or lipid coating to mask particles and evade the immune system (Knowles et al., 2018). Developing vectors that are capable of transferring exogenous DNA into specific bacteria is one of the most significant obstacles that must be overcome when employing the CRISPR/Cas9 technology in the production of antimicrobials (Lino et al., 2018). The CRISPR/Cas9 scheme can be delivered in a variety of ways depending on the technology used. A few examples of these approaches are polymer-derivatized CRISPR nanoparticles, conjugative plasmids, and phages (Verma et al., 2019). Phages are infectious agents that attack bacteria. They do this by attaching themselves to specific receptors on the surface of the bacteria and then injecting their genome into the cytoplasm. The CRISPR/Cas9 system has been implemented with both of these phage vector variants thus far. One phagemid created a temperate/virulent phage. The effort that aimed to use phagemids in the delivery of a customized CRISPR/Cas9 system into several bacterial species, like *E. coli* or *S. aureus*, has indeed been effective, and selective death has been seen (Knowles et al., 2018). Phagemids, on the other hand, cannot construct the full viral vector assembly without the assistance of helper phages. This is the primary disadvantage of using phagemids (Beisel, Gomaa, & Barrangou, 2014). The other type of phages, known as virulent as well as temperate phages, was utilized because of the superior bactericidal properties that they possessed in comparison to other types of phages. When designing phages, one of the primary challenges that must be addressed is the potential for disrupting

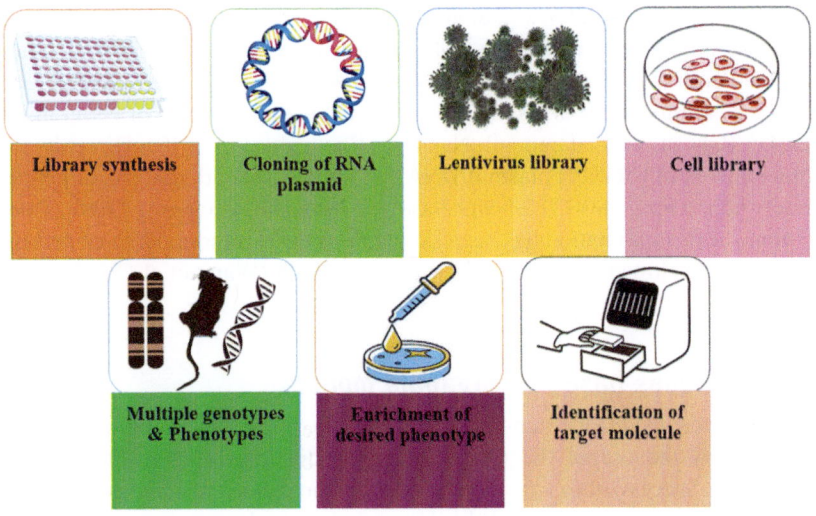

FIG. 3

The representation of genome-wide screening facilitated by CRISPR/cas9 that enables systematic examination of genes at unprecedented scales.

viral assembly and replication. This can happen when a long DNA fragment, such as a CRISPR/Cas9 system, is introduced into a phage, which can cause problems with transport and genome packaging (Doerflinger, Forsyth, Ebert, Pellegrini, & Herold, 2017). Genome-wide screening is facilitated by CRISPR/cas9 which enables systematic examination of genes at unprecedented scales (Fig. 3). The Cas9 and RNA libraries provide rapid genetic screening for gain or loss of functional mutations. As a first step, a plasmid library of Cas9 guide RNAs is constructed through the synthesis and cloning of guide RNAs intended to target coding and noncoding sections of the genome. Cells of interest are transduced with a lentivirus containing this library. Cas9 selection is followed by selective pressure, enriching cells for the desired phenotype. The phenotype causal locus can be pinpointed by sequencing the guide RNAs in this enriched cell population (Zhang, 2019).

4 Techniques utilizing CRISPR/Cas9
4.1 Mouse model techniques

Using CRISPR/Cas9 induced mutagenesis, five genes (Tet1, 2, 3, Sry, and Uty-8 variants) have been changed sequentially in mouse embryonic stem cells (ES cells) to facilitate the creation of multi-gene cancer models (Mou, Kennedy, Anderson, Yin, & Xue, 2015). Cas9 mRNA and sgRNAs were injected directly into the mouse zygote to cause mutations in the Tet1 and Tet2 genes, and the resulting mutant mice had an 80% survival rate (Wang et al., 2013). The generation of pulmonary adeno-carcinomas in animal studies by employing Cre-dependent somatic stimulation of oncogenic Kras (G12D) in association with CRISPR/Cas9-driven genome editing allows for the functional evaluation of genes that were suspected of being involved in cancer (Ghosh, Venkataramani, Nandi, & Bhattacharjee, 2019). Direct hydrodynamic infusion of CRISPR plasmid DNA encoding Cas9 and sgRNAs targeting the tumour suppressor genes like phosphatase and tensin homologue (PTEN) as well as p53 alone or in combination has been shown to produce tumours in mice and is currently being exploited as a model for cancer research (Wang et al., 2013).

4.2 Techniques based on organoid models

In human intestinal epithelium-derived organoids, numerous mutations have been introduced by the use of the CRISPR/Cas9-mediated editing technique (Baliou et al., 2018). Successful *in vitro* emulation of colon cancer is achieved by selecting and implanting isogenic organoids having variations in the tumour suppressor genes APC, SMAD4, and TP53 as well as oncogenes KRAS and/or PIK3Ca (Ghosh, Venkataramani, et al., 2019). Mismatch repair deficient colorectal cancer has been successfully modelled by CRISPR/Cas9-mediated gene deletion in colon organoids (Baliou et al., 2018). Editing the genome of PDAC driver genes using CRISPR–Cas9 in pancreatic tumour organoids demonstrates that the Wnt protein niche occurs independently during the process of carcinogenesis (Drost et al., 2017).

4.3 Techniques based on cell lines

Specific, inducible, and allele-specific suppression of protein function is now possible through the use of CRISPR/Cas9 and TALEN-induced knock-in using inactivating degron tags (Degron KI) which have improved knowledge of the pharmacological effects of EZH2 and PI3K compounds in cancer cell lines (Zhou et al., 2015). The genomic rearrangements CD74-ROS1 translocation, EML4-ALK, and KIF5B-RET inversion sequences, which are the main causes of lung cancer, have been studied using CRISPR/Cas9 (Choi & Meyerson, 2014). Additionally, cell lines with chromosomal translocations have been created using CRISPR/Cas9-mediated genome engineering to provide details on early pathological events in acute myeloid leukaemia (AML) as well as Ewing's sarcoma (Ghosh, Venkataramani, et al., 2019). Additionally, cell lines have been created using CRISPR/Cas9-mediated genome editing, which was then proceeded by a second *in vivo* screen to find new AML treatment targets (Zhou et al., 2015). Besides this, the CRISPR-mediated alteration of the HER2 gene in cell lines from breast cancer indicated a unique method of anti-cancer effects of HER2 designed to target CRISPR/Cas9, which is a potential replacement for the therapeutic medication named Herceptin (Kesik-Brodacka, 2018).

4.4 Techniques based on targeting miRNA

MicroRNAs (miRNAs) are a type of tiny, non-coding RNA. MicroRNAs (miRNAs) regulate gene expression at both the transcriptional and post-transcriptional stages by binding to the 3′-untranslated region of their target genes (Jiang, Meng, Yang, & Luo, 2020). The expression of target genes (such as those involved in the cell cycle, apoptosis, tumour cell metastasis, and chemotherapy resistance) is tightly regulated by miRNAs, which have recently been revealed to have strong ties to the emergence and progression of malignancies (Hernandez, Sanchez-Jimenez, Melguizo, Prados, & Rama, 2018). The targeted editing of microRNAs in tumour cells as well as the knocking down of microRNAs that promote the occurrence and development of tumours can prevent tumour growth and counteract chemotherapy resistance, hence improving the effectiveness of treatment for tumours (Baranwal & Alahari, 2010). A stable gene phenotype was observed in the transplanted tumour tissue 2 weeks after the subcutaneous administration of the colorectal cancer cell line HT-29 with knocked-out miR-17 in nude mice. This finding demonstrates that CRISPR/Cas9 can be used to establish a stable phenotype of targeted miRNA deletion. In human colorectal HCT116 cells, the results revealed that the levels of miR-17, miR-200c, and miR-141 had dramatically decreased. This suggests that CRISPR/Cas9 can successfully target particular miRNAs and reduce the levels of expression of certain miRNAs (Choi & Meyerson, 2014). The use of CRISPR/Cas9 to target and knock down certain miRNAs could have beneficial implications in cancer therapy (Jiang et al., 2020).

4.5 CRISPR/Cas9 in clinical trails

The CRISPR/Cas9 system for altering cancer genes was initially tested on individuals with aggressive forms of lung cancer (NCT02793856). During this clinical experiment, the immune cells from the recipient's blood were extracted, and then *ex vivo* CRISPR/Cas9 alteration was performed. This rendered the PD-1 protein inactive (Zhang, Tee, Wang, Huang, & Yang, 2015). Genome editing using *in vivo* CRISPR/Cas9 is currently being performed and shows promise, but it has not yet been used in clinical trials (Pursey, Sünderhauf, Gaze, Westra, & van Houte, 2018).

5 Challenges in the field of CRISPR/Cas9 system

Although CRISPR/Cas9 is a promising new tool for eliminating or re-sensitizing resistant bacteria, its efficacy in the complex human microbial community of billions of cells per gram of matrix and hundreds of species remains to be determined (Luther, Lee, Nagaraj, Scaletti, & Rotello, 2018). The identification of plasmids or even mobile genetic components harbouring varied resistance genes within a particular species is challenging, making it even more difficult to evaluate the real-world setting (Lino et al., 2018). Predicting how a whole population would react to a change presents another significant challenge when employing CRISPR/Cas9-based antimicrobials. It could have unintended consequences, such as the proliferation of a certain population due to the deletion of a certain plasmid, and set off a chain reaction that ultimately results in the proliferation of pathogenic species (Pursey et al., 2018). Increased susceptibility to infection with *Clostridioides difficile* in the stomach, for instance, has been connected to changes in microbial diversity and metabolite levels, which in turn have been linked to illnesses like diabetes (Harrison et al., 2014). In purified or mixed cultures, CRISPR/Cas9 engineering enables precise separation of closely related strains. With conventional antibiotics, bacteriophages, selected markers, or even other control approaches, selective and programmable microbial eradication is nearly impossible (Mahfouz, Piatek, & Stewart, 2014). Although developing "smart" antibiotics to combat multi-resistant organisms is important, distribution and off-target editing continue to be obstacles. The eradication of microbial communities through CRISPR/Cas9 remains controversial, and its ramifications are unclear (Luther et al., 2018). Transportation and overcoming regulatory hurdles are two further issues that need to be addressed to fully implement CRISPR/Cas9 systems. Resistance genes are widespread in bacteria, although they are encoded in different sites depending on the type of bacteria. However, while phages are effective vectors, delivering CRISPR/Cas9 systems through them is difficult because of phage restricted host ranges. Conjugative plasmids, which may be transmitted across bacteria, are another vector for delivery, but they have their own set of drawbacks, including a narrow host range, low conjugation effectiveness, and a high barrier to absorption (Redman et al., 2016). Though, the effectiveness of several of these newly introduced vectors for delivery, such as silica- and lipid-based particle formation remains to be determined, despite numerous research investigating them (Mitchell et al., 2021). Several suggestions have been made that can be implemented to prevent the future exploitation of safety flaws in genome editing, such as increasing the amount of research done to decrease mosaicism and off-target effects (Lino et al., 2018). Genome-editing technologies could be disseminated using relatively risk-free vector systems, like safe virus systems. Attempting to counteract the consequences of gene drive by simultaneously developing their opposite. In biosafety-critical genome-editing research, developing proportional biosafety risk classification and implementing suitable containment measures are essential. Furthermore, upgrading current standards and guidance on biosafety as well as biosecurity or offering international guidelines can be helpful at the governance level. Increased oversight is necessary because of the additional risks posed to biosafety by the use of genome-editing technologies in agricultural breeding programmes for plants and animals (Serajian et al., 2021).

6 Conclusion

Genetic engineering of microorganisms has been studied for decades. Many sectors have found a use for ZFNs, TALENs, and CRISPR/Cas9. One potential future solution to the problem of antibiotic resistance is the use of CRISPR/Cas9 for fatal self-targeting, selective eradication of selected bacterial strains, or targeting of antibiotic resistance as well as virulence genes. In addition, CRISPR/Cas9 with nuclease activity disabled provides a novel strategy for regulating bacterial gene expression. However, several obstacles remain, including safety concerns, an uptick in escape mutants, the possibility of off-target alterations, and the inefficiencies of delivery techniques. Despite the obstacles, CRISPR/Cas9-based techniques have been useful in the fight against infectious diseases. Before this technology can be used as a treatment method and the benefits of these preclinical trials may be realized in the clinic, more research is needed to establish its safety. The evolution of a prokaryotic immune defence system into a robust gene-editing platform highlights the importance of continuing to invest in research to create innovative solutions to unforeseen obstacles. Researchers have recently also developed a coronavirus quick detection method using the CRISPR/Cas9 system, providing nations with hope for a solution to this problem.

References

Aghamiri, S., Talaei, S., Ghavidel, A. A., Zandsalimi, F., Masoumi, S., Hafshejani, N. H., et al. (2020). Nanoparticles-mediated CRISPR/Cas9 delivery: Recent advances in cancer treatment. *Journal of Drug Delivery Science and Technology, 56*, 101533.

Alves-Bezerra, M., Furey, N., Johnson, C. G., & Bissig, K.-D. (2019). Using CRISPR/Cas9 to model human liver disease. *JHEP Reports, 1*(5), 392–402.

Baliou, S., Adamaki, M., Kyriakopoulos, A. M., Spandidos, D. A., Panayiotidis, M., Christodoulou, I., et al. (2018). CRISPR therapeutic tools for complex genetic disorders and cancer. *International Journal of Oncology, 53*(2), 443–468.

Baranwal, S., & Alahari, S. K. (2010). miRNA control of tumor cell invasion and metastasis. *International Journal of Cancer, 126*(6), 1283–1290.

Beisel, C. L., Gomaa, A. A., & Barrangou, R. (2014). A CRISPR design for next-generation antimicrobials. *Genome Biology, 15*(11), 1–4.

Carnes, E. C., Lopez, D. M., Donegan, N. P., Cheung, A., Gresham, H., Timmins, G. S., et al. (2010). Confinement-induced quorum sensing of individual Staphylococcus aureus bacteria. *Nature Chemical Biology, 6*(1), 41–45.

Choi, P. S., & Meyerson, M. (2014). Targeted genomic rearrangements using CRISPR/Cas technology. *Nature Communications, 5*(1), 1–6.

Cui, Y., Xu, J., Cheng, M., Liao, X., & Peng, S. (2018). Review of CRISPR/Cas9 sgRNA design tools. *Interdisciplinary Sciences: Computational Life Sciences, 10*(2), 455–465.

Doerflinger, M., Forsyth, W., Ebert, G., Pellegrini, M., & Herold, M. J. (2017). CRISPR/Cas9—The ultimate weapon to battle infectious diseases? *Cellular Microbiology, 19*(2), e12693.

Donohoue, P. D., Barrangou, R., & May, A. P. (2018). Advances in industrial biotechnology using CRISPR-Cas systems. *Trends in Biotechnology, 36*(2), 134–146.

Drost, J., Van Boxtel, R., Blokzijl, F., Mizutani, T., Sasaki, N., Sasselli, V., et al. (2017). Use of CRISPR-modified human stem cell organoids to study the origin of mutational signatures in cancer. *Science, 358*(6360), 234–238.

Ferrari, G., Thrasher, A. J., & Aiuti, A. (2021). Gene therapy using haematopoietic stem and progenitor cells. *Nature Reviews Genetics, 22*(4), 216–234.

Gaj, T., Gersbach, C. A., & Barbas, C. F., III. (2013). ZFN, TALEN, and CRISPR/Cas-based methods for genome engineering. *Trends in Biotechnology, 31*(7), 397–405.

Ghosh, C., Sarkar, P., Issa, R., & Haldar, J. (2019). Alternatives to conventional antibiotics in the era of antimicrobial resistance. *Trends in Microbiology, 27*(4), 323–338.

Ghosh, D., Venkataramani, P., Nandi, S., & Bhattacharjee, S. (2019). CRISPR-Cas9 a boon or bane: The bumpy road ahead to cancer therapeutics. *Cancer Cell International, 19*(1), 1–10. https://doi.org/10.1186/s12935-019-0726-0.

Gilbert, L. A., Horlbeck, M. A., Adamson, B., Villalta, J. E., Chen, Y., Whitehead, E. H., et al. (2014). Genome-scale CRISPR-mediated control of gene repression and activation. *Cell, 159*(3), 647–661.

Givens, B. E., Naguib, Y. W., Geary, S. M., Devor, E. J., & Salem, A. K. (2018). Nanoparticle-based delivery of CRISPR/Cas9 genome-editing therapeutics. *The AAPS Journal, 20*(6), 1–22.

Harrison, M. M., Jenkins, B. V., O'Connor-Giles, K. M., & Wildonger, J. (2014). A CRISPR view of development. *Genes & Development, 28*(17), 1859–1872.

He, Y.-Z., Kuang, X., Long, T.-F., Li, G., Ren, H., He, B., et al. (2022). Re-engineering a mobile-CRISPR/Cas9 system for antimicrobial resistance gene curing and immunization in Escherichia coli. *Journal of Antimicrobial Chemotherapy, 77*(1), 74–82.

Hernandez, R., Sanchez-Jimenez, E., Melguizo, C., Prados, J., & Rama, A. R. (2018). Down-regulated microRNAs in the colorectal cancer: Diagnostic and therapeutic perspectives. *BMB Reports, 51*(11), 563.

Jiang, F., & Doudna, J. A. (2017). CRISPR-Cas9 structures and mechanisms. *Annual Review of Biophysics, 46*(May), 505–529. https://doi.org/10.1146/annurev-biophys-062215-010822.

Jiang, C., Meng, L., Yang, B., & Luo, X. (2020). Application of CRISPR/Cas9 gene editing technique in the study of cancer treatment. *Clinical Genetics, 97*(1), 73–88. https://doi.org/10.1111/cge.13589.

Kesik-Brodacka, M. (2018). Progress in biopharmaceutical development. *Biotechnology and Applied Biochemistry, 65*(3), 306–322.

Kholosy, W. M., Visscher, M., Ogink, K., Buttstedt, H., Griffin, K., Beier, A., et al. (2021). Simple, fast and efficient iTOP-mediated delivery of CRISPR/Cas9 RNP in difficult-to-transduce human cells including primary T cells. *Journal of Biotechnology, 338*, 71–80.

Knowles, B. R., Yang, D., Wagner, P., Maclaughlin, S., Higgins, M. J., & Molino, P. J. (2018). Zwitterion functionalized silica nanoparticle coatings: The effect of particle size on protein, bacteria, and fungal spore adhesion. *Langmuir, 35*(5), 1335–1345.

Lino, C. A., Harper, J. C., Carney, J. P., & Timlin, J. A. (2018). Delivering CRISPR: A review of the challenges and approaches. *Drug Delivery, 25*(1), 1234–1257.

Luther, D. C., Lee, Y. W., Nagaraj, H., Scaletti, F., & Rotello, V. M. (2018). Delivery approaches for CRISPR/Cas9 therapeutics in vivo: Advances and challenges. *Expert Opinion on Drug Delivery, 15*(9), 905–913.

Mahfouz, M. M., Piatek, A., & Stewart, C. N., Jr. (2014). Genome engineering via TALENs and CRISPR/Cas9 systems: Challenges and perspectives. *Plant Biotechnology Journal, 12*(8), 1006–1014.

Mellor, S. M. (2022). The utilization of CRISPR/Cas9 in monogenic disorders authors. *Spectra Undergraduate Research Journal, 2*(2), 3.

Mitchell, M. J., Billingsley, M. M., Haley, R. M., Wechsler, M. E., Peppas, N. A., & Langer, R. (2021). Engineering precision nanoparticles for drug delivery. *Nature Reviews Drug Discovery, 20*(2), 101–124.

Mou, H., Kennedy, Z., Anderson, D. G., Yin, H., & Xue, W. (2015). Precision cancer mouse models through genome editing with CRISPR-Cas9. *Genome Medicine, 7*(1), 1–11.

Pavel-Dinu, M., Wiebking, V., Dejene, B. T., Srifa, W., Mantri, S., Nicolas, C. E., et al. (2019). Gene correction for SCID-X1 in long-term hematopoietic stem cells. *Nature Communications, 10*(1), 1–15.

Pursey, E., Sünderhauf, D., Gaze, W. H., Westra, E. R., & van Houte, S. (2018). CRISPR-Cas antimicrobials: Challenges and future prospects. *PLoS Pathogens, 14*(6), e1006990.

Redman, M., King, A., Watson, C., & King, D. (2016). What is CRISPR/Cas9? *Archives of Disease in Childhood-Education and Practice, 101*(4), 213–215.

Safari, F., Farajnia, S., Arya, M., Zarredar, H., & Nasrolahi, A. (2018). CRISPR and personalized Treg therapy: New insights into the treatment of rheumatoid arthritis. *Immunopharmacology and Immunotoxicology, 40*(3), 201–211.

Sashital, D. G., Wiedenheft, B., & Doudna, J. A. (2012). Mechanism of foreign DNA selection in a bacterial adaptive immune system. *Molecular Cell, 46*(5), 606–615.

Serajian, S., Ahmadpour, E., Rodrigues Oliveira, S. M., de Lourdes Pereira, M., & Heidarzadeh, S. (2021). Crispr-cas technology: Emerging applications in clinical microbiology and infectious diseases. *Pharmaceuticals, 14*(11), 1–23. https://doi.org/10.3390/ph14111171.

Shukla, A., Jani, N., Polra, M., Kamath, A., & Patel, D. (2021). CRISPR: The multidrug resistance endgame? *Molecular Biotechnology, 63*(8), 676–685.

Suzuki, K., Tsunekawa, Y., Hernandez-Benitez, R., Wu, J., Zhu, J., Kim, E. J., et al. (2016). In vivo genome editing via CRISPR/Cas9 mediated homology-independent targeted integration. *Nature, 540*(7631), 144–149.

Tian, P., Wang, J., Shen, X., Rey, J. F., Yuan, Q., & Yan, Y. (2017). Fundamental CRISPR-Cas9 tools and current applications in microbial systems. *Synthetic and Systems Biotechnology, 2*(3), 219–225.

Verma, R., Sahu, R., Singh, D. D., & Egbo, T. E. (2019). A CRISPR/Cas9 based polymeric nanoparticles to treat/inhibit microbial infections. *Seminars in Cell & Developmental Biology, 96*, 44–52.

Vigouroux, A., & Bikard, D. (2020). CRISPR tools to control gene expression in Bacteria. *Microbiology and Molecular Biology Reviews, 84*(2). https://doi.org/10.1128/MMBR.00077-19.

Wan, F., Draz, M. S., Gu, M., Yu, W., Ruan, Z., & Luo, Q. (2021). Novel strategy to combat antibiotic resistance: A sight into the combination of CRISPR/Cas9 and nanoparticles. *Pharmaceutics, 13*(3), 352.

Wang, H., Yang, H., Shivalila, C. S., Dawlaty, M. M., Cheng, A. W., Zhang, F., et al. (2013). One-step generation of mice carrying mutations in multiple genes by CRISPR/Cas-mediated genome engineering. *Cell, 153*(4), 910–918.

Watson, J. D., & Crick, F. H. C. (1953). The structure of DNA. *Cold Spring Harbor Symposia on Quantitative Biology, 18*, 123–131.

Zhang, F. (2019). Development of CRISPR-Cas systems for genome editing and beyond. *Quarterly Reviews of Biophysics, 52*. https://doi.org/10.1017/s0033583519000052.

Zhang, X.-H., Tee, L. Y., Wang, X.-G., Huang, Q.-S., & Yang, S.-H. (2015). Off-target effects in CRISPR/Cas9-mediated genome engineering. *Molecular Therapy--Nucleic Acids, 4*, e264.

Zhen, S., Hua, L., Liu, Y. H., Gao, L. C., Fu, J., Wan, D. Y., et al. (2015). Harnessing the clustered regularly interspaced short palindromic repeat (CRISPR)/CRISPR-associated Cas9 system to disrupt the hepatitis B virus. *Gene Therapy, 22*(5), 404–412.

Zhou, Q., Derti, A., Ruddy, D., Rakiec, D., Kao, I., Lira, M., et al. (2015). A chemical genetics approach for the functional assessment of novel cancer Genes. *Cancer Research, 75*(10), 1949–1958.

Genome engineering in *Aspergillus niger*

Hongzhi Dong[a,b] **and Li Pan**[c,d,*]

[a]*Key Laboratory for Northern Urban Agriculture of Ministry of Agriculture and Rural Affairs, Beijing University of Agriculture, Beijing, China*
[b]*College of Bioscience and Resources Environment, Beijing University of Agriculture, Beijing, China*
[c]*School of Biology and Biological Engineering, South China University of Technology, Guangzhou, China*
[d]*Guangdong Provincial Key Laboratory of Fermentation and Enzyme Engineering, South China University of Technology, Guangzhou, China*
[*]*Corresponding author: e-mail address: btlipan@scut.edu.cn*

Abbreviations

amyA	gene of alpha-amylase
argB	gene of ornithine carbamoyltransferase
ble	bleomycin resistance gene
CRISPR	clustered regularly interspaced short palindromic repeats
CRISPR-HDR	CRISPR/Cas9 homologous direct repair
dDNA	donor DNA
glaA	gene of glucoamylase
goxC	gene of glucose oxidase
gRNA	guide RNA
hygB	hygromycin B resistance gene
NHEJ	non-homologous end joining
ORF	opening reading frame
pTef	promoter of tef1 from *A. nidulans*
pyrG	gene of orotidine-5′-phosphate decarboxylase
sgRNA	single-guide RNA
tRNA	transfer RNA
tTef	terminator of tef1 from *A. nidulans*

1 Introduction

Aspergillus niger, an organism that is generally recognized as safe, has been widely used in the food industry (Meyer, 2008). To optimize enzyme fermentation and investigate the regulatory mechanisms of secondary metabolite biosynthesis, classical genome editing methods are required to construct mutants (Meyer, 2008). However, there are several difficulties associated with constructing classical knock-out cassettes and recycling selection markers (Niu, Arentshorst, Seelinger, Ram, & Ouedraogo, 2016). CRISPR/Cas9 is considered the most effective tool for gene editing and has been broadly applied across species (Dicarlo et al., 2013; Liu, Chen, Jiang, Zhou, & Zou, 2015; Ma et al., 2015; Nødvig, Nielsen, Kogle, & Mortensen, 2015). The CRISPR/Cas9 system is used to insert and delete genes with the repair of non-homologous end joining (NHEJ) or donor DNA (dDNA)-harbouring homologies near the double-stranded break that it creates (Peters et al., 2016; Pohl, Kiel, Driessen, Bovenberg, & Nygård, 2016; Ran et al., 2013; Zeng et al., 2015).

The CRISPR/Cas9 system has been developed for filamentous fungi such as *Trichoderma reesei*, *Aspergillus oryzae*, *Aspergillus fumigatus*, and *Aspergillus carbonarius* (Fuller, Chen, Loros, & Dunlap, 2015; Katayama et al., 2016; Liu, Perkins, Petriello, & Hennig, 2015; Weyda et al., 2017; Zhang, Meng, Wei, & Lu, 2016). In *A. niger*, based on the CRISPR/Cas9 technique, Nødvig and colleagues successfully mutated the polyketide synthase gene *fwnA* using a series of pFC plasmids and then flanking transfer RNA (tRNA) genes to express multiple single-guide RNAs (sgRNAs) in one expression cassette to edit multiple sites (Nødvig et al., 2015). Zheng found that the 5S rRNA gene promoter can be used as a sgRNA promoter and used the CRISPR/Cas9 system to delete a 48-kb gene cluster (Zheng et al., 2018). Song tested 37 tRNA promoters to drive the transcription of sgRNA and finally concluded that 36 of these tRNAs, including a tRNA gene plus 100 base pairs of upstream sequence, were reliable and efficient in the CRISPR/Cas9 system (up to 97% efficiency) (Song, Ouedraogo, Kolbusz, Nguyen, & Tsang, 2018). Sarkari used CRISPR/Cas9 together with the Golden Gate cloning technique to integrate an aconitic acid expression cassette into the genome of *A. niger* (Sarkari et al., 2017).

In the present study, we developed a toolbox called the CRISPR/Cas9 homologous direct repair (CRISPR-HDR) system and tested this system with dDNAs and *in vitro* synthesized sgRNAs in *A. niger*. Using the developed CRISPR-HDR toolbox, we separately deleted 0-, 2-, 10-, and even two 50-kb fragments of secondary metabolism gene clusters. We then modified the plasmid pFC330 by adding a tandem sgRNA expression cassette to transcribe two sgRNAs (*amyA* site and *glaA* site). By co-transforming the modified pFC330 with dDNAs to *A. niger*, we obtained single-, double-, and triple-site deleted strains with efficiencies of 31.8%, 54.5%, and 13.6%, respectively. Using this optimized CRISPR-HDR toolbox, we even successfully inserted the glucose oxidase gene *goxC* at the loci of *amyA* and *glaA* simultaneously and increased the production of GoxC to 869.86 U/mL, which was fourfold higher than that of the control strain.

2 Materials

2.1 Nucleotide preparation or construction

(1) pFC series plasmids, include pFC330, pFC331, pFC332, and pFC333 (Nødvig et al., 2015);
(2) pMD19 and pUC57 plasmids were used for the *de novo* construct;
(3) Prime STAR, used for high-fidelity PCR enzyme, was purchased from TAKARA;
(4) Synthesis of primers and Sanger sequencing were ordered from Tianyi Huiyuan Co.;
(5) HiPure PCR Pure Maxi Kit for PCR product purification was purchased from Magen Co.;
(6) HiScribe™ T7 Quick High Yield RNA Synthesis Kit, which was used for *in vitro* synthesis of RNA, was purchased from NEB;
(7) Qubit, used to measure the concentration and purity of nucleotides, was purchased from Thermo Fisher Scientific.

2.2 Manipulation of strains

(1) *A. niger* strain CBS513.88 was purchased from the CBS Fungal Conservation Center, Netherlands;
(2) The formulation of DPY medium was as follows: 5 g of yeast extract, 10 g of peptone, 20 g of glucose, 5 g of KH_2PO_4, and 0.5 g of $MgSO_4 \cdot 7H_2O$ per litre, autoclaving at 115 °C, 20 min;
(3) The formulation of CD medium was as follows: 10 g of glucose, 3 g of $NaNO_3$, 2 g of KCl, 0.5 g of $MgSO_4 \cdot 7H_2O$, 1 g of $K_2HPO_4 \cdot 3H_2O$, 0.1 g of $FeSO_4 \cdot 7H_2O$, and 20 g of agar per litre; pH 5.5, autoclaving at 115 °C, 20 min;
(4) The formulation of the fermentation medium was as follows: 50 g of liquefied starch (New Probe, Guangzhou), 30 g of maltose syrup (New Probe), 20 g of soy powder (New Probe), 20 g of glucose per litre, autoclaving at 115 °C, 20 min;
(5) The details of STC solution: 50 mM $CaCl_2$, 1 M sorbitol, 10 mM Tris-Cl (pH 7.5), pH adjusted to 5.6 with HCl.

2.3 Measurement of enzyme activity

(1) The standard glucose oxidase was purchased from New Probe;
(2) The enzyme diluent contained 0.1 mol/L disodium hydrogen phosphate and 0.05 mol/L citric acid, pH 6.2;
(3) The formula of buffer I was as follows: 0.1 g linanisamine was dissolved into 10 mL of methanol, stored at 4 °C, and used right after it was ready;
(4) The formula of buffer II was: 18 g D-glucose powder was dissolved into 100 mL distilled water;
(5) The formula of buffer III was as follows: Horseradish peroxidase (Beyotime) was diluted to 90 U/mL, and used immediately;

(6) The formula of buffer IV, named linanisamine working solution, was as follows: Dissolve 0.1 mL of buffer I in enzyme diluent;

(7) The microplate reader was purchased from Thermo Fisher Scientific.

2.4 Detection of secondary metabolism

(1) The formula of the WATM medium was as follows: 2.0 g/L yeast extract, 3.0 g/L peptone, 2.0 g/L dextrose, 30.0 g/L sucrose, 5.0 g/L corn steep solids, 2.0 g/L NaNO$_3$, 1.0 g/L K$_2$HPO$_4$·3H$_2$O, 0.5 g/L MgSO$_4$, 0.2 g/L KCl, 0.01 g/L FeSO$_4$·7H$_2$O, 20.0 g/L agar, pH 7.0;

(2) The HPLC system comprised an Agilent 1200 HPLC system equipped with a column (Luna 5u C18 250*4.60 mm 5-μm Phenomenex, USA) and a Diode Array Detector (DAD, Agilent) and an ultra-high resolution TOF mass detector (Bruker maXis, Germany);

(3) Solution A of HPLC was 0.1% (v/v) formic acid solution;

(4) Solution B of HPLC was an absolute acetonitrile solution.

3 Methods

The protocol includes the following aspects: preparation of nucleic acids (Cas9 protein plasmids, sgRNA, and donors), the transformation of host strains, detection of Cas9 protein, and evaluation of the yield of target products.

3.1 Choice of appropriate Cas9 protein expression plasmids

Cas9 protein was expressed using the pFC series plasmid (pFC330, pFC331, pFC332, and pFC333) as mentioned in Nødvig's research (Nødvig et al., 2015), which contains all the expression elements of protein Cas9, which contains the promoter (pTef), the cDNA of Cas9, the nuclear localization signal, and the terminator (Ttef) of *Aspergillus nidulans*. The only distinction of the four plasmids is the resistance marker: *pyrG* of pFC330, *argB* of pFC331, *hygB* of pFC332, and *ble* of pFC333.

3.2 Construction of sgRNAs

sgRNAs could obtained by two strategies: expression through a plasmid harbouring a complete set of expression elements (promoter, ORF and terminator) *in vivo,* or synthesis *in vitro*.

3.2.1 sgRNAs expression in vivo through plasmids

The sgRNA expression cassette pAfU6-fwnA consisted of the 400 bp of *A. fumigatus* U6 promoter (pAfU6) (Zhang et al., 2016), 20 bp of *fwnA* protospacer, a sgRNA scaffold (76 bp), and 138 bp of the *A. oryzae* U6 terminator (tAoU6) (Katayama et al., 2016). These were usually constructed using an overlap PCR method as follows:

(1) Design primers for overlap PCR (Fig. 1A).

(2) Using F-U6p and R-U6p1 as primers, the genome of *A. fumigatus* as the template, the fragment S1 was obtained;

(3) Using F-U6p and R-U6p2 as primers, fragment S1 as template, the fragment S2 was obtained;

(4) Using F-U6t and R-U6t as primers, the genome of *A. oryzae* as a template and the fragment of AoU6 terminator was obtained;

(5) Using F-U6p and R-U6t as primers, fragment S2 and AoU6 as the template, the whole cassette of sgRNA of the target site was obtained by fusion PCR.

(6) Ligate T-Vector pMD19 and sgRNA cassette obtained in step 5, using a vector ligation kit, then transform *Escherichia coli* DH5α and screen by Ampicillin resistance, to obtain the positive sgRNA expression plasmid.

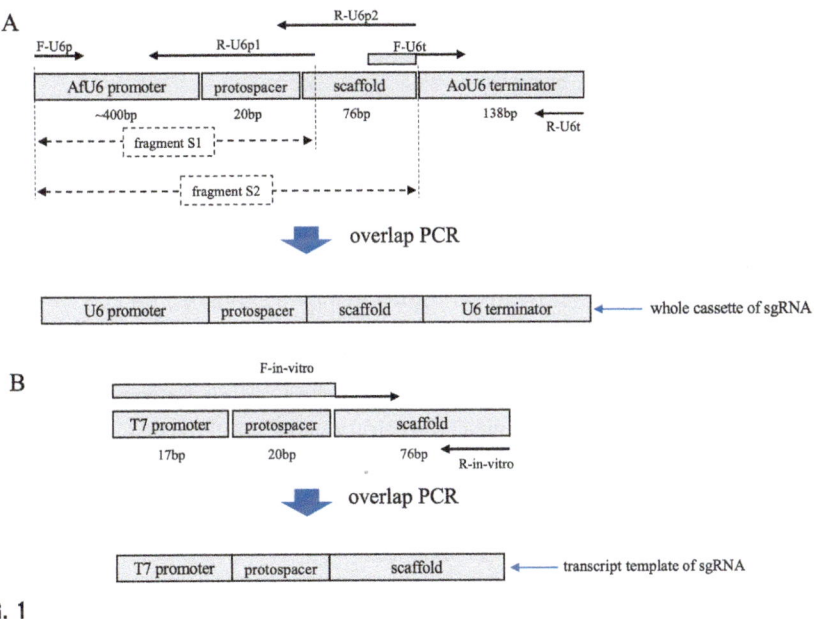

FIG. 1

Schematic diagram of the construction of transcription template of sgRNA.

3.2.2 sgRNAs synthesis in vitro

For the *in vitro* synthesis of gRNA, a minimal T7 promoter (18 bp), a variable pro-tospacer (20 bp), and a gRNA scaffold (76 bp) were used as templates. The workflow was as follows:

(1) Design primers for overlap PCR (Fig. 1B);
(2) Amplify the plasmid (named pUC57-sgscaffold) that contains the sgRNA scaffold by Tianyi Huiyuan Co.
(3) Using F-*in-vitro* and R-*in-vitro* as the primers, and pUC57-sgscaffold as the template, the transcript template of sgRNA was obtained by PCR.
(4) Purify the template of sgRNA using a HiPure PCR Pure Maxi Kit;
(5) *In vitro* transcription of the template through HiScribe™ T7 Quick High Yield RNA Synthesis Kit (NEB);
(6) Purify the transcribed RNAs using phenol/chloroform extraction with ethanol precipitation and then dissolve them in RNase-free water;
(7) Authenticate the quantity and quality of sgRNAs using gel electrophoresis and Qubit.

3.3 Preparation of donor DNAs

The donor DNA contained three components in order: the upstream homologous arm of the target site, the resistance gene, and the downstream homologous arm of the target site.

3.3.1 Construction of donor DNAs with short homologous arms (39 bp)

(1) Design primers to amplify the resistance marker;
(2) Using F-mini and R-mini as the primers, the fragment containing a screening marker as the template, and the donor of the mini-arm was obtained by PCR.

3.3.2 Construction of donor DNAs with long homologous arms (500–2000 bp)

(1) Using F-up and R-up as the primers, the genome of the host strain as the template, the upstream fragment of the donor was obtained by PCR (Fig. 2B);
(2) Using F-down and R-down as the primers, the genome of the host strain as the template, the downstream fragment of the donor was obtained by PCR;
(3) Using F-marker and R-marker as the primers, the fragment containing a screening marker as the template, the selection marker of the donor was obtained by PCR;
(4) Using F-up and R-down as the primers, the fragments of upstream, downstream, and selection marker as the template, the tandem fragment of donor contain norm-arm (regular length) was obtained.
(5) Ligate T-Vector pMD19 and the tandem fragments using a vector ligation kit, then transform *E. coli* DH5α to obtain the recombinant vector.
(6) Using the pre-selected endonuclease, linearize the vector above to get the donor DNA.

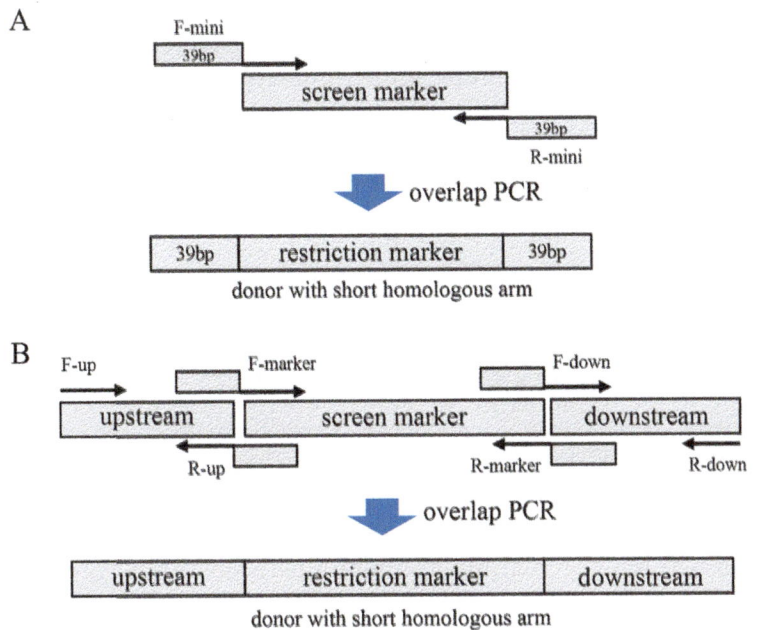

FIG. 2

Schematic diagram of the construction of donor DNA.

3.4 Transformation of host strains and verification of positive transformants

A. niger was transformed using the polyethylene PEG/CaCl$_2$-protoplast transformation method.

(1) Innoculate 1E5 spores of the host strain of *A. niger* into liquid DPY medium and cultivate in a shaking incubator with 200 rpm, 30 °C;

(2) Harvest the thallus to obtain enough biomass for biotransformation;

(3) Filter the thallus using a filter flask and weigh the biomass. Digest each 2 g of thallus with 10 mL of Yatalase (TAKARA);

(4) Mix the thallus and the fresh Yatalase solution, lyse the thallus with Yatalase at 30 °C, 100 rpm flask incubation for 2–3 h, till a large number of protoplasts are produced as seen by microscopic examination;

(5) Prepare several solid CD medium plates;

(6) Filter the protoplast solution with Miracloth (Merck) and collect the protoplasts into a 50-mL tube, and then centrifuge at 900 × *g*, 10 min, then discard the supernatant and add 20 mL of STC buffer (see Section 2.2), centrifuge at 900 × *g*, 10 min, repeat once again to get the pure protoplasts resuspended in 3 mL STC buffer;

(7) Divide the protoplasts into 5-mL tubes, equally for the experimental group, positive control, and negative control, and add 1 mL PEG/CaCl$_2$ buffer, corresponding screening drugs, and DNA. Invert several times to mix them sufficiently;

(8) Pour the mixture onto the solid CD medium plate, and shake the plate gently to distribute the mixture evenly over the solid plate;

(9) Place the transformation plate in an incubator of 30 °C, till the transformants gradually emerge (usually 3–5 days).

3.5 Detecting the yield of target products and evaluation of the genome edit effect or efficiency

Evaluation of the genome edit may be dependent on the purpose. The procedures for product detection are described below:

3.5.1 Detecting the activity of glucose oxidase

(1) Cultivate positive transformants in fermentation medium for 3 days, then harvest the thallus and centrifuge at 5000 × *g* for 10 min to collect the supernatant and enzyme.

(2) Definition of enzyme activity: The amount of enzyme that resolves 1 μmol glucose to gluconolactone and hydrogen peroxide in 1 min, pH 5.1 and 35 °C, was defined as 1 U of enzyme activity;

(3) Dilute the standard sample to a gradient of dilutions (0.4, 0.6, 0.8, 1.0, 1.2, 1.6, 1.8, 2.0 U/mL), number the samples as No. 1–9;

(4) Inactivate the standard sample in boiling water, named No. 10;

(5) Prepare a glass tube of 10 mL, add 2.5 mL buffer I, 300 μL buffer II and 100 μL buffer III, incubate for 2 min at 37 °C;

(6) Add 100 μL sample of No. 1–10, incubate for 3 min at 37 °C;

(7) Add 2 mL sulphuric acid solution of 2 mol/L to terminate the reaction;

(8) Take the 200 μL mixture above to the coated wells and measure the absorption at 540 nm;

(9) Draw the standard curve as follows: Using No. 10 as the origin, the concentration of glucose oxidase as the x-axis, and the absorption at 540 nm as the y-axis. The standard equation is:

$$y = ax + b$$

(a and b are contants dependent on the standard curve)

(10) Inoculate the positive transformants into 50 mL liquid DPY medium (in 250-mL shake flask), with 1E5 spores, and cultivate in a shake incubator with 200 rpm, 30 °C for 7 days;

(11) Take 5 mL fermentation broth of positive transformants and centrifuge (8000 rpm, 5 min) to get the supernatant (or enzyme sample) to measure the enzyme activity;

(12) Prepare a glass tube of 10 mL, add 2.5 mL buffer I, 300 μL buffer II and 100 μL buffer III, incubate for 2 min at 37 °C;

(13) Add 100 μL enzyme samples or inactivated enzyme samples, and incubate for 3 min at 37 °C;

(14) Add 2 mL sulphuric acid solution of 2 mol/L to terminate the reaction;

(15) Put 200 μL mixture above into a microplate and measure the absorption at 540 nm through a microplate reader;

(16) Dilute the enzyme samples to an appropriate reading in the range of 0.2–0.4, for readings in this range are approximately considered to conform to the linear relationship in the standard curve.

(17) Calculate the enzyme activity according to the following formula:

$$Y = X(measurement) * n + b$$

(Y, the enzyme activity of sample; X, the reading of instrument; n, the dillution ratio)

3.5.2 Secondary metabolism detection

Secondary metabolites were extracted as described (Wang et al., 2018) and analysed by HPLC-TOF-ESI/MS using a previously described method (Wang et al., 2018).

(1) Inoculate *A. niger* transformants onto WATM solid medium in a glass dish with a diameter of 20 cm, and cultivate for 7 days at 25 °C;

(2) Put 100 mL mixture (99 mL ethyl acetate and 1 mL formic acid) onto the solid medium and incubate for 24 h;

(3) Suction the extract solution with a 50-mL syringe and put it into two 50-mL tubes;

(4) Centrifuge at 8000 rpm, 4 °C for 5 min and collect the supernatant as the sample;

(5) Dry the sample in a rotary evaporator at 38 °C;

(6) Dissolve the dried sample in MeOH:ddH$_2$O (9:1) and vortex for 30 s till the sample is totally dissolved;

(7) Analyse the sample by HPLC-TOF-ESI/MS. Solution A and Solution B were used as the mobile phases in the gradient elution at 0.8 mL/min with a time course of increasing solution B of 10–100% B for 0–15 min and 100–10% B for 15–30 min. Mass spectra in positive ion mode were recorded in 30 min.

4 Notes

(1) For the preparation of buffer II for the measurement of enzyme activity of glucose oxidase, the powder of glucose must be devoid of water. It could be dried off through treatment at 105 °C for 3 h;

(2) For the construction of the transcription template of sgRNA, the primer of R-U6p1 and R-U6p2 were designed to contain the sequence of 20 bp of protospacer; overlap regions in the primers should be within the range of 15–25 bp (Fig. 1);

(3) For the construction of donor DNA, the primers of F-mini and R-mini (Fig. 2) should contain the 39 bp of homologous arms, and the length of homologous arms was dependent on the purpose: homologous arms as short as 39 bp (mini-arm) could simplify the flowsheet, but the conventional length of the homologous arm (norm-arm) leads to greater integration efficiency.

(4) For the construction of donor DNA, the screening marker must be used in the host strain to be gene-edited.

(5) For the construction of the donor DNA, the endonuclease is not permanent and should satisfy the following conditions: cannot exist in the donor (do not destroy the tandem structure of the donor); should be as inexpensive as possible; obtained easily.

(6) For the strategy of multiple genes knockout: There are one copy of *glaA* and two copies of *amyA* in *A. niger*. We used two sgRNAs (one target to *glaA* and another one target to *amyA*) to knockout three target sites. There are 22 transformants and the knockout rates are: 100% of *glaA* (22/22) and 54.5% & 13.6% of *amyA*(7/22 of single and 3/22 of double, means 54.5% and 13.6%).

References

Dicarlo, J. E., Norville, J. E., Mali, P., Rios, X., Aach, J., & Church, G. M. (2013). Genome engineering in Saccharomyces cerevisiae using CRISPR-Cas systems. *Nucleic Acids Research, 41*(7).

Fuller, K. K., Chen, S., Loros, J. J., & Dunlap, J. C. (2015). Development of the CRISPR/Cas9 system for targeted gene disruption in Aspergillus fumigatus. *Eukaryotic Cell, 14*(11), 1073–1080. Available from: http://www.ncbi.nlm.nih.gov/pubmed/26318395.

Katayama, T., Tanaka, Y., Okabe, T., Nakamura, H., Fujii, W., Kitamoto, K., et al. (2016). Development of a genome editing technique using the CRISPR/Cas9 system in the industrial filamentous fungus Aspergillus oryzae. *Biotechnology Letters, 38*(4), 637–642. Available from: http://www.ncbi.nlm.nih.gov/pubmed/26687199.

Liu, R., Chen, L., Jiang, Y., Zhou, Z., & Zou, G. (2015). Efficient genome editing in filamentous fungus Trichoderma reesei using the CRISPR/Cas9 system. *Cell Discovery, 1*, 15007.

Liu, D., Perkins, J. T., Petriello, M. C., & Hennig, B. (2015). Exposure to coplanar PCBs induces endothelial cell inflammation through epigenetic regulation of NF-κB subunit p65. *Toxicology and Applied Pharmacology, 289*(3), 457–465.

Ma, X., Zhang, Q., Zhu, Q., Liu, W., Chen, Y., Qiu, R., et al. (2015). A robust CRISPR/Cas9 system for convenient, high-efficiency multiplex genome editing in monocot and dicot plants. *Molecular Plant, 8*(8), 1274–1284.

Meyer, V. (2008). Genetic engineering of filamentous fungi—Progress, obstacles and future trends. *Biotechnology Advances, 26*(2), 177–185.

Niu, J., Arentshorst, M., Seelinger, F., Ram, A. F. J., & Ouedraogo, J. P. (2016). A set of iso-genic auxotrophic strains for constructing multiple gene deletion mutants and parasexual crossings in Aspergillus niger. *Archives of Microbiology*, *198*(9), 861–868.

Nødvig, C. S., Nielsen, J. B., Kogle, M. E., & Mortensen, U. H. (2015). A CRISPR-Cas9 system for genetic engineering of filamentous fungi. *PLoS One*, *10*(7), e0133085. Available from: http://www.ncbi.nlm.nih.gov/pubmed/26177455.

Peters, J. M., Colavin, A., Shi, H., Czarny, T. L., Larson, M. H., Wong, S., et al. (2016). A comprehensive, CRISPR-based functional analysis of essential genes in bacteria. *Cell*, *165*(6), 1493–1506.

Pohl, C., Kiel, J. A. K. W., Driessen, A. J. M., Bovenberg, R. A. L., & Nygård, Y. (2016). CRISPR/Cas9 based genome editing of Penicillium chrysogenum. *ACS Synthetic Biology*, *5*(7), 754–764. Available from: http://www.ncbi.nlm.nih.gov/pubmed/27072635.

Ran, F. A., Hsu, P. D., Wright, J., Agarwala, V., Scott, D. A., & Zhang, F. (2013). Genome engineering using the CRISPR-Cas9 system. *Nature Protocols*, *8*(11), 2281–2308.

Sarkari, P., Marx, H., Blumhoff, M. L., Mattanovich, D., Sauer, M., & Steiger, M. G. (2017). An efficient tool for metabolic pathway construction and gene integration for Aspergillus niger. *Bioresource Technology*, *245*(Pt. B), 1327–1333. Available from: http://www.ncbi.nlm.nih.gov/pubmed/28533066.

Song, L., Ouedraogo, J. P., Kolbusz, M., Nguyen, T. T. M., & Tsang, A. (2018). Efficient genome editing using tRNA promoter-driven CRISPR/Cas9 gRNA in Aspergillus niger. *PLoS One*, *13*(8), 1–17.

Wang, B., Lv, Y., Li, X., Lin, Y., Deng, H., & Pan, L. (2018). Profiling of secondary metab-olite gene clusters regulated by LaeA in Aspergillus niger FGSC A1279 based on genome sequencing and transcriptome analysis. *Research in Microbiology*, *169*(2), 67–77.

Weyda, I., Yang, L., Vang, J., Ahring, B. K., Lübeck, M., & Lübeck, P. S. (2017). A comparison of Agrobacterium-mediated transformation and protoplast-mediated trans-formation with CRISPR-Cas9 and bipartite gene targeting substrates, as effective gene tar-geting tools for Aspergillus carbonarius. *Journal of Microbiological Methods*, *135*, 26–34. Available from: http://www.ncbi.nlm.nih.gov/pubmed/28159628.

Zeng, H., Wen, S., Xu, W., He, Z., Zhai, G., Liu, Y., et al. (2015). Highly efficient editing of the actinorhodin polyketide chain length factor gene in Streptomyces coelicolor M145 using CRISPR/Cas9-CodA(sm) combined system. *Applied Microbiology and Biotechnol-ogy*, *99*(24), 10575–10585. Available from: http://www.ncbi.nlm.nih.gov/pubmed/26318449.

Zhang, C., Meng, X., Wei, X., & Lu, L. (2016). Highly efficient CRISPR mutagenesis by microhomology-mediated end joining in Aspergillus fumigatus. *Fungal Genetics and Bi-ology*, *86*, 47–57.

Zheng, X., Zheng, P., Zhang, K., Cairns, T. C., Meyer, V., Sun, J., et al. (2018). 5S rRNA promoter for guide RNA expression enabled highly efficient CRISPR/Cas9 genome edit-ing in Aspergillus niger. *ACS Synthetic Biology*, *8*(7), 1568–1574. Available from: http://www.ncbi.nlm.nih.gov/pubmed/29687998.

Transformation IV

Natural transformation as a tool in *Acinetobacter baylyi*: Evolution by amplification of gene copy number

Isabel Pardo[a], Stacy R. Bedore[b], Melissa P. Tumen-Velasquez[b,†],
Chantel V. Duscent-Maitland[b], Alyssa C. Baugh[b],
Suvi Santala[c], and Ellen L. Neidle[b,*]

[a]*Centro de Investigaciones Biológicas Margarita Salas (CIB), Spanish National Research
Council (CSIC), Madrid, Spain*
[b]*Department of Microbiology, University of Georgia, Athens, GA, United States*
[c]*Faculty of Engineering and Natural Sciences, Tampere University, Tampere, Finland*
*Corresponding author: e-mail address: eneidle@uga.edu

Abbreviations

Ab[R]	antibiotic resistance
ALE	adaptive laboratory evolution
Cit	citrate
cPCR	colony PCR
EASy	Evolution by Amplification and Synthetic biology
GDA	gene duplication and amplification
gDNA	genomic DNA
GOI	gene(s) of interest
Gua	guaiacol
Km	kanamycin
qPCR	quantitative PCR
SBF	synthetic bridging fragment
Sm	streptomycin
Sp	spectinomycin
TPA	terephthalic acid

†Current affiliation: Biosciences Division, Oak Ridge National Laboratory, Oak Ridge, TN, United States.

Methods in Microbiology, Volume 52, ISSN 0580-9517, https://doi.org/10.1016/bs.mim.2023.01.001

1 Introduction

Adaptive laboratory evolution (ALE) is a widespread and powerful tool to engineer microorganisms with enhanced tolerance, increased production capabilities, or new phenotypes (Dragosits & Mattanovich, 2013). By growing the cells under an appropriate selective pressure for multiple generations, the desired property can be attained due to the accumulation of beneficial mutations that lead to improved activities (i.e., optimization) or novel functions (i.e., innovation) (Barrick & Lenski, 2013). However, in the timescales typically employed in experimental evolution, the low frequency of spontaneous, beneficial mutations may impose constraints that are difficult to overcome. Specifically, in the early stages of selection, gene/protein expression may be insufficient to sustain growth. Low expression could be problematic if selection depends on the weak side activity of an enzyme, if toxic intermediates accumulate, and/or if non-native proteins are unstable. In nature, gene duplication and amplification (GDA) facilitate adaptation to new or changing environments by increasing expression via transient changes in gene dosage. This amplification is a reversible mechanism that allows the reduction of gene copy number once selective pressure is removed or other beneficial mutations are propagated throughout the population (Andersson & Hughes, 2009; Elliott, Cuff, & Neidle, 2013; Seaton et al., 2012).

Although gene duplications tend to occur significantly more frequently than point mutations (Anderson & Roth, 1981; Seaton et al., 2012), the probability remains low that a sufficient number of cells in the starting population of a culture will randomly have a duplication in the correct location to be selectively beneficial in ALE (Herrmann et al., 2022). However, if a tandem duplication encompassing the region of interest does exist, it provides long stretches of identical sequence that allow further amplification mediated by homologous recombination. Gene copy number then fluctuates stochastically, and if increased dosage is advantageous, cells will be selected that, on average, have the same level of increased expression (Reams & Neidle, 2003) (Fig. 1A). A paradigmatic example of the relevance of GDA in experimental evolution is the emergence of the aerobic citrate-utilizing (Cit$^+$) phenotype in *Escherichia coli* during a long-term evolution experiment (Blount, Barrick, Davidson, & Lenski, 2012). This event, which took place after >31,000 generations (15 years), was due to the GDA of *citT*, which encodes a citrate: succinate antiporter. Later works purposefully evolving *E. coli* for citrate utilization found that Cit$^+$ clones obtained after at least 30 days of selection also possessed multiple copies of *citT* (Van Hofwegen, Hovde, & Minnich, 2016).

To induce GDA in the laboratory at a precise position, the locus of interest may be artificially duplicated using DNA manipulation techniques (Stoudenmire et al., 2017) (Fig. 1B). Recently, we developed a method named Evolution by Amplification and Synthetic Biology (EASy) in *A. baylyi* ADP1 (Tumen-Velasquez et al., 2018). A schematic overview of this method is shown in Fig. 2. EASy takes advantage of the exceptional natural competence and efficient homologous recombination of *A. baylyi*. In a first step, the gene(s) of interest (GOI) and an antibiotic resistance (Ab^R) selection marker are integrated in the chromosome of *A. baylyi*. Next,

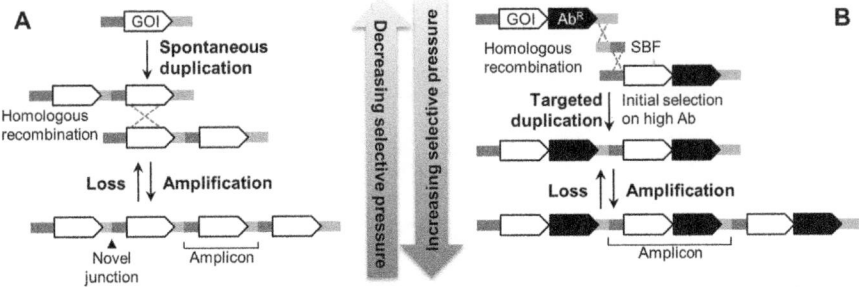

FIG. 1

Overview of natural and EASy-mediated gene duplication and amplification (GDA). (A) In natural GDA, the infrequent duplication of the gene of interest (GOI), by homologous or illegitimate recombination, in a few individuals in a population provides a stretch of sequence identity for subsequent homologous recombination. With increasing selective pressure, mutants with higher gene dosage are selected. (B) With EASy, the initial duplication of a targeted chromosomal segment is induced by transforming with a synthetic bridging fragment (SBF). High antibiotic concentrations allow the rapid selection of mutants with multiple copies of the antibiotic selection marker (Ab^R), which is present in the amplicon delimited by the SBF. Then, growth under selective conditions without antibiotic drives further GDA of the GOI.

a specific region of the chromosome, encompassing the GOI and the Ab^R marker, is amplified by transforming cells with a synthetic DNA bridging fragment (SBF). The SBF provides sufficient sequence identity for initial duplication by homologous recombination, and it precisely delimits the chromosomal segment that is duplicated, i.e., the amplicon. By including the Ab^R marker within the amplicon, mutants with multiple copies can be easily selected in the presence of high antibiotic concentrations. This step bypasses a potential bottleneck in the emergence of amplification mutants when direct selection for the desired phenotype is too demanding to the cells. Afterwards, the amplification mutants can be transferred to the selective conditions for ALE for the emergence of the desired phenotype. At this point, high antibiotic concentrations are removed from the medium. Thus, the only selective pressure driving GDA is the target condition for ALE (e.g., growth on a non-native substrate). Since chromosomal duplications are inherently unstable, beneficial mutations that result in a growth advantage will be selected during ALE and lead to a decrease in the average copy number of the amplicon in the evolving population. These beneficial mutations may be found in the GOI and/or in other regions of the chromosome.

We successfully used EASy to enable the catabolism of non-native aromatic carbon sources in *A. baylyi*, namely guaiacol and terephthalic acid (TPA) (Pardo et al., 2020; Tumen-Velasquez et al., 2018). Similarly, after the functional replacement of a native aromatic catabolic pathway with a non-native pathway, EASy was used to increase thermotolerance (Bedore, 2021). *A. baylyi* can utilize a broad range of aromatic compounds through the β-ketoadipate pathway, by which many different substrates are converted by peripheral routes into either catechol or protocatechuic acid (Fuchs, Boll, & Heider, 2011; Seaton & Neidle, 2018). The aromatic rings of these compounds are then cleaved by intradiol dioxygenases before metabolites

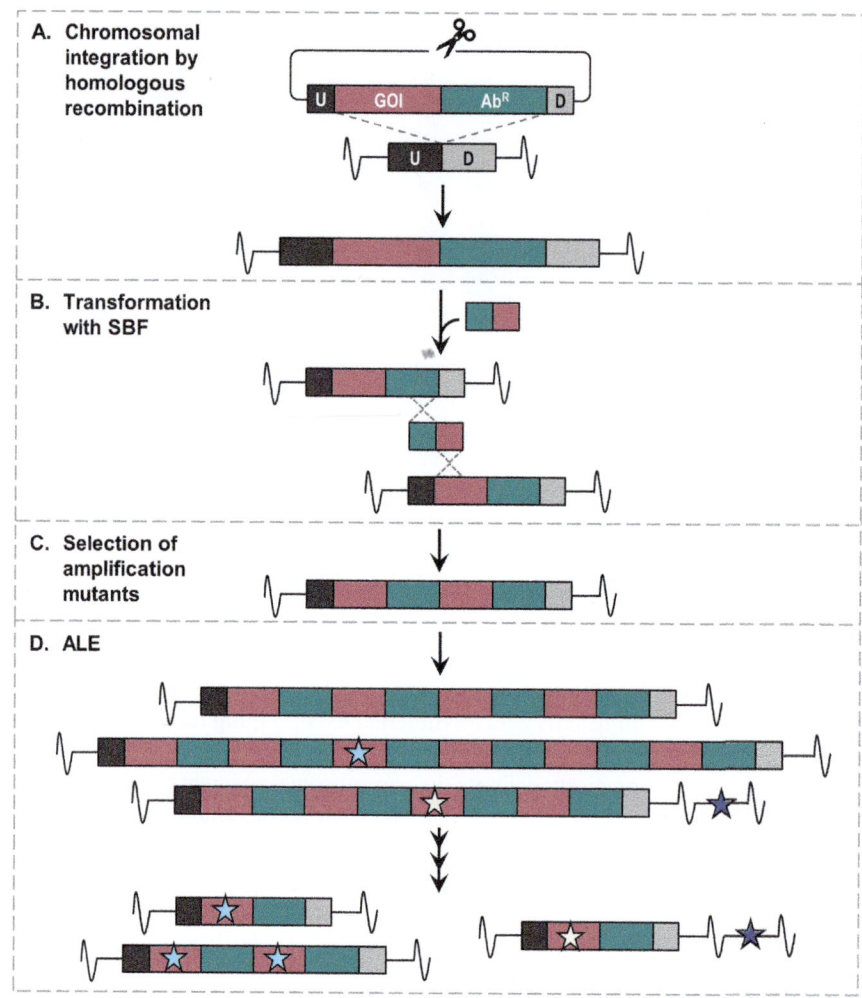

FIG. 2

Overview of the EASy method. (A) The gene of interest (GOI) and the antibiotic selection marker (Ab^R), flanked by sequences identical to the upstream (U) and downstream (D) regions of the targeted locus, are integrated in the *Acinetobacter baylyi* chromosome by homologous recombination. Alternatively, a native gene or chromosomal region can be targeted for amplification by simply introducing the Ab^R in an adjacent position. (B) Mutants are transformed with a synthetic bridging fragment (SBF), which delimits gene duplication and amplification to the desired chromosomal segment. (C) The amplification mutants that have incorporated the SBF are selected in the presence of high antibiotic concentrations. (D) The amplification mutants are subjected to adaptive laboratory evolution (ALE) under the desired selective conditions, in which antibiotic selection is removed. During ALE, the tandem array of amplicon copies may expand to sustain growth under the selective conditions. Beneficial mutations (shown as stars) may appear in the gene of interest or elsewhere in the chromosome. As the beneficial mutations accumulate, the copy number of the amplicon is reduced. Finally, alleles holding beneficial mutations are selected and fixed throughout the population.

are channelled into central metabolism. To enable *A. baylyi* to utilize guaiacol, we introduced into its chromosome the *gcoAB* genes from *Amycolatopsis* sp. ATCC 39116, which encode a guaiacol *O*-demethylase for the conversion of guaiacol to catechol (Fig. 3A). EASy-mediated GDA of *gcoAB* gave rise to mutants capable of growing on guaiacol as the sole carbon source (Gua+ phenotype). After ~1000 generations of ALE, Gua+ mutants with a single copy of *gcoAB* were obtained. These clones presented either point mutations in the heterologous *gcoAB* genes or translational fusions between *gcoAB* and *catA* (encoding the catechol dioxygenase). Other mutations in different loci were likewise selected (Tumen-Velasquez et al., 2018). In the second example, EASy was performed on strains harbouring the TPA catabolic genes *tph* from *Comamonas* sp. E6, which encode the necessary activities for the conversion of TPA to protocatechuic acid. To allow uptake of TPA into the cell, transport genes (*tpiAB*) were also integrated into the chromosome of *A. baylyi*. These genes, encoding the transmembrane proteins of the tripartite TPA transporter from *Comamonas* sp. E6, were maintained in single copy to avoid potential toxic effects resulting from overexpression (Fig. 3B). In these studies, mutants with a single copy of the *tph* genes were not obtained. Moreover, no mutations in the *tph* genes were identified in eight evolved strains. Nonetheless, EASy allowed the identification of native *A. baylyi* transporters that had evolved for improved uptake of TPA. We further demonstrated that the heterologous *Comamonas* transporter is not necessary for *A. baylyi* to grow on TPA. This outcome would not have been possible without prior amplification of the foreign catabolic genes, a step that was necessary for the emergence of the Tpa+ phenotype (Pardo et al., 2020).

In this chapter, we provide some tips and best practices for performing EASy experiments with *A. baylyi*, particularly focused on the evolution of heterologous catabolic pathways. However, we note that this method is amenable to many variations, and we have successfully applied it in separate laboratories using slightly different approaches. We anticipate that EASy can be adapted for a range of applications.

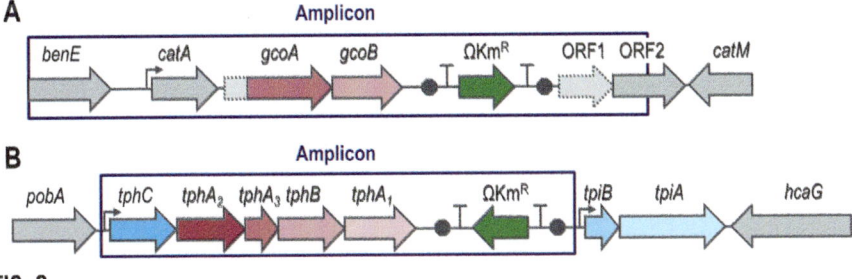

FIG. 3

Schematic representation of the chromosomal integrations in *Acinetobacter baylyi* for EASy of (A) guaiacol and (B) TPA utilization. The native *A. baylyi* genes are shown in grey. The foreign genes for guaiacol and TPA utilization are shown in pink (catabolic genes) and blue (transport genes). Promoters driving the expression of the heterologous genes are indicated with angled arrows. The ΩKm^R cassette used for antibiotic selection contains a kanamycin resistance gene (green) flanked by transcription and translation terminating signals, respectively shown as "T"s and black octagons. The open-reading frame disruption caused by insertion of *gcoAB* in panel (A) is indicated with dashed lines. The EASy amplicons are bound by purple boxes.

2 General considerations

2.1 Design of the amplicon and synthetic bridging fragment

To design an EASy experiment, the first step is to determine the precise chromosomal region to be amplified, i.e., the amplicon. In most experiments, this design involves the chromosomal insertion of target DNA that will be amplified, although in some experiments, we have amplified a region of the wild-type chromosome (Stoudenmire et al., 2017). As depicted in Fig. 1A, the endpoint of the amplicon is created by a novel junction that joins two chromosomal regions that are not normally adjacent. This novel junction is engineered by the choice of DNA segments connected on the SBF (Fig. 1B). When used as donor DNA, this SBF promotes homologous recombination with chromosomal DNA to duplicate the desired region. Also essential to the experimental design is inclusion of an Ab^R marker in the chromosome within the intended amplicon. This marker provides a means to select and maintain increased gene dosage after transforming with the SBF.

The simplest way to construct the parent strain is to insert the GOI (or larger region of interest) into the chromosome together with the Ab^R marker (Fig. 2). We have used Ω fragments encoding either resistance to kanamycin (Km^R) or streptomycin/spectinomycin (Sm^R/Sp^R) (Eraso & Kaplan, 1994; Prentki & Krisch, 1984). In these fragments, the Ab^R marker is flanked by transcription and translation termination signals (Fig. 3). This configuration ensures that the insertion of the Ab^R marker does not interfere with the expression of upstream and downstream genes, independent of the orientation. However, any DNA construction that isolates the expression of an Ab^R marker from the rest of the amplicon is expected to work.

An important consideration when designing the amplicon is the potential toxicity caused by overexpression of certain genes. For example, when engineering TPA transport and catabolism in *A. baylyi*, we found that high expression levels of genes encoding transmembrane proteins were lethal to the cells, even when in single copy in the chromosome (Pardo et al., 2020). Therefore, we decided to insert them as a separate transcription unit that was not included in the amplicon (Fig. 3B). The architecture of the amplicon can also be tailored to different gene expression strategies. For EASy of guaiacol utilization, the *gcoAB* genes were inserted as transcriptional fusions directly downstream of *catA*, so that they were under control of the *cat* promoter (Fig. 3A). On the other hand, the *tph* genes were placed under control of the *tac* promoter (de Boer, Comstock, & Vasser, 1983), a strong constitutive promoter in *A. baylyi*, so that they were independently transcribed during EASy TPA consumption (Fig. 3B). Synthetic promoter sequences, of varying strength, for use in strain ADP1 have recently been characterized (Biggs et al., 2020).

Once the desired amplicon is planned, the SBF is designed accordingly to delimit GDA to the chromosomal segment of interest. For this purpose, the last ~1000 bp of the 3′-end of the amplicon are fused to the first ~1000 bp of the 5′-end, in tandem. This length provides a sufficiently long stretch of sequence identity to enable homologous recombination in *A. baylyi*. Again, this step is amenable to variations. The SBF can be designed to limit the amplicon to the foreign DNA sequence (Fig. 4, amplicon 1), or to include the native upstream and downstream sequences of the insertion locus (Fig. 4, amplicon 2). We have successfully used both approaches.

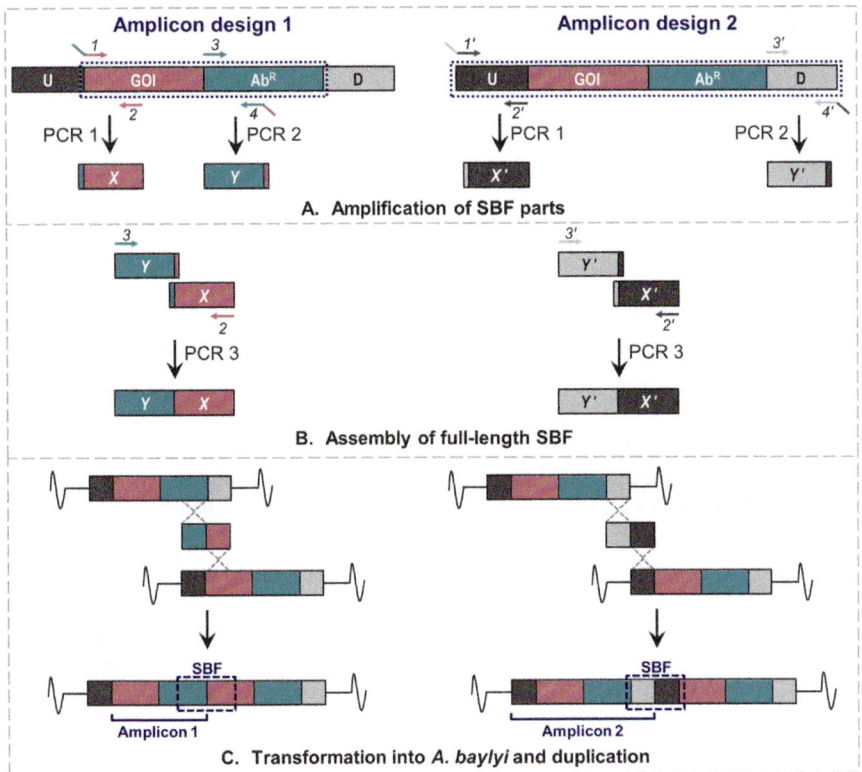

FIG. 4

Workflow for the construction of the synthetic bridging fragment (SBF) for EASy. (A) In a first step, the target chromosomal segment that will be amplified is determined (i.e., amplicon, bound by a dotted line). In the figure, amplicon design 1 (left side) is limited to the gene of interest (GOI) and the antibiotic selection marker (Ab^R), whereas amplicon design 2 (right side) includes the upstream (U) and downstream (D) chromosomal regions of the targeted insertion locus. Once the amplicon is designed, the first and last ~1000 bp of the target amplicon are amplified by PCR. The forward primer to amplify the 5'-end of the amplicon (primer *1*) is designed to have a ~20 bp overhang identical to the 3'-end of the target amplicon. Likewise, the reverse primer to amplify the 3'-end of the target amplicon has a ~20 bp overhang that is the reverse complement of the 5'-end of the amplicon (primer *4*). (B) The full-length SBF is assembled in 15 cycles of a primer-less PCR by sequence overlap extension, using the PCR products form the first round of PCR as self-priming templates (fragments *X* and *Y* in panel A). Then, primers *2* and *3* are added to amplify the full-length SBF in another 15 cycles of PCR. (C) The SBF is transformed into the recipient strain to precisely delimit GDA to the target chromosomal region, with the integrated SBF forming part of the novel junction (bound by dashed lines). Depending on the design of the SBF, the resulting amplicon will be different (indicated with brackets).

2.2 Chromosomal integration by natural transformation

Having designed the DNA cassette that will be integrated into the chromosome of *A. baylyi*, it needs to be flanked by upstream and downstream sequences identical to the chromosomal locus targeted for insertion (Figs. 2 and 6). These regions of identity can be as small as 300 bp (de Berardinis et al., 2008), although longer stretches of identical sequences increase the efficiency of homologous recombination and facilitate the integration of larger DNA fragments. The construction of the DNA cassette flanked by the homology regions can be performed using any DNA assembly technique (e.g., restriction cloning, sequence overlap extension PCR, Gibson assembly, Golden Gate assembly, etc.). Since linear DNA fragments are used for the integration into the chromosome, the assembly products can be directly transformed into *A. baylyi* or sub-cloned into a plasmid for sequence verification and maintenance. If the latter option is used, a simple digestion with a restriction enzyme to linearize the plasmid backbone needs to be performed before transformation. In all cases, the genotype of the resulting transformant should be confirmed.

As noted above, *A. baylyi* has a remarkable competency for natural transformation, which allows using many different variations of the same two-step protocol: (1) incubation of *A. baylyi* with transforming DNA; and (2) selection or screening to identify desired transformants (Biggs et al., 2020; Metzgar et al., 2004; Santala, Karp, & Santala, 2016; Suárez et al., 2020; Suárez, Renda, Dasgupta, & Barrick, 2017). The incubation step can be done in liquid or solid medium, which can be either minimal or rich. The same protocol can be used for replicative plasmids or linear DNA fragments. For chromosomal integrations, unpurified linear DNA fragments can be added directly from assembly or digestion reactions, and even genomic DNA from lysed cells can be used. Bedore et al. (2023) describe in detail the protocols that we routinely use in the laboratory, although these may be entirely customized. Any one of the transformation protocols given can be used throughout the different steps of EASy.

2.3 Selection of amplification mutants

As noted in the introductory section, the SBF provides sufficient sequence identity for initial duplication of the target chromosomal segment. Further amplification of the duplicated region can occur and be maintained if selectively advantageous. Therefore, following transformation with the SBF, amplification mutants are selected in the presence of high antibiotic concentration. This selection is used to maintain a tandem array of multiple copies of the amplicon. Under the conditions described in the Procedures section, we use a concentration of 1 g/L Km or Sm/Sp (using a combination of both Sm and Sp) which inhibits growth of mutants carrying a single copy of the Ab^R gene. Nevertheless, we have found that this concentration may need to be optimized when using resistance genes from different sources or growing on different media (unpublished data). In particular, the ΩKm^R fragment we typically use (positions 10,499–12,705 in GenBank MN266288.1) harbours the *nptII* gene from Tn5, whereas the $\Omega Sm/Sp^R$ fragment (GenBank M60473.1) harbours the *aadA* gene from plasmid R100 (Fellay, Frey, & Krisch, 1987; Prentki & Krisch, 1984). Researchers implementing EASy in their laboratory should first optimize the selection conditions that only permit growth of the amplification mutants. Spontaneous drug-resistant mutants are not observed. Typically, 5–10 copies of the drug marker are selected by these antibiotic concentrations.

Once colonies are obtained that are resistant to high antibiotic concentrations, GDA can be checked by colony PCR (cPCR) or quantitative real-time PCR (qPCR). In the first approach, a pair of primers that specifically amplify the novel junction created by the SBF can be used (e.g., primers used for PCR 3 in Fig. 4). Although this method is not able to quantify copy number, it is a simple and fast way to screen possible amplification mutants. To quantify gene copy number, qPCR is used instead. Primers targeting any region of the amplicon can be employed. However, primers that specifically bind the Ab^R marker have the advantage that they can be re-used in different EASy experiments to give reproducible estimations of the gene copy number. In previous work, we have shown that the estimated copy number is virtually the same using primers binding the Ab^R marker or the GOI (Pardo et al., 2020). Clones confirmed to have multiple copies of the amplicon can then be transferred to the selective medium for ALE. The antibiotic is omitted at this point so that gene dosage of the amplicon can vary and be selected based on the desired phenotype.

In a variation of this approach, two different chromosomal regions of the same parent strain can be amplified. This approach was recently demonstrated using the $\Omega Sm/Sp^R$ marker to select amplification in one chromosomal region and the ΩKm^R marker in a different region (unpublished data). In this case, the two distinct regions were manipulated sequentially in the construction of a synthetic metabolic pathway. The goal is to build complex pathways in a modular fashion. This approach is designed to alleviate problems caused by imbalances between the synthesis and degradation of metabolic intermediates that might be toxic. After moving the doubly-amplified populations to selective growth medium, without antibiotics, the new pathway may evolve to reduce the accumulation of toxic metabolites via variation in the gene dosage of different parts of the pathway.

2.4 Interpreting gene copy number estimations and obtaining single-copy mutants

GDA is a highly dynamic process, where gene copy number rapidly fluctuates, and copy number decreases in the absence of selection. Therefore, genomic DNA extracted from cells growing under the selective conditions should be used for an accurate estimation of the gene copy number by qPCR. It is important to note that this number represents the average for all the cells in a population. On occasions, the gene copy number for an evolving population will stabilize at 2–4 copies, but a significant proportion of individual cells will carry a single copy of the amplicon. In Section 4.4, we provide two approaches to isolate single-copy mutants. The first approach is based on the screening of isolated colonies from the evolving population for the absence of the SBF, indicative of a single copy of the amplicon (Section 4.4.1). The second approach consists of replacing the tandem repeats of the amplicon with a single copy of the GOI (Section 4.4.2). An overview of this allelic replacement method is shown in Fig. 5. In a first step, the tandem amplicon array in an evolved isolate is replaced with a selectable/counter-selectable marker, such as the *sacB*-Km^R cassette (Jones & Williams, 2003). The counter-selectable marker allows the isolation of mutants in a second transformation step, in which a single copy of the non-evolved amplicon is re-introduced. It should be noted that if the original GDA was selected with one Ab^R marker (such as Km^R) a different marker (such as Sm/Sp^R) should be used in combination with the *sacB* counter-selectable marker.

Once mutants have been isolated in which the evolved amplicons have been replaced with a single copy of the original version of this DNA, further analyses are possible (Fig. 5B). If these mutants grow in the selective medium, then the mutations accrued throughout the chromosome are sufficient to sustain growth with a single copy of the non-evolved GOI. If such mutants do not grow in the selective medium, they can be transformed with individual copies of the evolved GOI alleles, amplified from the multi-copy population by PCR. Beneficial mutations introduced by allelic replacement can be selected and identified (Fig. 5C). We note that recombination between different alleles of the GOI may occur during PCR or transformation. Therefore, mutations that were present in separate copies of the gene may be combined in the same allele. Likewise, two mutations that are sufficiently distant in the same copy of the GOI may be segregated. By plating cells transformed with the evolved alleles on solid ALE selective medium, faster-growing clones that acquire a single copy of an allele with beneficial mutations can be identified and isolated for sequencing. However, given enough time, spontaneous amplification mutants may also arise. The novel junction may occur anywhere in the chromosome—even several kbp apart from the GOI. In this case, screening for the absence of the SBF does not assure that mutants with the desired phenotype will present a single copy of the GOI. Similarly, a PCR using primers that bind the upstream and downstream regions flanking the insertion locus of the amplicon may still give a band resembling that of the single-copy parent strain before GDA. Therefore, the presence of the GOI in single copy needs to be confirmed by qPCR or whole-genome sequencing. Any method of whole-genome sequencing should be sufficient for such purposes.

The final goal of EASy is to isolate beneficial mutations that improve cell fitness under the selective conditions. Ideally, the epistatic effect of the mutations accrued during ALE allows growth of evolved mutants with a single copy of the amplicon. These mutations are identified with whole-genome sequencing of isolated evolved clones and mapping to the reference genome sequence with appropriate software (Deatherage & Barrick, 2014). The latest annotated genome sequence for *A. baylyi* APD1 is deposited in GenBank under accession number NC_005966. It is possible, however, that evolved mutants with a single copy of the amplicon cannot be obtained, despite following the protocols mentioned above. For example, it may not be possible to remove and reintroduce amplicon copies as illustrated in Fig. 5. One issue that may arise with ALE experiments is a reduction in the transformability

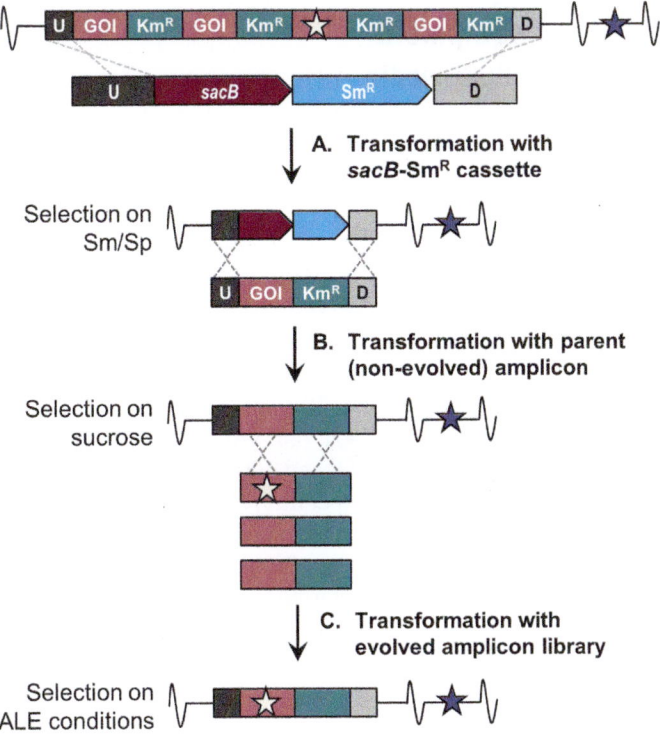

FIG. 5

Overview of the allelic replacement strategy to obtain single-copy mutants from EASy experiments. (A) An evolved isolate from an EASy experiment carries four copies of the amplicon, which encompasses the gene of interest (GOI) and a kanamycin selection marker (Km^R). One of the copies of the gene of interest presents a beneficial mutation (white star). A second beneficial mutation has been selected elsewhere in the chromosome (purple star). The multi-copy amplicon array is replaced by a DNA cassette carrying a streptomycin/spectinomycin selection marker (Sm^R) and a *sacB* counter-selection marker, flanked by sequences identical to upstream (U) and downstream (D) chromosomal regions for homologous recombination. Transformants are selected on Sm/Sp plates. (B) Mutants are then transformed with the parent, non-evolved amplicon. Transformants are selected on medium with sucrose. (C) Mutants obtained in step B are then transformed with single copies of the amplicon derived from the initial EASy isolate. Transformants that incorporate mutated alleles that provide a growth benefit are selected on the same medium used for ALE.

of ADP1. This problem can be avoided using a transposon-free strain, ADP1-ISx (Suárez et al., 2017). The choice of parent strain may depend on the goal of the experiment. In some cases, we prefer to use ADP1, in which the insertion sequences are intact, because transposition generates beneficial genomic mutations during ALE. Regardless of whether a single copy of the GOI is obtained, valuable information can be gained from whole-genome sequencing data, as we discuss in the Section 5.

3 Material and equipment

3.1 Strains and culture media

- *A. baylyi* ADP1 (from culture collections ATCC 33305 or DSM 24193). A transposon-free derivative of ADP1 is also available, ADP1-ISx (Suárez et al., 2017).
- Chemically-competent *E. coli* cells to be used as hosts for plasmid construction and maintenance (any host strain). Typical strains used for this purpose include *E. coli* strains DH5α and XL1-Blue (Agilent Technologies).
- Metals 44 solution (per litre, 2.5 g EDTA; 10.95 g $ZnSO_4 \cdot 7H_2O$; 5 g $FeSO_4 \cdot 7H_2O$; 1.54 g $MnSO_4 \cdot 7H_2O$; 392 mg $CuSO_4 \cdot 5H_2O$; 250 mg $Co(NO_3)_2 \cdot 6H_2O$; and 177 mg $Na_2B_4O_7 \cdot 10H_2O$)
- Concentrated base solution (per litre, 20 g nitriloacetic acid dissolved in 600 mL H_2O with 14.6 g KOH; 28.9 g $MgSO_4$; 6.67 g $CaCl_2 \cdot 2H_2O$; 18.5 mg $Mo_7O_{24} \cdot 4H_2O$; 198 mg $FeSO_4 \cdot 7H_2O$; and 100 mL of Metals 44 solution)
- Minimal medium for *Acinetobacter* (per litre, 25 mL 0.5 M KH_2PO_4; 25 mL 0.5 M Na_2HPO_4; 10 mL 10% $(NH_4)_2SO_4$; 1 mL concentrated base) supplemented with an appropriate carbon source (e.g., 20 mM pyruvate, succinate, or acetate). Although organic acids tend to be preferred to sugars, glucose (25–50 mM) can be used. A longer lag phase will be observed with glucose. In addition, casein amino acids are good carbon sources and can be added in concentration 0.2% (with or without other carbon sources). For plates, 1.5% agar is used.
- YTS agar plates for sucrose counterselection (autoclave 3 g yeast extract, 6 g tryptone, 11 g agar in 300 mL water; then add 300 mL filter-sterilized 50% sucrose)

3.2 Reagents for DNA manipulation

- Custom PCR primers
- Cloning vector, such as pUC18 or pUC19 (Yanisch-Perron, Vieira, & Messing, 1985)
- Plasmid carrying the ΩKm^R or $\Omega Sm/Sp^R$ fragment, e.g., pHP45Ω-Km (Fellay et al., 1987), pUI1637, pUI1638 (Eraso & Kaplan, 1994)
- High-fidelity DNA polymerase and additional components (e.g., NEB Phusion)
- Taq polymerase and reagents for colony PCR (e.g., Bioline MyTaq HS Red Mix, NEB LongAmp Taq DNA polymerase)
- NEBuilder HiFi DNA assembly master mix (NEB)
- Restriction enzymes
- DNA ligase (e.g., NEB Quick ligation kit, T4 DNA ligase)
- TaqMan probes
- TaqMan Gene Expression Master Mix (Thermo Scientific)
- Agarose
- TAE (Tris Acetate EDTA) buffer, final working solution for electrophoresis running buffer (1×): 40 mM Tris, 20 mM acetic acid, and 0.4 mM EDTA
- Plasmid purification kit (e.g., Zymo Research ZR Plasmid miniprep kit)

- PCR purification kit (e.g., Zymo Research DNA Clean & Concentrator kit)
- Genomic DNA (gDNA) extraction kit (e.g., Zymo Research Quick-DNA Miniprep Plus kit)

3.3 Equipment

- Orbital shaker
- Incubator
- DNA electrophoresis chamber
- UV or blue-light transilluminator
- NanoDrop or comparable DNA quantification instrument
- Thermocycler
- Real-time PCR cycler
- Spectrophotometer or device for following cell growth

---

4 Experimental procedures
4.1 Construction of the amplicon and chromosomal integration

As indicated in Section 2, different DNA manipulation techniques can be used to assemble the cassette that will be integrated into the *A. baylyi* chromosome. Here, we describe a procedure using NEBuilder HiFi DNA assembly for plasmid cloning as described in Pardo et al., 2020 (Fig. 6):

1. Design PCR primers for the amplification of the gene of interest (including 5′ and 3′ untranslated regions) and the upstream and downstream homology regions (~1 kbp each) of the targeted insertion locus. These oligos should have 15–20 nt overhangs complementary to each other and the cloning vector for subsequent assembly. Additionally, the reverse primer amplifying the gene of interest and the forward primer amplifying the downstream homology region should include a restriction site for the insertion of the ΩKm^R fragment in Step 7.
2. Amplify the upstream and downstream homology regions using *A. baylyi* genomic DNA (gDNA) or whole cells as template with a high-fidelity DNA polymerase (e.g., Phusion) following the manufacturer's indications. If whole cells are used, include an initial denaturation step at 95 °C for 5 min in the PCR program. Follow the same protocol to amplify the gene of interest from gDNA of a heterologous host, plasmid, or synthetic DNA.
3. Linearize the cloning vector (e.g., pUC19) by PCR or digestion with restriction enzymes.
4. Purify the PCR and/or digestion products and measure DNA concentration with NanoDrop (or comparable instrument for quantification).
5. Prepare NEBuilder HiFi DNA assembly reaction including the upstream and downstream homology regions, the gene of interest, and the linearized cloning vector following the manufacturer's indications.

6. Transform an aliquot of the assembly reactions into chemically-competent *E. coli* and select on LB agar with the appropriate antibiotic.
7. Purify plasmids by miniprep and verify correct assembly by sequencing or other methods.
8. Digest the plasmid carrying the ΩKm^R fragment (e.g., pUI1637) with a restriction enzyme whose target sequence is present on the inverted flanking polylinkers. Purify the ΩKm^R fragment in an agarose gel (2.4 kbp). Digest the plasmid carrying the gene of interest (recipient) with the same enzyme.
9. Prepare a T4 ligase reaction with the recipient plasmid and the ΩKm^R fragment.
10. Transform an aliquot of the ligation reactions into chemically-competent *E. coli* cells and select on LB agar +50 mg/L Km.
11. Purify plasmids by miniprep. Optionally, the orientation of the ΩKm^R fragment can be checked by PCR or diagnostic restriction digestion.
12. Digest the resulting plasmid with a restriction enzyme cutting in the vector backbone for transformation into *A. baylyi*. No purification of the DNA or heat inactivation of the enzyme is necessary.
13. Transform DNA into *A. baylyi* using any of the protocols provided in Bedore et al. (2023).

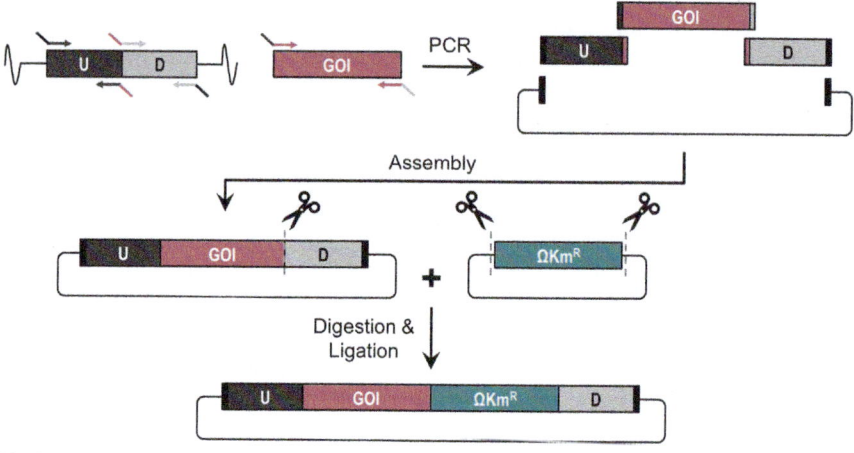

FIG. 6

Workflow for the construction of the amplicon in a replicative plasmid for subsequent integration in *Acinetobacter baylyi*. The gene of interest (GOI) and the flanking upstream (U) and downstream (D) chromosomal regions for homologous recombination are amplified by PCR, using primers with appropriate 15–20 nt overhangs for NEBuilder HiFi DNA assembly into a cloning vector. The resulting plasmid (recipient) and the donor plasmid carrying the ΩKm^R fragment are digested with the same restriction enzyme. Then, the ΩKm^R fragment is ligated into the recipient plasmid between the GOI and the downstream homology region.

4.2 Construction of the SBF and amplification of gene copy number

Once the desired DNA cassette has been integrated into the *A. baylyi* chromosome, the resulting mutant (i.e., parent strain) is transformed with the SBF to facilitate gene duplication. Here, we describe the construction of the SBF using sequence overlap extension PCR (Fig. 4), although other DNA assembly techniques can be used (or can be commercially synthesized). For future experiments, it is convenient to construct a plasmid carrying the SBF that can be stored and used as needed.

1. Amplify the first ~1000 bp of the target amplicon using a forward primer with a 15–20 bp overhang identical to the 3′-end of the amplicon, and a standard reverse primer. Use a high-fidelity DNA polymerase and the plasmid or mutant strain carrying the DNA cassette constructed in Section 4.1 as template.
2. Similarly, amplify the last ~1000 bp of the target amplicon using a standard forward primer and a reverse primer with 15–20 bp overhang that is the reverse complement of the 5′-end.
3. Purify the PCR and/or digestion products and measure DNA concentration.
4. Prepare a primer-less PCR mix using ~30 ng of each of the of the purified DNA fragments and amplify for 15 cycles. The 30–40 bp stretch of sequence identity generated by the overlapping primers should have a melting temperature of 55–65 °C. Then, add the standard forward and reverse primers used in steps 1 and 2, respectively (i.e., primers *3* and *4* in Fig. 4). Amplify the full-length SBF for another 15 cycles.
5. Check correct assembly of the full-length SBF in an agarose gel (expected size of ~2 kbp). If needed, purify the correct band by excision from the agarose gel.
6. Transform the SBF fragment into the parent *A. baylyi* strain containing a single copy of the region targeted for amplification. For selection of amplification mutants, plate cells on minimal medium containing 1 g/L Km.
7. We recommend streak purification of colonies at least three times on 1 g/L Km plates to rule out false positive clones. Often, colonies grow poorly on the first transfer.
8. Confirm integration of the SBF in the high-antibiotic resistant colonies by PCR, using the same primers used for amplification of the SBF in step 4. Amplification mutants should be maintained under selective pressure, such as in the presence of high antibiotic concentration, to prevent undesired decrease of the amplicon copy.

4.3 Adaptive laboratory evolution and monitoring of gene copy number over time

Selection of the amplification mutants in high antibiotic concentrations ensures obtaining colonies in 1–2 days, which can be quickly checked for increased gene dosage by colony PCR or qPCR. The confirmed amplification clones can then be transferred to the desired solid selective medium without antibiotic, e.g., minimal medium with a non-native carbon source. In this way, the only selective pressure is the target condition for adaptive laboratory evolution (ALE). In our experience, it can take up to 2 weeks for colonies to appear on the selective agar plates, although growth will be faster when re-streaked on a second selective plate. These clones can then be used to initiate an ALE experiment, which we describe below following a serial passage method. Generally, as in any ALE experiment, it is recommended to start with a moderate selective pressure and gradually increase it over time, e.g., substrate concentration, temperature, dilution, transfer rate. Optionally, the changes of gene copy number over time can be monitored by qPCR. Here we describe a method using fluorescent probes, e.g., 6FAM-MGBNFQ labelled TaqMan™ probes (Thermo Scientific). Fluorescent DNA-binding dyes such as SYBR Green can also be used, but we find reduced background noise when using probes. The qPCR protocol described below specifies the DNA concentrations that we normally use and that give suitable amplification curves to estimate gene copy number. We note that we have been able to detect copy numbers above 100 in an EASy experiment.

4.3.1 Adaptive laboratory evolution by serial transfer

1. Inoculate sterile test tubes containing at least 2 mL liquid selective medium with the amplification clone(s). Use a heavy inoculum for this first transfer to liquid medium. Incubate under the appropriate conditions.
2. Once stationary phase is reached, transfer an aliquot of the culture to fresh medium, diluting cells at least 100-fold. Repeat this operation as many times as needed until the desired phenotype is reached or no further improvements are observed.
3. While performing the serial transfers, periodically remove an aliquot of the culture for preservation of the cells as glycerol stocks and extraction of genomic DNA for the quantitation of the average gene copy number of the evolving population (e.g., once a week).
4. Extract genomic DNA with a molecular biology kit (e.g., Quick-DNA Miniprep Plus kit from Zymo Research) following the manufacturer's instructions and measure DNA concentration with NanoDrop. Purified DNA can be stored at $-20\,°C$ or $-80\,°C$ for long-term storage.

4.3.2 Gene copy number analysis by quantitative PCR

1. Design primer pairs and probes targeting the amplicon and a reference gene that is present in single copy in the chromosome using an appropriate software. In Table 1, we provide the sequences used for qPCR of the Tn5 *nptII* gene present in the ΩKm^R cassette with *rpoA* as reference gene.
2. Prepare separate master mixes (without DNA) for quantitation of the *nptII* (Km^R) and *rpoA* genes with the TaqMan™ Gene Expression Master Mix. The final reaction mixes should contain each primer and probe at concentration of 0.2 μM.
3. Dispense the master mixes into separate wells in a real-time PCR 96-well plate, with at least 3 technical replicates per sample and target. Add DNA to a final

Table 1 Primer and probe sequences for qPCR.

Target	Forward primer	Reverse primer	Probe
nptII[a]	gcgttggctaccc gtgata	ggaagcggtcagcccatt	tgaagagcttggcggc
rpoA[b]	gctcgacgccttc tatttcaa	tttacgtcgcattctatt gtcttctt	tcaaccacagcagcgc caggc

[a]Sequence within the ΩKm[R] cassette, used as a measure of amplicon gene dosage.
[b]Sequence within a housekeeping gene, used as a measure of a single-copy chromosomal gene.

concentration of 0.05 ng/μL to each well such that the final reaction volume is 20 μL (1 ng DNA per well). Genomic DNA from the parent strain containing a single copy of the amplicon should be included as control.

4. Using the genomic DNA from the parent strain as template, prepare a calibration curve with increasing amounts of DNA. We typically use a 5-point curve from 0.02 to 12.5 ng gDNA per well, prepared as 5-fold serial dilutions.

5. Cover the plate with an optically-clear adhesive film appropriate for real-time PCR and run a program with the following cycles: 50 °C for 2 min; 95 °C for 10 min; 40 cycles of 95 °C for 15 s and 60 °C for 1 min.

6. Using appropriate software, quantify the amount of *nptII* and *rpoA* for each sample. The *nptII:rpoA* ratio is considered to be equivalent to the relative gene copy number of the amplicon. The ratio for the parent strain should be equal to one.

4.4 Obtaining single-copy mutants from EASy

The ultimate goal of EASy is to obtain mutants with a single copy of the amplicon thanks to the accumulation of beneficial mutations. These mutations may appear in the GOI and/or in other chromosomal loci that provide a growth benefit to the cell when combined. On occasions, a single copy of the amplicon will not be reached following ALE. Nevertheless, single-copy mutants can sometimes be obtained by screening or by allelic replacement.

4.4.1 Isolation and screening by colony PCR

The gene copy number that results from qPCR analysis is an average of the cells that make up the population. Therefore, in a population with a copy number of 4 or less some cells will present a single copy of the amplicon. These can be isolated using the following procedure:

1. Dilute and plate cells from the evolving population on the same selective medium used for ALE with agar. To obtain well isolated colonies, it is generally sufficient to plate 100 μL of a 10^{-6} dilution from a saturated culture ($OD_{600} \sim 2$).

2. Screen isolates for the absence of the SBF by colony PCR (cPCR), using the same pair of primers as in Section 4.2, step 7. Only colonies that still possess multiple copies of the amplicon should give a product of the expected size. However, since the qPCR-determined copy number represents the average in a population, colonies need to be screened to identify those with a single copy of the region of interest. If the average copy number is 2 or higher, we recommend screening at least 100 colonies.

3. Confirm the phenotype of the SBF-negative isolates by streak purification (at least three times) on selective medium to ensure a pure clonal population. We also recommend genotypic and phenotypic confirmation of the presence of the Ab^R gene that was initially present in the parent strain prior to amplification and evolution (Section 4.1).

4. Extract genomic DNA from the selected isolates and repeat the PCR to check for the presence of the SBF. Simultaneously, perform additional PCRs using primers that can distinguish alternative conformations. For example, with a primer pair, each of which binds outside of the amplicon, only clones that harbour a single copy of the amplicon should give an appropriately sized product. However, experimental design may need to be tailored to the specific situation.

5. Clones that are confirmed to have a single copy of the amplicon can be further analysed by Sanger sequencing of the GOI or whole-genome sequencing to identify beneficial mutations.

4.4.2 Allelic replacement in evolved populations

This method consists of a series of transformations to replace the tandem amplicon array from an isolated clone of the evolved population with a single copy of the amplicon. This method facilitates analysis of the fitness contribution of mutations in the GOI (or in other regions of the chromosome). However, the success of this method may be limited by the known loss of natural competency of *A. baylyi* during ALE (Renda, Dasgupta, Leon, & Barrick, 2015).

1. Transform an isolated clone from the evolved population with a DNA cassette carrying a Sm/Sp^R selectable marker and a *sacB* counter-selectable marker. This DNA cassette should be flanked by at least 1000 bp of sequence identical to the upstream and downstream regions of where the tandem array of the amplicon is inserted. The same DNA assembly strategy as the one used for the initial integration of the amplicon can be used (Section 4.1).

2. Select mutants on agar plates containing 25 mg/L Sm/Sp. Lower transformation efficiency may occur due to the loss of natural competency during ALE.

3. Confirm Sm/Sp^R colonies for correct allelic replacement by cPCR. Also confirm sensitivity to Km and the inability to grow on the ALE selective medium.

4. Transform mutants from step 3 with a single copy of the non-evolved amplicon, as in Section 4.1. Select for sucrose-resistant transformants on YTS. Since sucrose selection can have high background growth, re-streak colonies on fresh YTS plates at least three times before confirming by cPCR.

5. Amplify individual copies of the amplicon from the evolved isolate by PCR, using primers that bind the ends of the amplicon. Alternatively, amplify only the GOI.
6. Transform cells containing a single copy of the non-evolved allele with the PCR products obtained in step 5. These PCR products can be integrated by direct recombination with the allele present in the chromosome. Include a no DNA control.
7. Select transformants on plates containing the selective medium used for ALE. Colonies growing faster than the control are expected to have acquired evolved alleles with beneficial mutations.
8. Confirm that mutants growing on the selective medium present a single copy of the gene of interest by qPCR.

5 Summary and concluding remarks

In this chapter, we provide general guidelines to engineer new phenotypes in *A. baylyi* via controlled GDA followed by ALE. Although we focus on expansion of the catabolic capabilities of *A. baylyi*, the EASy method has potential for use in any application in which the desired phenotype can be linked to improved growth. For example, EASy might facilitate the isolation of mutations that increase detoxifying activities or that modify the substrate preference of promiscuous enzymes. The principle of EASy is to compensate for initial low gene expression or enzymatic activities with increased gene dosage. Indeed, the single-copy parent strains used in our EASy experiments were unable to grow under selective conditions, making it impossible to use direct selection in ALE without increased gene dosage. Furthermore, we did not observe spontaneous GDA in the parent strains. With EASy, one can target the amplification of the GOI by transforming with a SBF. Since direct selection for the desired phenotype may be too demanding to the cells, GDA is first induced in the presence of high antibiotic concentrations. Growth under these conditions is achieved by introducing an Ab^R marker in the target chromosomal segment, i.e., the amplicon. With this strategy, multi-copy mutants can be obtained within 1–2 days to be transferred to ALE conditions.

The EASy method is highly tuneable and can be adapted to various standard DNA manipulation techniques. This flexibility is enabled by the natural competence and genetic malleability of *A. baylyi*. These same host properties can be exploited to screen beneficial mutations acquired from ALE by reverse engineering (Bedore et al., 2023; Luo et al., 2022). However, EASy presents some limitations. Importantly, there is no "counter-selective" pressure to drive the decrease of gene copy number, other than the inherent instability of the tandem amplicon array. Therefore, it is possible that a single-copy of the amplicon is not fixed in the evolving population after several generations of ALE. In some cases, when the average copy number is 2–4, it is possible to obtain single-copy mutants, as we have described in Section 4.4.2. In others, average gene copy numbers above 10 make this task more difficult, as we observed for EASy of TPA utilization.

While selection for high-level gene dosage using antibiotic selection is convenient, there are situations where it may be preferable to use direct selection after transformation with the SBF. Our initial EASy studies to achieve growth on guaiacol and TPA did not allow direct selection. However, recent studies using selection for growth without antibiotic markers have been more successful (unpublished results). By monitoring growth in liquid cultures, the positive effect of gene amplification on growth has been observed without first using antibiotic selection. While this approach is still under development, one benefit could be that gene dosage will more closely match that needed for the target selection. Thus, if selection initially demands low-level amplification, the problem of incomplete allele segregation may be reduced.

Regardless of problems that result if a haploid state is not obtained at the end of EASy experiments, valuable information can still be obtained from the whole-genome sequence data of isolates that maintain multiple amplicons. Beneficial mutations that are selected elsewhere in the chromosome can be identified and reverse engineered into the single-copy parent strain to evaluate their phenotypic effect. For example, during EASy of TPA utilization, we identified mutations that improved TPA uptake in *A. baylyi*. When these mutations were engineered into the single-copy parent strain, we observed spontaneous GDA of the *tph* catabolic genes that led to a Tpa$^+$ phenotype, in contrast to the unmodified parent strain. This result proved that uptake of TPA was the first limiting step to obtain TPA-utilizing mutants. The EASy-mediated or spontaneous GDA can be identified with whole-genome sequencing by an increased coverage of certain regions of the chromosome, usually delimited by short sequence stretches with a high density of polymorphisms—indicative of novel junctions. For the analysis of whole-genome sequencing data, we refer the reader to the open-source *breseq* computational pipeline, which we have found is optimal for the detection of novel junctions formed during GDA (Deatherage & Barrick, 2014).

In conclusion, EASy is a powerful tool to engineer improved cell catalysts through laboratory evolution. Additionally, EASy can be applied in more fundamental research addressing gene regulation or the dynamics of GDA in a microbial population. In all, the simplicity and versatility of the EASy method in combination with the natural competence of *A. baylyi* make this tool accessible to most microbiology laboratories.

Acknowledgements

I. Pardo wishes to thank the Spanish National Research Council, Reina Sofía Foundation, and Primafrío Foundation for funding under agreement no. 20210510. Research at the University of Georgia in the United States, described in this chapter, was funded by grants from the National Science Foundation (MCB2225858) and the U.S. Department of Energy, Office of Science, Office of Biological and Environmental Research, Genomic Science Program (DE-SC0022220). S. Santala would like to thank the Novo Nordisk Foundation (grant NNF21OC0067758) and the Academy of Finland (grant no. 334822 and 347204).

References

Anderson, P., & Roth, J. (1981). Spontaneous tandem genetic duplications in *Salmonella typhimurium* arise by unequal recombination between rRNA (*rrn*) cistrons. *Proceedings of the National Academy of Sciences of the United States of America*, 78(5), 3113–3117. https://doi.org/10.1073/PNAS.78.5.3113.

Andersson, D. I., & Hughes, D. (2009). Gene amplification and adaptive evolution in bacteria. *Annual Review of Genetics*, 43(1), 167–195. https://doi.org/10.1146/annurev-genet-102108-134805.

Barrick, J. E., & Lenski, R. E. (2013). Genome dynamics during experimental evolution. *Nature Reviews Genetics*, 14(12), 827–839. https://doi.org/10.1038/nrg3564.

Bedore, S. R. (2021). *Metabolic expansion in Acinetobacter baylyi ADP1 for enhanced aromatic compounds catabolism* (Doctoral Dissertation). University of Georgia.

Bedore, S. R., Neidle, E. L., Pardo, I., Luo, J., Baugh, A. C., Duscent-Maitland, C. V., et al. (2023). Natural transformation as a tool in *Acinetobacter Baylyi*: Streamlined engineering and mutational analysis. *Methods in Microbiology*, 52.

Biggs, B. W., Bedore, S. R., Arvay, E., Huang, S., Subramanian, H., & Tyo, K. E. J. (2020). Development of a genetic toolset for the highly engineerable and metabolically versatile *Acinetobacter baylyi* ADP1. *Nucleic Acids Research*, 48(9), 5169–5182. https://doi.org/10.1093/nar/gkaa167.

Blount, Z. D., Barrick, J. E., Davidson, C. J., & Lenski, R. E. (2012). Genomic analysis of a key innovation in an experimental *Escherichia coli* population. *Nature*, 489(7417), 513–518. https://doi.org/10.1038/nature11514.

de Berardinis, V., Vallenet, D., Castelli, V., Besnard, M., Pinet, A., & Weissenbach, J. (2008). A complete collection of single-gene deletion mutants of *Acinetobacter baylyi* ADP1. *Molecular Systems Biology*, 4(1), 174. https://doi.org/10.1038/msb.2008.10.

de Boer, H. A., Comstock, L. J., & Vasser, M. (1983). The *tac* promoter: A functional hybrid derived from the *trp* and *lac* promoters. *Proceedings of the National Academy of Sciences*, 80(1), 21–25. https://doi.org/10.1073/pnas.80.1.21.

Deatherage, D. E., & Barrick, J. E. (2014). Identification of mutations in laboratory-evolved microbes from next-generation sequencing data using *breseq*. In L. Sun, & W. Shou (Eds.), *Vol. 1151. Engineering and analyzing multicellular systems. Methods in Molecular Biology* (pp. 165–188). New York: Springer. https://doi.org/10.1007/978-1-4939-0554-6_12.

Dragosits, M., & Mattanovich, D. (2013). Adaptive laboratory evolution—Principles and applications for biotechnology. *Microbial Cell Factories*, 12(1), 64. https://doi.org/10.1186/1475-2859-12-64.

Elliott, K. T., Cuff, L. E., & Neidle, E. L. (2013). Copy number change: Evolving views on gene amplification. *Future Microbiology*, 8(7), 887–899. https://doi.org/10.2217/fmb.13.53.

Eraso, J. M., & Kaplan, S. (1994). *prrA*, a putative response regulator involved in oxygen regulation of photosynthesis gene expression in *Rhodobacter sphaeroides*. *Journal of Bacteriology*, 176(1), 32–43.

Fellay, R., Frey, J., & Krisch, H. (1987). Interposon mutagenesis of soil and water bacteria: A family of DNA fragments designed for in vitro insertional mutagenesis of gram-negative bacteria. *Gene*, 52(2–3), 147–154. https://doi.org/10.1016/0378-1119(87)90041-2.

Fuchs, G., Boll, M., & Heider, J. (2011). Microbial degradation of aromatic compounds—From one strategy to four. *Nature Reviews Microbiology*, 9(11), 803–816. https://doi.org/10.1038/nrmicro2652.

Herrmann, J. A., Koprowska, A., Winters, T. J., Villanueva, N., Nikityuk, V. D., & Reams, A. B. (2022). Gene amplification mutations originate prior to selective stress in *Acinetobacter baylyi*. *G3 Genes Genomes Genetics*, 2022. https://doi.org/10.1093/g3journal/jkac327.

Jones, R. M., & Williams, P. A. (2003). Mutational analysis of the critical bases involved in activation of the AreR-regulated sigma54-dependent promoter in *Acinetobacter* sp. strain ADP1. *Applied and Environmental Microbiology*, *69*(9), 5627–5635. https://doi.org/10.1128/AEM.69.9.5627-5635.2003.

Luo, J., McIntyre, E. A., Bedore, S. R., Santala, V., Neidle, E. L., & Santala, S. (2022). Characterization of highly ferulate-tolerant *Acinetobacter baylyi* ADP1 isolates by a rapid reverse engineering method. *Applied and Environmental Microbiology*, *88*(2). https://doi.org/10.1128/AEM.01780-21.

Metzgar, D., Bacher, J. M., Pezo, V., Reader, J., Doring, V., & Crecy-Lagard, V. (2004). *Acinetobacter* sp. ADP1: An ideal model organism for genetic analysis and genome engineering. *Nucleic Acids Research*, *32*(19), 5780–5790. https://doi.org/10.1093/nar/gkh881.

Pardo, I., Jha, R. K., Bermel, R. E., Bratti, F., Gaddis, M., & Johnson, C. W. (2020). Gene amplification, laboratory evolution, and biosensor screening reveal MucK as a terephthalic acid transporter in *Acinetobacter baylyi* ADP1. *Metabolic Engineering*, *62*, 260–274. https://doi.org/10.1016/j.ymben.2020.09.009.

Prentki, P., & Krisch, H. M. (1984). In vitro insertional mutagenesis with a selectable DNA fragment. *Gene*, *29*(3), 303–313. https://doi.org/10.1016/0378-1119(84)90059-3.

Reams, A. B., & Neidle, E. L. (2003). Genome plasticity in *Acinetobacter*: New degradative capabilities acquired by the spontaneous amplification of large chromosomal segments. *Molecular Microbiology*, *47*(5), 1291–1304. https://doi.org/10.1046/j.1365-2958.2003.03342.x.

Renda, B. A., Dasgupta, A., Leon, D., & Barrick, J. E. (2015). Genome instability mediates the loss of key traits by *Acinetobacter baylyi* ADP1 during laboratory evolution. *Journal of Bacteriology*, *197*(5), 872–881. https://doi.org/10.1128/JB.02263-14.

Santala, V., Karp, M., & Santala, S. (2016). Bioluminescence-based system for rapid detection of natural transformation. *FEMS Microbiology Letters*, *363*(13), 125. https://doi.org/10.1093/FEMSLE/FNW125.

Seaton, S. C., Elliott, K. T., Cuff, L. E., Laniohan, N. S., Patel, P. R., & Neidle, E. L. (2012). Genome-wide selection for increased copy number in *Acinetobacter baylyi* ADP1: Locus and context-dependent variation in gene amplification. *Molecular Microbiology*, *83*(3), 520–535. https://doi.org/10.1111/j.1365-2958.2011.07945.x.

Seaton, S. C., & Neidle, E. L. (2018). Using aerobic pathways for aromatic compound degradation to engineer lignin metabolism. In G. T. Beckham (Ed.), *Lignin valorization: Emerging approaches* (pp. 252–289). The Royal Society of Chemistry. https://doi.org/10.1039/9781788010351-00252.

Stoudenmire, J. L., Schmidt, A. L., Tumen-Velasquez, M. P., Elliott, K. T., Laniohan, N. S., & Karls, A. C. (2017). Malonate degradation in *Acinetobacter baylyi* ADP1: Operon organization and regulation by MdcR. *Microbiology*, *163*(5), 789–803. https://doi.org/10.1099/mic.0.000462.

Suárez, G. A., Dugan, K. R., Renda, B. A., Leonard, S. P., Gangavarapu, L. S., & Barrick, J. E. (2020). Rapid and assured genetic engineering methods applied to *Acinetobacter baylyi* ADP1 genome streamlining. *Nucleic Acids Research*, *48*(8), 4585–4600. https://doi.org/10.1093/nar/gkaa204.

Suárez, G. A., Renda, B. A., Dasgupta, A., & Barrick, J. E. (2017). Reduced mutation rate and increased transformability of transposon-free *Acinetobacter baylyi* ADP1-ISx. *Applied and Environmental Microbiology*, *83*(17), e01017–e01025. https://doi.org/10.1128/AEM.01025-17.

Tumen-Velasquez, M., Johnson, C. W., Ahmed, A., Dominick, G., Fulk, E. M., & Neidle, E. L. (2018). Accelerating pathway evolution by increasing the gene dosage of chromosomal segments. *Proceedings of the National Academy of Sciences*, *115*(27), 7105–7110. https://doi.org/10.1073/pnas.1803745115.

Van Hofwegen, D. J., Hovde, C. J., & Minnich, S. A. (2016). Rapid evolution of citrate utilization by *Escherichia coli* by direct selection requires *citT* and *dctA*. *Journal of Bacteriology*, *198*(7), 1022–1034. https://doi.org/10.1128/JB.00831-15.

Yanisch-Perron, C., Vieira, J., & Messing, J. (1985). Improved M13 phage cloning vectors and host strains: Nucleotide sequences of the M13mp18 and pUC19 vectors. *Gene*, *33*(1), 103–119. https://doi.org/10.1016/0378-1119(85)90120-9.

Natural transformation as a tool in *Acinetobacter baylyi*: Streamlined engineering and mutational analysis

Stacy R. Bedore[a,‡], Ellen L. Neidle[a,‡], Isabel Pardo[b], Jin Luo[c], Alyssa C. Baugh[a], Chantel V. Duscent-Maitland[a], Melissa P. Tumen-Velasquez[a,†], Ville Santala[c], and Suvi Santala[c,*]

[a]*Department of Microbiology, University of Georgia, Athens, GA, United States*
[b]*Centro de Investigaciones Biológicas Margarita Salas (CIB), Spanish National Research Council (CSIC), Madrid, Spain*
[c]*Faculty of Engineering and Natural Sciences, Tampere University, Tampere, Finland*
*Corresponding author: e-mail address: suvi.santala@tuni.fi

Abbreviations

Ap	ampicillin
CEMENT	combinatorial evaluation of mutations examined by natural transformation
Cm	chloramphenicol
EASy	evolution by amplification and synthetic biology
Gm	gentamicin
Km	kanamycin
LB	lysogeny broth (also known as Luria-Bertani medium)
MM	minimal medium
Pob	*p*-hydroxybenzoate
RAMSES	rapid advantageous mutation screening and selection
Sm	streptomycin
Sp	spectinomycin

[†]Current Affiliation: Biosciences Division, Oak Ridge National Laboratory, Oak Ridge, TN, United States
[‡]Equal contributions

Methods in Microbiology, Volume 52, ISSN 0580-9517, https://doi.org/10.1016/bs.mim.2023.01.002

1 Introduction

Acinetobacter baylyi ADP1 has a remarkably efficient system of natural competence and homologous recombination that is ideal for a wide variety of experiments. Several publications highlight these features and describe new toolkits and optimized methods for this bacterium (Biggs et al., 2020; De Berardinis, Durot, Weissenbach, & Salanoubat, 2009; Elliott & Neidle, 2011; Metzgar et al., 2004; Santala & Santala, 2021; Young, Parke, & Ornston, 2005). Our goal is to complement these publications by illustrating the power of simple transformation techniques that tolerate large variations in the amount and type of donor DNA, the state of the recipient cells, and the incubation conditions. The technical simplicity of such transformation makes it possible to use and modify methods for diverse applications. The provision of easy-to-follow protocols is intended to facilitate the increased adoption of strain ADP1 as a model organism.

In the 1980s, we demonstrated the utility of allelic replacement to reveal mutations present on relatively small DNA fragments in *A. baylyi* (previously designated *A. calcoaceticus*) (Neidle & Ornston, 1986). At that time, there were few molecular biology resources for bacteria other than *Escherichia coli*. Chromosomal alteration by natural transformation enabled the modification of a defined genomic segment to be displayed as a specific phenotype. Additionally, this approach obviated the need for expression vectors, which were not yet available for *A. baylyi*. Decades later, the range of available expression vectors and DNA constructs is nearly limitless. Nevertheless, the ability to analyse chromosomal change via transformation-based methods in *A. baylyi* continues to push the boundaries of current techniques.

A recently developed method, called rapid advantageous mutation screening and selection (RAMSES), enhances adaptive laboratory evolution techniques by exploiting natural transformation (Luo, Mcintyre, et al., 2022). This method addresses two common issues that can impede the practical utility of evolved strains selected for desired phenotypes. First, such strains typically carry multiple mutations. Determining the significance of these mutations, alone or in combination, is usually laborious and time consuming. Second, introducing these mutations into new strains with different genetic backgrounds can be complicated, especially if multiple changes are needed. Both problems can be overcome using experiments in which PCR-generated fragments carrying mutations from the adapted strain are added directly to growing cultures. Under selective conditions, these specific mutations get incorporated into the chromosome by allelic replacement. Cells with a competitive advantage emerge over time, and whole genome sequence analysis of the competitive 'winners' reveal beneficial mutations. This approach makes reverse engineering easy and informative. Moreover, the construction of new strains for specific applications can be readily accomplished with minimal effort (Luo, Mcintyre, et al., 2022).

In addition to new applications, natural transformation remains a powerful tool for inexpensive and rapid studies that can be done by students at all levels. A demonstration of genetic principles using *A. baylyi* natural competence was developed as a college level laboratory exercise many years ago, using a protocol that is still in use (Earnest & Rosenbaum, 1993). In a laboratory course at the University of Georgia (U.S.A.), research using *A. baylyi* transformation permits students, with relatively little research experience, to conduct authentic projects that test novel hypotheses and yield publishable metabolic insights (Bedore et al., 2022; Stoudenmire et al., 2017).

A. baylyi is not unique in being naturally transformable, but it is unusual in being highly competent throughout its growth phase and early into stationary phase. This feature of nearly continuous competency, combined with an active homologous recombination machinery, makes the genetic manipulation of ADP1 unrivalled for ease and efficiency. For example, in a comparison of *A. baylyi* and *Bacillus subtilis* as hosts for a method based on recombinational capture of PCR products, *A. baylyi* was found to be approximately 100-fold more efficient than *B. subtilis* (Melnikov & Youngman, 1999). To date, there is no evidence of fundamental differences in recombination or mutational processes compared to other bacteria. For example, although we tout the plasticity of strain ADP1 and exploit gene amplification, the frequency of spontaneous duplication in *A. baylyi* is comparable to that in *E. coli* (Seaton et al., 2012).

One factor affecting allelic replacement, and its corresponding utility, could be a low level of DNA restriction and/or degradation following transformation. Small fragments and DNA from diverse sources are readily integrated into the genome following transformation. Recent attempts to improve recombinational frequencies by inactivating the single restriction modification system of strain ADP1 failed to have an effect, regardless of the source of DNA (Jiang et al., 2020). Thus, while future studies are needed to improve understanding of some aspects of natural transformation/recombination, the practical ease of genetic manipulation results in the ongoing expansion of metabolic engineering and synthetic biology applications for *A. baylyi* (Jiang et al., 2020; Luo, Efimova, Volke, Santala, & Santala, 2022; Suárez et al., 2020).

2 General considerations

2.1 Convenience and optimization

Successful transformation of *A. baylyi* strains can be accomplished with large variations in methodology. In all cases, the ease of genome editing derives in large part from the ability to use linear donor DNA. Whether this donor DNA consists of PCR products, linearized plasmids, or genomic DNA, recombination with the recipient chromosome yields viable transformants only when allelic replacement occurs (Fig. 1A). In contrast, many systems depend on transformation with circular plasmids, wherein lowered rates of success (or other complications) may result because single recombination events integrate the entire plasmid in the genome. As illustrated in Fig. 1, a wide range of chromosomal changes can be made by allelic replacement,

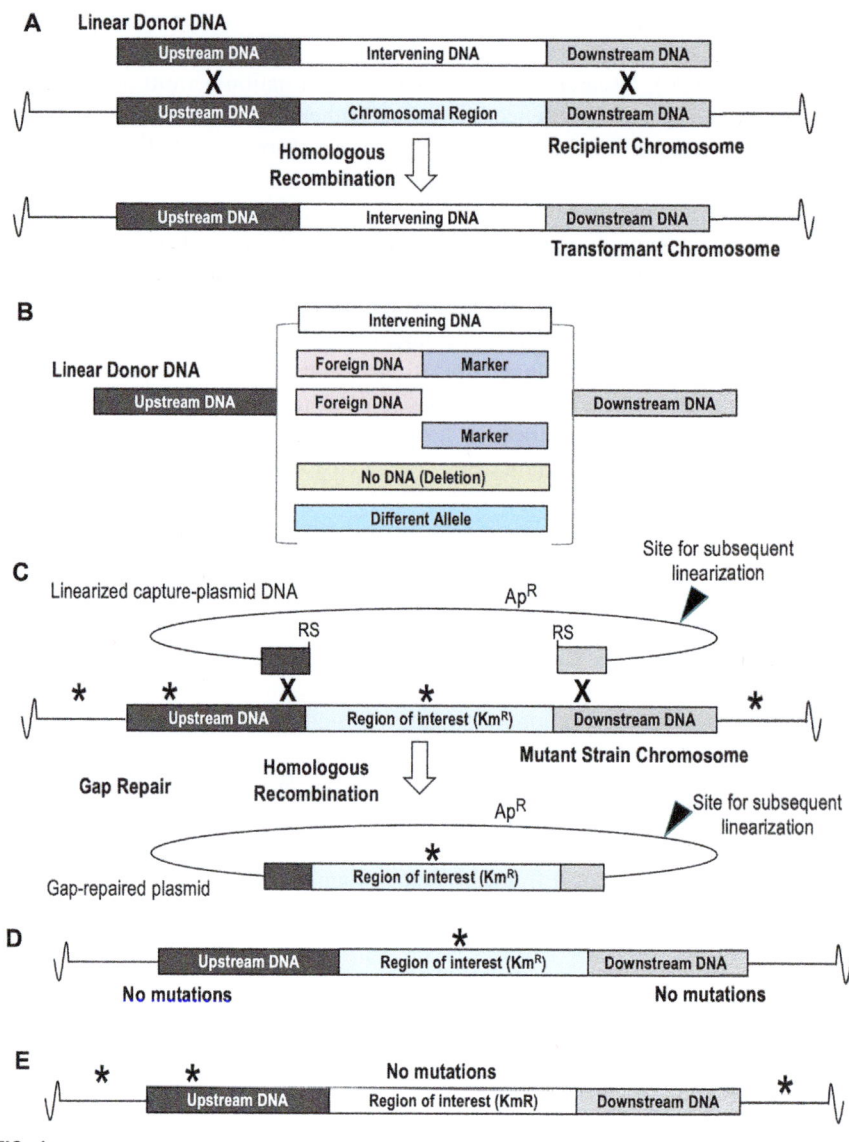

FIG. 1

Allelic replacement techniques in *A. baylyi*. (A) Homologous recombination events (X) will replace a chromosomal region with intervening DNA of the donor. A targeted insertion will result if there is no DNA between the regions of homology in the recipient. (B) The type of intervening donor DNA can vary and may include a selectable and/or counter-selectable marker. (C) A recombinational gap-repair method can clone a chromosomal region of interest from a strain with multiple mutations (*). A 'capture plasmid' is used as donor DNA after being cleaved at a unique engineered restriction site (RS). Homologous recombination including large and small insertions, deletions, and/or the targeted replacement of specific genes or chromosomal regions. Markers can be used for selection and/or counter selection. In circumstances where selection for a specific replacement is not possible, it is feasible to screen for the desired transformants. Typically, it is sufficient to screen several hundred colonies from a nonselective plate following transformation. PCR or other methods can be used to identify the correct transformant(s) by checking individual or pooled isolates.

Since experiments can be done using different methods, the exact protocol chosen may depend on convenience. The care taken to optimize experimental procedures should be guided by the specific scientific goals. For example, in the case of simple allelic replacement with a selectable marker (Fig. 1), the highest possible efficiency of transformation and recombination should not be needed. For this purpose, cells may be used in almost any growth phase, and donor DNA may require very small stretches of sequence identity to promote homologous recombination with the chromosome. In contrast, if the isolation of the desired transformant requires screening, it may be worthwhile to obtain the highest allelic replacement efficiency that is possible. Throughout this article, and its companion (Pardo et al., 2023), we indicate where variation in methodology may be tolerated. Moreover, genomic context can affect what conditions are optimal for a specific purpose in ways that are not predictable (Seaton et al., 2012). Thus, the use of various experimental protocols and/or trying multiple approaches in parallel can help ensure success. While extended passaging can result in reduced transformability (Suárez, Renda, Dasgupta, & Barrick, 2017), such reduction is not usually problematic for bacteria stored temporarily on benchtops or in refrigerators. However, it is important to maintain glycerol stocks (at $-80\,°C$) for long-term storage and to use such stocks if there is any evidence that transformation efficiencies are low.

As in all transformation protocols, important issues include: (1) preparation of the recipient cells and donor DNA, (2) introduction of DNA into cells and growth conditions following transformation, and (3) the isolation and confirmation of

---

(X) generates a 'gap-repaired' plasmid that can be propagated in *E. coli*. Subsequent linearization of this resulting plasmid facilitates transfer of the captured DNA into the chromosome of a new recipient strain. To use the plasmid as donor DNA, a restriction site is employed that is outside the *A. baylyi* DNA region (black triangle). (D) Allelic replacement of the analogous region of a recipient with no other genomic mutations (than those of an original strain being investigated) can be selected with a drug-resistance marker (Km^r) in the region of interest. The resulting transformant can be used to assess the significance of mutation(s) in the region of interest. (E) Comparable allelic-replacement steps to those described in (D) can be used to introduce unmutated DNA into the chromosome of the mutant that has multiple chromosomal mutations. Allelic replacement of the region of interest allows analysis of the effects of genomic mutations that reside outside the target region.

desired transformants. These issues are discussed below. In addition, examples of transformation assays are provided to illustrate a range of applications that are uniquely possible with this bacterial system.

2.2 Preparation of recipient cells and donor DNA

Different culture conditions for *A. baylyi* may be used for transformation experiments. Systematic studies to establish a standardized workflow determined that high transformation efficiency resulted from growth at 30°C, with liquid cultures aerated by shaking at 250 rpm (Biggs et al., 2020). In those studies, cells used for natural transformation were grown in rich medium to make procedures as simple as possible. Although it was shown that transformation efficiency decreased with growth at 37°C, some laboratories successfully and routinely grow *A. baylyi* at 37°C rather than at 30°C. The higher temperature is convenient for the use of common equipment for culturing *A. baylyi* and *E. coli*. The transformation efficiency at 37°C is sufficient for nearly all applications. Moreover, minimal medium rather than rich medium may be used to prepare competent *A. baylyi* cultures. Growth in minimal medium with a single carbon source may reduce contamination (especially in undergraduate laboratory courses) and, in some cases, a higher transformation efficiency is obtained compared to growth in rich medium.

To prepare cells for transformation, a starting culture of *A. baylyi* is typically grown overnight. Such cultures are diluted, DNA is added, and additional growth is allowed for approximately 3–6h, as illustrated by the rightward arrow in Fig. 2A (Biggs et al., 2020). Using an alternative protocol, the overnight culture may simply be supplemented with an additional aliquot of a carbon source to allow the resumption of growth, as illustrated in Fig. 2 by the downward arrow leading to panels B and C. In this case, cells require a short period of growth (approximately 30 min) prior to spreading competent cells on the surface of a plate with selective or non-selective growth medium. Although this approach may not yield maximal competency, the high concentration of recipient cells usually enables good results. To obtain higher transformation efficiencies, cell growth may be monitored to ensure that cultures are in late exponential phase prior to spreading competent cells on plates, and/or a cell concentration step may be used for a further increase in the number of recipient cells.

Genomic change in *A. baylyi* can be accomplished with diverse types of donor DNA. As described elsewhere, replicative plasmids may be introduced into recipient cells (Biggs et al., 2020). Here we focus on transformations with linear donor DNA to modify the chromosome via homologous recombination, a type of genome editing that is uniquely efficient in *A. baylyi*. This type of allelic exchange requires only that there is sequence identity between the donor and recipient DNA in regions upstream and downstream of the targeted change (Fig. 1). A simple way to prepare DNA from a donor *A. baylyi* strain involves heat-based lysis followed by filtration to remove whole cells. This method has the advantage of providing large stretches of DNA for homologous recombination. This method is typically used to combine mutations

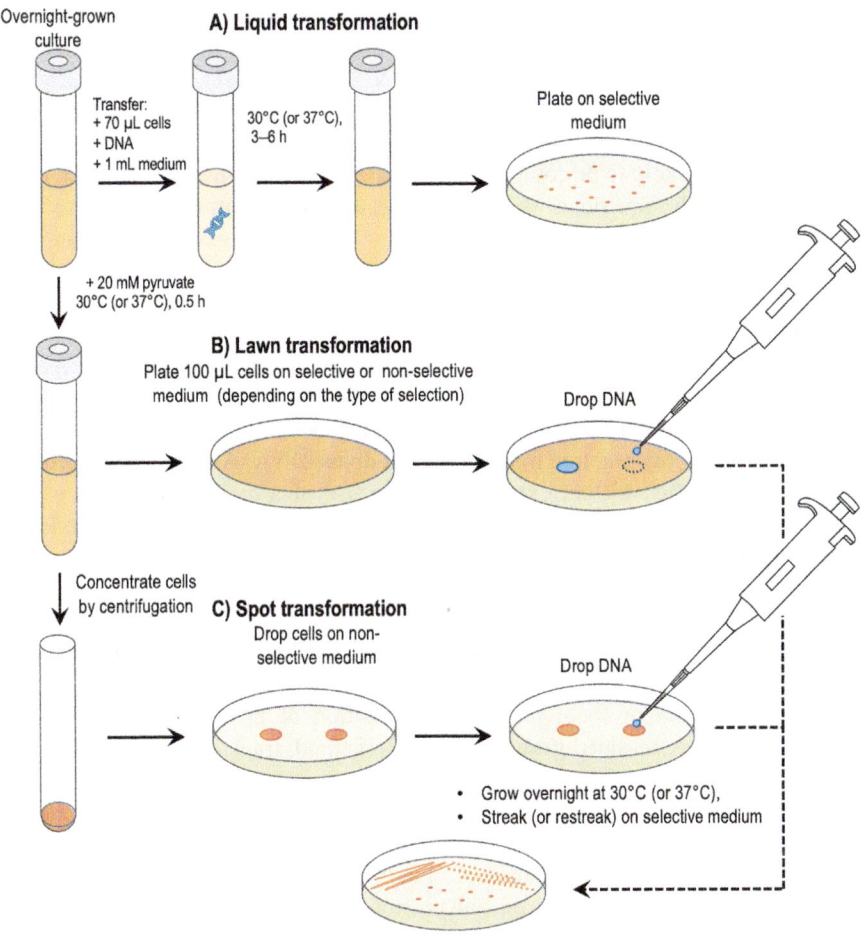

FIG. 2

Different methods for transforming *A. baylyi*. (A) Liquid transformation is highly efficient. (B) Lawn transformation can be done on selective medium if recipient cells can survive (see Figs 3–5). For selections based on antibiotic resistance, a recovery period on non-selective medium is needed. (C) Spot transformations on a non-selective medium allows different concentrations and proportions of DNA and cells to be mixed prior to selection or screening. Variations and combinations of these methods are well tolerated. Modifications may be tailored to specific needs, convenience, or available resources. If no selection is available, screening of individual or pooled transformants will typically identify allelic replacement (with linear donor DNA) at a frequency of approximately 10^{-2} to 10^{-3}.

in different *A. baylyi* strains. In this case, the genomic location of all known mutations should be considered, as multiple allelic exchanges may take place, and several mutations can be co-transformed (Gerischer & Ornston, 2001). Localized or whole genome sequencing of the resulting transformant is recommended to confirm the genotype. Similarly, purified genomic DNA may be used to transform the recipient. If the results are likely to be complex, it may be preferable to use fragments of donor DNA to define the positions where homologous recombination will occur. When plasmid DNA is used, it should be linearized by restriction digestion with conditions that promote complete cutting (e.g. use of multiple enzymes, optimization of the time of restriction enzyme digestion, the amount of DNA, and the amount of enzyme). Uncut plasmid DNA can result in the chromosomal integration of the entire plasmid via a single recombination event (Palmen, Vosman, Buijsman, Breek, & Hellingwerf, 1993). Thus, transformants should always be evaluated to ensure allelic replacement rather than plasmid integration. PCR products from any type of single or multi-piece assembly reaction may be used as donor DNA. Although PCR products need not be purified by gel extraction or the use of a clean-up procedure, such purification steps can simplify experiments by preventing the formation of undesired transformants and/or by improving transformation efficiencies.

2.3 Introduction of DNA into cells, and growth conditions following transformation

One transformation protocol involves the addition of DNA to diluted recipient cells followed by continued growth. With this method, transformants can be plated directly on selective medium after several hours of incubation (Fig. 2A). Alternatively, cells and DNA can be mixed in different ways on plates (Fig. 2B and C). The lawn transformation method (Fig. 2B) enables rapid assays to study a range of topics related to the significance of mutations. In ways that are not possible in other organisms, specific genotypic and phenotypic effects can be readily correlated (Young et al., 2005). These rapid, and powerful genetic tests can streamline subsequent experimental strategies.

Several examples are used to demonstrate how lawn transformation assays can be used (1) to screen libraries of mutated alleles, (2) to evaluate the effects of chromosomal context, and (3) to evaluate multiple mutations (alone and in combination) that arise during adaptive laboratory evolution. In all these examples, a culture sample is spread across the surface of a selective plate containing a medium insufficient for this recipient to grow. In marked spots, cell-free linear DNA is dropped directly on the plate surface. When these selections involve a change in the ability to use a carbon source (or other selection that does not severely inhibit the recipient), rather than acquisition of antibiotic resistance, there is no need for a recovery period involving growth on non-selective medium. Transformation and homologous recombination take place within cells on the plate surface. When allelic replacement alters the phenotype appropriately, growth of transformants is observed in regions corresponding to the position(s) where DNA was dropped.

In the first example (Fig. 3), an auxotrophic mutant, which lacks a critical enzyme, is unable to grow on minimal medium without amino acid supplementation. DNA encoding a different (non-homologous) enzyme can compensate for the missing *A. baylyi* enzyme and confer growth. The compensating DNA encodes a variant (positive control) of an *Escherichia coli* enzyme whose native sequence does not

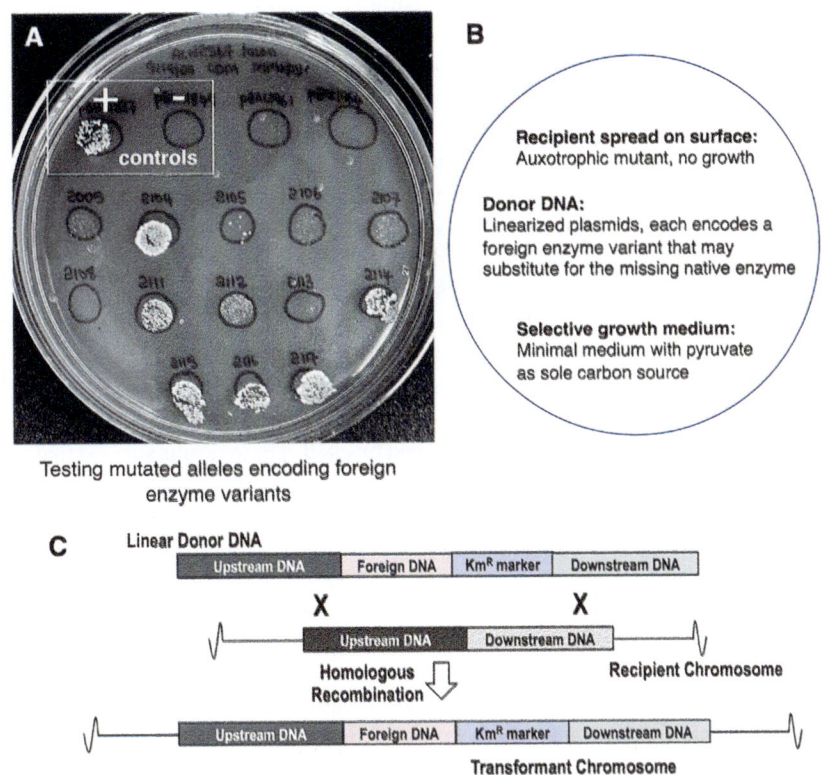

FIG. 3

Lawn transformation assay on selective medium. (A) A recipient strain, which is unable to grow on the agar plate medium, was spread across the surface. Cell-free linear DNA dropped in marked spots can transform the recipient and replace the corresponding chromosomal region via homologous recombination. Growth of transformants indicates that allelic replacement enables expression of a foreign enzyme to compensate for a missing enzyme in the auxotrophic recipient. (B) Details corresponding to the experimental results shown in (A). (C) Schematic illustration of the allelic replacement. In the experiment shown, selection depends on prototrophy without antibiotics. An alternative way to assess transformants involves antibiotic resistance, using a marker adjacent to the foreign enzyme gene and a method that includes a recovery period on non-selective medium (Fig. 2). If Km^R-transformants are selected, their growth can be evaluated without of any prior selection for prototrophy.

confer growth (negative control). Alleles encoding different variants of this *E. coli* enzyme can be tested in separate spots. The results not only indicate which variants permit prototrophic growth, but the time needed for colonies/confluent growth to appear in each spot provides preliminary information that correlates with the growth rate of the mutant. This type of rapid analysis enables the identification of mutants with desired phenotypes. Such data can hone subsequent mutagenic strategies and focus further analyses on the most promising candidates.

In the second example (Fig. 4), a similar assay was used to understand a mutation that arose during experimental evolution. In this case, a foreign pathway was introduced into a mutant that was unable to grow on *p*-hydroxybenzoate (Pob) as the carbon source. Isolates that grew on this substrate (Pob$^+$ derivatives), were obtained with Evolution by Amplification and Synthetic biology (EASy) (Tumen-Velasquez et al., 2018), a method that is described more fully in our companion chapter (Pardo et al., 2023). One critical mutation for this acquired phenotype inactivates a native ADP1 enzyme (here termed EnzB) that has functional overlap to an enzyme encoded by the foreign pathway (here termed EnzA). Since these enzymes have significant sequence differences, the preference for EnzA could relate to enzyme activity. However, using the transformation assay (Fig. 4), it was shown that either enzyme enables Pob$^+$ growth if the corresponding gene is in the chromosome within the region of the foreign DNA. In contrast, neither enzyme was sufficient for growth if its gene was in the position normally occupied by *enzB*. These results suggest that some type of regulation related to proximity of the associated genes and/or proteins affects metabolic function. This example illustrates the use of chromosomal alteration by natural transformation and allelic replacement to investigate chromosomal context, an issue that is typically difficult to study in other organisms.

A third example of a transformation assay (Fig. 5) illustrates a novel approach to assess the significance of mutational combinations. This plate-based assay is essentially a variation of the RAMSES method described above (Luo, Mcintyre, et al., 2022). To distinguish these methods, we designate the plate-based technique by a different acronym, CEMENT, which refers to combinatorial evaluation of mutations examined by natural transformation. Strains derived from laboratory evolution experiments are typically found to carry multiple mutations. However, to determine which mutations are required for the selected phenotype can be difficult. In the example shown, a Pob$^-$ parent strain (unable to use Pob as the carbon source), evolved to yield a Pob$^+$ derivative. Some of the approximately 12 mutations in various locations of the chromosome that were revealed by whole genome sequencing are depicted by asterisks (Fig. 5A). In this case, the parent strain was spread on selective growth medium with Pob as the carbon source. Individual linear fragments were obtained that each carry one mutation from the Pob$^+$ isolate as well as surrounding regions of sequence identity to allow allelic replacement of the corresponding region of the parent strain. After testing various mutations, alone and in combination, the assay revealed that a combination of three specific mutations allowed growth on the selective Pob plate. We have not yet determined the limit of how many individual

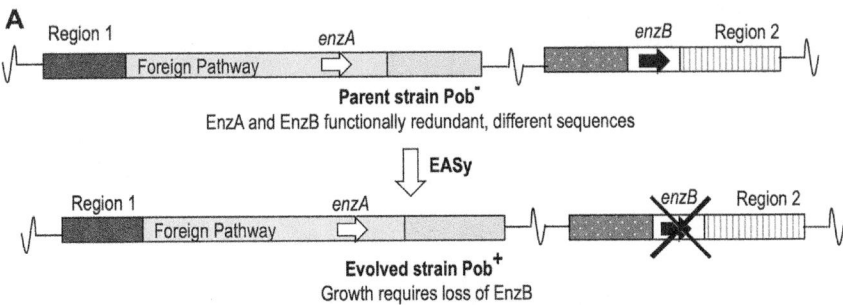

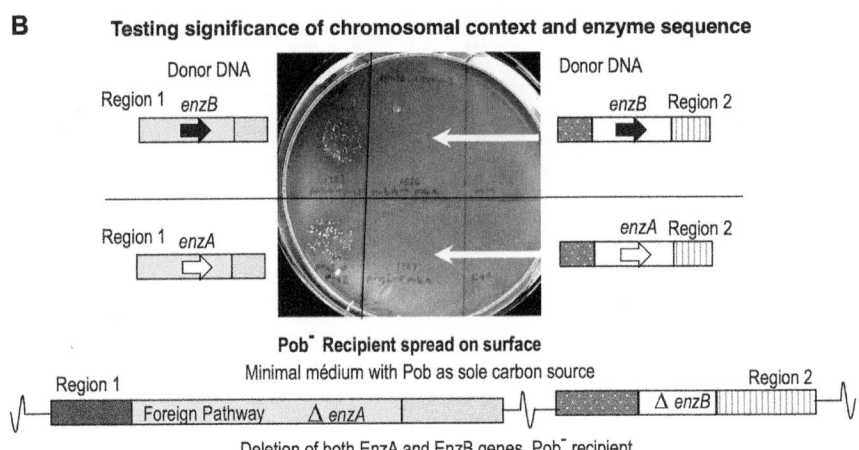

FIG. 4

Lawn transformation assay to investigate mutations in a Pob$^+$ evolved strain derived by the EASy method from a Pob$^-$ parent strain. (A) Schematic representation of chromosomal configurations in the parent and evolved strains. The acquired phenotype of the evolved mutant required both expression of a foreign pathway, including a gene (*enzA*), integrated in the chromosome (region 1), and additional mutations such as the inactivation of a native *A. baylyi* gene (*enzB*) in a different chromosomal locus (region 2). (B) A new Pob$^-$ strain was generated by deleting both *enzA* and *enzB*, which encode enzymes that catalyse the same reaction despite some significant differences in their sequences. Linear DNA fragments were generated in which each gene (*enzA* or *enzB*) resides within sequences corresponding to region 1 or region 2. The Pob$^-$ deletion strain was used as the recipient in a lawn transformation assay with cell-free donor DNA for each of the four configurations (*enzA* in region 1, *enzA* in region 2, *enzB* in region 1, and *enzB* in region 2). Transformants grew on Pob when either gene could be inserted in region 1, whereas neither conferred growth in region 2. These results suggest that genomic context, rather than the enzyme sequence is most important for a Pob$^+$ phenotype.

A

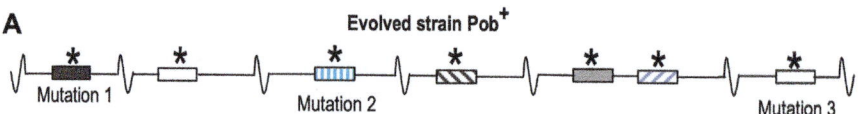

B

Testing significance of different mutations for Pob⁺ phenotype

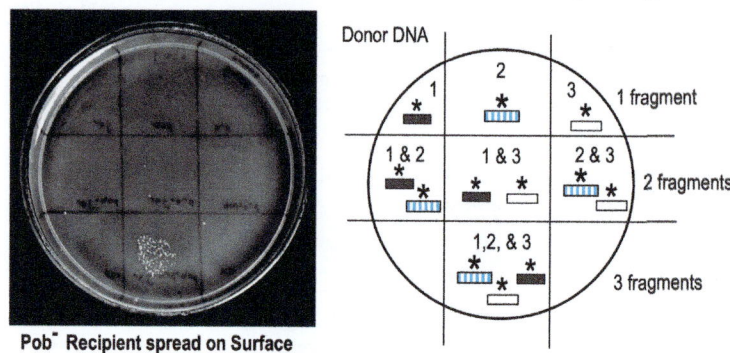

Pob⁻ Recipient spread on Surface
Minimal medium, Pob (sole carbon source)

FIG. 5

Combinatorial evaluation of mutations examined by natural transformation (CEMENT).
(A) Schematic representation of the chromosomal configuration of a Pob⁺ mutant that
evolved from a Pob⁻ parent strain. Whole genome sequencing revealed multiple mutations
(*) throughout the chromosome. Linear DNA fragments were generated that each
corresponded to a genomic region (rectangle). (B) Cell-free linear fragments, each carrying a
known mutation from the evolved strain, were used as donor DNA in a lawn transformation
assay. The recipient strain, spread on the surface of a selective plate with Pob as the
carbon source, was the Pob⁻ parent from which the evolved mutant was derived. Multiple
fragments were tested, and a combination of three specific fragments (mutations) proved
sufficient to transform the recipient to grow on the Pob plate. Growth was observed only
when all three fragments were dropped in the same spot (bottom center). The diagram at the
right indicates the regions of the plate where the DNA fragments were dropped. Each
was tested individually and as a pair with one other fragment as well as in the three-fragment
combination. The results highlight the importance of this combination of mutations
for Pob⁺ growth.

mutations can be efficiently transformed in this fashion. Nevertheless, this approach,
together with RAMSES, can streamline mutational analyses as well as the construction of strains with multiple defined mutations.

2.4 Isolation and confirmation of desired transformants

Transformants generated by any of the methods shown in Fig. 2 should be confirmed
after streak purification. Such confirmation can be accomplished using PCR,
screening for drug resistance, localized DNA sequencing and/or whole genome

sequencing. When linearized plasmids are used as the donor DNA, it is important to ensure that chromosomal changes result from allelic replacement rather than plasmid integration. Another issue to consider is whether the isolation method could lead to the selection of unanticipated mutations. For example, in the experiments shown in Figs 3–5, transformants are selected directly for growth on a given substrate, which could result in the isolation of spontaneous mutations that are not introduced by the donor DNA. While such mutations may be rare, it is important to consider all possibilities. For example, in Fig. 3, the introduction of a mutated allele could lead to slow growth that increases the chance of selecting spontaneous mutations that increase the growth rate. To test the phenotypes of transformants that have never been exposed to the selective carbon source, an alternative approach would be to select allelic replacement using expression of the adjacent antibiotic resistance marker. When using drug resistance, the recipients should be allowed to grow on non-selective medium to allow time for recovery and expression before selection in the presence of the antibiotic.

3 Material and equipment
3.1 Strains and culture media

- *Acinetobacter baylyi* ADP1 (from culture collections ATCC 33305 or DSM 24193); genome sequence in GenBank under accession number NC_005966. A transposon-free derivative of ADP1 is also available, ADP1-ISx (Suárez et al., 2017).
- Chemically competent *Escherichia coli* cells, to be used as hosts for plasmid construction and maintenance (any host strain). Typical strains used for this purpose include *E. coli* strains DH5α and XLI Blue (Agilent Technologies).
- Minimal medium (MM) for *A. baylyi* (per litre, 25 mL 0.5 M KH_2PO_4; 25 mL 0.5 M Na_2HPO_4; 10 mL 10% $(NH_4)_2SO_4$; 1 mL concentrated base) supplemented with an appropriate carbon source (e.g. 20 mM pyruvate or 10 mM succinate). Common carbon sources such as glucose and acetate (25–50 mM) also work well, although a longer lag phase is observed with glucose. In addition, casein amino acids serve as an excellent carbon source (especially if strains overexpress certain proteins) and can be added in concentration 0.2% (with or without other carbon sources). For solid media, 1.5% agar is used.
  - Concentrated base solution (per litre, 20 g nitriloacetic acid dissolved in 600 mL H_2O with 14.6 g KOH; 28.9 g $MgSO_4$; 6.67 g $CaCl_2 \cdot 2H_2O$; 18.5 mg $Mo_7O_{24} \cdot 4H_2O$; 198 mg $FeSO_4 \cdot 7H_2O$; and 100 mL of Metals 44 solution).
  - Metals 44 solution (per litre, 2.5 g EDTA; 10.95 g $ZnSO_4 \cdot 7H_2O$; 5 g $FeSO_4 \cdot 7H_2O$; 1.54 g $MnSO_4 \cdot 7H_2O$; 392 mg $CuSO_4 \cdot 5H_2O$; 250 mg $Co(NO_3)_2 \cdot 6H_2O$; and 177 mg $Na_2B_4O_7 \cdot 10H_2O$).

- Minimal medium, different formulations: alternative recipes may also be used for *A. baylyi*. Examples of alternative defined media are described elsewhere (Biggs et al., 2020; Hartmans, Smits, Van Der Werf, Volkering, & De Bont, 1989).
- Rich medium, such as Lysogeny Broth (LB, also known as Luria Bertani medium), (per litre, 10 g Bacto-tryptone, 5 g yeast extract and 10 g NaCl); Note: an alternative composition with only 1 g per litre NaCl may be used to improve *A. baylyi* growth.
- Supplements added as needed. Antibiotics are used at the following final concentrations: kanamycin (Km), $25 \, \mu g \, mL^{-1}$; spectinomycin (Sp) and streptomycin (Sm), $12–15 \, \mu g \, mL^{-1}$ (each); ampicillin (Ap), $150 \, \mu g \, mL^{-1}$, chloramphenicol (Cm), $25–50 \, \mu g \, mL^{-1}$, and gentamicin (Gm), $15 \, \mu g \, mL^{-1}$. Note: ADP1 has some natural resistance to Ap. If testing *A. baylyi* transformants, check the resistance level of the recipient in comparison to the transformant, and, if needed, the drug concentration can be increased to discriminate between natural and acquired resistance.
- YTS agar plates for sucrose counterselection (autoclave 3 g yeast extract, 6 g tryptone, 11 g agar in 300 mL water; then add 300 mL filter-sterilized 50% sucrose).

3.2 Buffers and reagents for DNA preparation and manipulation

- Gel electrophoresis methods may be used with any standard agarose and buffers.
- Lysis buffer can be used to generate genomic donor DNA (0.05% sodium dodecyl sulphate [SDS] in 0.15 M NaCl-0.015 M citrate, trisodium salt).
- Molecular biology reagents include standard materials for PCR, restriction digestion, cloning, and DNA isolation. Such materials depend on the experiment and may include: PCR primers, cloning vectors, plasmids carrying the ΩKm^R or $\Omega Sm/Sp^R$ fragment, e.g., pHP45Ω-Km (Fellay, Frey, & Krisch, 1987), pUI1637, and pUI1638 (Eraso & Kaplan, 1994), DNA polymerase and PCR components (such as buffer and dNTPs), restriction enzymes, and T4 DNA ligase.
- Purification methods/kits: donor DNAs need not be purified. However, in some cases, experimental strategies or efficiencies can be improved by using a commercial clean-up kit or by excising and purifying a specific DNA band from an agarose gel. Any purification method or commercial kit may be used.
- Salt Sodium Citrate (SSC) buffer is used for the heat-based cell lysis method: 0.15 M NaCl and 0.015 M citrate, trisodium salt.
- Syringe filter units can be used to ensure that whole cells are removed from donor DNA (disposable, polyethersulfone (PES) membrane; pore size $0.2 \, \mu m$, diameter 13 mm).

3.3 **Equipment**

- Bacterial growth will require standard equipment, such as incubators and a method for the aeration of liquid cultures (such as an orbital shaker or a roller drum for test tubes).
- Gel electrophoresis equipment will be used for DNA analysis, requiring an electrophoresis chamber and a power supply. To visualize DNA in the gels, a UV transilluminator is needed.
- Quantification of DNA can be done using any equipment for any method (such as a spectrophotometer, fluorometer, or specialized quantification machine). For most purposes, DNA quantification is not required.
- A thermocycler will be needed for PCR.
- For RAMSES, a microplate reader is needed (e.g. Tecan Spark multimode microplate reader, Tecan, Switzerland)

4 **Experimental procedures**
4.1 **Design and preparation of donor DNA**

Using allelic replacement to modify the chromosome of an *A. baylyi* recipient strain requires the donor DNA to carry sequences identical to the chromosome in regions upstream and downstream of the targeted modification (Fig. 1). All methods of obtaining cell-free genomic DNA from *A. baylyi* donor strains will produce linear fragments that have long sequence stretches matching the recipient chromosome. These long stretches of adjacent DNA promote homologous recombination and increase the efficiency of allelic replacement. Using smaller fragments lowers this efficiency but can be advantageous by defining the chromosomal change and enabling the design/synthesis of mutations.

The design of PCR products or plasmids should take into consideration the DNA needed for recombination. Typically, approximately 500 bp to 2 kbp of sequence identity between donor and recipient DNA on either side of the target modification enables high-efficiency allelic replacement (Biggs et al., 2020). However, much smaller regions of sequence identity can be sufficient. For example, in one study, the introduction of a point mutation was evaluated using different donor DNA fragments (Gerischer & Ornston, 1995). All fragments had 385 bp of identity with the chromosome in the region downstream of the mutation. In the region upstream of the mutation, 122 bp was sufficient for generating frequent transformants. Surprisingly, transformants were also obtained, albeit at low frequency, when the donor DNA in the region upstream of the mutation had only 4 bp of identity with the recipient chromosome. Suitable constructs may be made by any molecular biology method (restriction cloning, PCR, any DNA assembly method, etc.) or by commercial synthesis. One convenient method for plasmid construction takes advantage of (in vivo) assembly in *E. coli* (Kostylev, Otwell, Richardson, & Suzuki, 2015).

4.1.1 Preparation of cell-free genomic DNA by a heat-based lysis method

Any method of purifying genomic DNA may be used. A simple method for generating donor DNA involves minor modifications of past protocols (Juni, 1972; Neidle & Ornston, 1986). In general, these lysates are highly effective as donor DNA for combining mutations from different strains.

1. Inoculate a 5 mL overnight liquid culture of a donor strain in minimal medium with a carbon source such as pyruvate (20 mM).
2. Harvest cells by centrifugation and suspend in 0.5 mL sterile prewarmed lysis buffer.
3. Incubate at 65 °C for 1–3 h.
4. Dilute 10-fold in SSC buffer and remove any remaining cells from 1 mL of the diluted lysate by filter sterilization (using a disposable unit).
5. The filtered lysate and different dilutions (10-fold and 100-fold) of it may be used. In some cases, using more dilute samples of the lysate can be helpful, perhaps by diluting SDS in the lysis buffer or inhibitors in the lysate. DNA concentration need not be determined.

4.1.2 PCR-products or linearized plasmids as donor DNA

Any method of PCR may be used to generate donor DNA. A preferred method for adding flanking DNA and/or cassettes with selectable or counter-selectable markers is (in vivo) assembly of DNA fragments in *E. coli* (Kostylev et al., 2015). Another good method is splicing by overlap-extension PCR (Horton, Cai, Ho, & Pease, 1990). If high fidelity PCR is desired, the following polymerases are good options: PrimeSTAR (Takara Biosciences), Phusion (NEB), and Q5 (NEB). However, low fidelity or mutagenic PCR conditions may be used deliberately to mutate the donor DNA (Kok, Young, & Ornston, 1999; Young, Kok, & Ornston, 2002). Any method of restriction digestion may be used to linearize a circular plasmid. The following points should be considered.

1. If plasmid DNA is used as the template for a PCR product, a small amount of circular template in the donor DNA may get integrated by a single recombination event in the transformant. Thus, using digestion conditions that promote complete cutting of the plasmid should be used, and transformants should be carefully characterized. It should be noted that most cloning vectors used in *E. coli* to generate donor alleles (such as pUC19 and other vectors with ColE1-based origins of replication) are not stably maintained in *A. baylyi* as independent replicons. However, they may persist in the cell for an undetermined amount of time (Gralton, Campbell, & Neidle, 1997).
2. Similarly, if genomic DNA is used as the template for PCR, a small amount of genomic template in the PCR product may be able to modify the chromosome of the recipient. While such issues occur infrequently, it is important to be aware of all possibilities and to note that very little DNA is needed for allelic replacement to occur.

4.2 Liquid transformation method (Fig. 2A)

The following transformation method works well with as little as 25 ng of DNA added to the culture (Biggs et al., 2020). However, while it is not critical to quantify the amount of DNA that is added to cells, a range of 25 ng to 2 µg DNA in a volume of 1–50 µL is typical.

1. Grow the recipient A. *baylyi* strain overnight in 5 mL liquid MM with a non-selective carbon source (e.g. 20 mM pyruvate), with aeration (such as shaking at 250 rpm) at 30 °C (or 37 °C). Alternatively, rich medium (LB) may be used.
2. Combine 70 µL of the recipient culture with 1 mL fresh growth medium and linear donor DNA in a culture tube. Prepare a similar culture without adding DNA to serve as a negative control.
3. Incubate culture and DNA for 3–6 h, with aeration (such as shaking at 250 rpm) at 30 °C (or 37 °C).
4. Spread 100 µL transformed cell culture onto a selective medium agar plate. Do the same for the control culture. Note: to obtain isolated transformants on the selective medium, the culture may need to be diluted or concentrated. Use the same dilutions or concentrations for the control culture.
5. Incubate plate(s) at 30 °C (or 37 °C). Depending on the selection, colonies typically appear within 1–3 days.

4.3 Lawn transformation (Fig. 2B)

Transformations can be done directly on selective growth medium when conditions are not lethal to the recipient. As shown in Figs 3–5, selection for growth on a new carbon source enables rapid assays for a range of metabolic investigations as well as for strain engineering. Such direct selection is not appropriate for antibiotic selection. For antibiotic selection, spread recipients to non-selective plates and allow growth/recovery before moving transformed cells to selective medium (after overnight incubation).

1. Grow the recipient A. *baylyi* strain overnight in 5 mL liquid MM with a non-selective carbon source (e.g. 20 mM pyruvate), with aeration (shaking at 250 rpm) at 30 °C (or 37 °C).
2. Add additional carbon source (100 µL 1 M pyruvate) and incubate with shaking for an additional 30 min to initiate cell growth.
3. Spread 100–200 µL culture onto solid selective medium (35 × 10 mm plate). Note: depending on the selection, it may be helpful to wash the cells and/or concentrate them by centrifugation and suspension in MM without a carbon source. Carryover of some growth substrate when spreading the recipient cells can sometimes be helpful by allowing cells to grow and increase the transformation efficiency. In other cases, such carryover can cause background growth that interferes with the experiment. While a negative control without

added DNA can help to identify the impact of carryover substrate, for lawn transformation assays the difference between growth in spots where DNA was dropped compared to the background growth (negative control) on the plate helps assess the effect of recipient growth and transformation by DNA. The background region and plating of cells without DNA can also provide information about reversion rates and/or spontaneous mutation rates of the parent culture. When possible, it is best to add donor DNA that should confer growth to serve as a positive control to assess the competency of the parent culture. Washing cells can be helpful in removing excess carbon source and can also be used to concentrate cells to increase the number of recipient cells. However, a disadvantage of washing the cells is that during the centrifugation and suspension steps the competency of the culture may be reduced. If one method proves insufficient, minor changes to the plating method can be beneficial.

4. Allow culture to dry on the plate such that the surface is not visibly wet. This step should not require more than approximately 1–2 min.

5. Mark on the plastic side of dish where DNA will be dropped and label appropriately. A small circle or region drawn on the plate is sufficient to demarcate the location of the DNA (and can be seen in Figs 3–5).

6. Drop 1–5 µL linear DNA (in isotonic solution) on top of the dried culture on the plate. Such drops typically contain 25–500 ng of DNA. For a negative control, use an isotonic solution without DNA. DNA that should not confer growth should also be used as a negative control.

7. Allow the DNA drop to air dry, until the surface is no longer visibly wet (several min).

8. Incubate at 30 °C (or 37 °C). Growth is typically observed in 1–3 days, depending on the experiment and the selection.

4.4 **Spot transformation** (Fig. 2C)

A. baylyi may also be transformed in small spots on the surface of a non-selective agar plate. Recipient cells and donor DNA can be mixed in drops and incubated together. One advantage of this method is that different amounts of cells and DNA can easily be combined in multiple spots on the same plate. Such variation may facilitate obtaining the desired transformants. Although altering the cell/ DNA ratio is typically not needed, the constraints of different experiments and selective conditions can be unpredictable. The initial steps are the same as for the lawn transformation method.

1. Grow the recipient *A. baylyi* strain overnight in 5 mL liquid MM with a non-selective carbon source (e.g. 20 mM pyruvate), with aeration (shaking at 250 rpm) at 30 °C (or 37 °C).

2. Add additional carbon source (100 µL 1 M pyruvate) and incubate with shaking for an additional 30 min to initiate cell growth.

3. To increase the concentration of recipient cells, the culture may be concentrated by centrifugation. The cell pellet can be suspended in a small amount of growth medium (approximately 100–500 µL MM, with or without carbon source).
4. Mark on the plastic side of dish where DNA will be dropped and label the spot(s) appropriately.
5. Drop samples of the recipient culture in spot(s) on the plate. Drops of different amounts may be used. However, procedures are simpler if spots do not mix/run along the plate surface. Volumes of 1–10 µL may be used regardless of the cell concentration.
6. Drop samples of the donor DNA (in isotonic solution) to mix with cells. A volume of 1–5 µL is appropriate (regardless of the DNA concentration). Typically, 25–500 ng of DNA will be added. If the volume is too large, the spot will spread. For a negative control, mix the same volume of liquid without DNA with cells.
7. Allow cell-DNA mixtures to air dry, until the spots will not spread when the plate is moved.
8. Incubate at 30 °C (or 37 °C) overnight before transferring cells to selective (or non-selective) medium by streak purification or suspending cells in a small amount of medium and spreading on an agar plate.

4.5 Isolation, screening, and genotypic/phenotypic confirmation of transformants

Regardless of the transformation method, the resulting isolates should be streak purified. Where possible, growth on selective medium will yield individual colonies that can be tested further. However, if selection is not possible, transformation efficiencies are usually high enough to obtain desired transformants by screening 100–300 colonies. Depending on the screening method, candidate colonies can be pooled to reduce the effort. For example, screening for a genotype by PCR can be done in the absence of a phenotypic screen. After individual transformants are identified, they are characterized further by one or more of the following steps.

1. Several colonies from the same isolate should be screened on different growth media (as appropriate) by patching each colony to multiple plates with a sterile toothpick (Bedore et al., 2022). Growth, antibiotic resistance and/or screening for counter-selectable markers, such as *sacB* (Jones & Williams, 2003) or *tdk* (Metzgar et al., 2004) help to confirm or detect phenotypic differences between the transformants and parent (recipient) strain. Drug resistance (typically ampicillin resistance encoded on the vector backbone) can identify transformants that result from plasmid integration rather than allelic replacement.
2. PCR with multiple sets of primers may be used to confirm expected genotypes. These tests can be done rapidly using colony PCR or with purified DNA templates. When possible, the primers used should bind the chromosome outside of the surrounding identical sequences included in the donor DNA to promote

homologous recombination (e.g. when using linearized plasmids or PCR products as donor DNA). In this way, allelic replacement in the targeted locus can be verified.

3. DNA sequencing of localized regions, and/or of the whole genome will identify genetic changes.

4.6 Method for rapid advantageous mutation screening and selection (RAMSES)

The agar plate assays described above, and shown in Figs 3–5, reveal the significance of mutations when transformants form colonies on selective media. However, such assays are unable to distinguish more subtle phenotypes that reflect growth differences under diverse selective conditions. The RAMSES method, introduced earlier, provides a powerful means to monitor the growth of transformants in a microtiter plate reader. As illustrated in Fig. 6, this method readily demonstrates which of the many genomic mutations, typically observed in a strain derived by laboratory evolution, are most beneficial for the desired phenotype (Luo, Mcintyre, et al., 2022). RAMSES allows seamless introduction of the individual mutated alleles (alone and in different combinations) into a recipient strain that does not grow under the selective conditions. Mutated PCR fragments, added directly to the cells, generate transformants that become enriched in the population when they have acquired selectively advantageous mutations. This method enables both the use of incremental concentrations of the selective compound, such as high concentrations of an aromatic compound, and the simultaneous comparison of multiple variations of donor DNA combinations. For RAMSES, either liquid transformation or spot transformation can be used. Here we describe the method based on spot transformation.

1. Analyse the whole genome sequencing data of any isolate(s) with an improved or novel phenotype to identify and choose the mutations to be studied.
2. Design PCR primers that cover the mutated areas in the genome. The amplified DNA fragments should each contain at least 500 bp of sequence identity with the recipient chromosome on each side of the mutated region to ensure successful allelic replacement.
3. Use genomic DNA of the mutant isolates as template for PCR to amplify the targeted region. To avoid unintended co-transformation of different mutated alleles, each PCR product should be purified using a gel extraction kit.
   We recommend not placing different PCR products in adjacent wells of the agarose gel to avoid cross-contamination.
4. Streak the recipient strain on LB agar and incubate at 30 °C overnight. For RAMSES, we recommend using a transposon-free recipient, such as *A. baylyi* ADP1-ISx (Suárez et al., 2017) or a strain derived from it. Such recipients have increased transformation efficiency and genomic stability to help avoid the emergence of undesired spontaneous mutations. However, ADP1 and

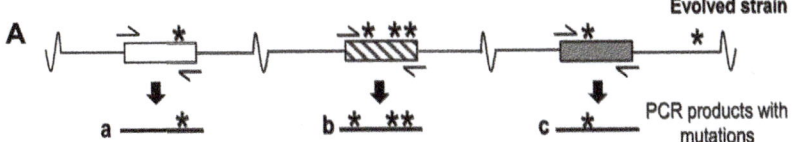

A

Individual and combinations of PCR products used to transform *A. baylyi* recipient

B

A. baylyi recipient in each well transformed with:

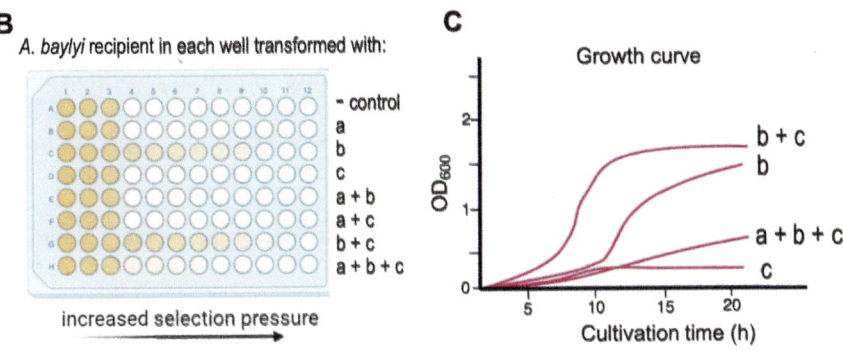

- control
a
b
c
a + b
a + c
b + c
a + b + c

increased selection pressure

C

Growth curve

b + c
b
a + b + c
c

OD$_{600}$

Cultivation time (h)

FIG. 6

Use of the RAMSES method to identify beneficial mutations in evolved strains. (A) Schematic representation of the chromosomal configuration of an evolved strain with multiple mutations (*). Using genomic DNA of the evolved strain as template, PCR with specific primers generates linear DNA fragments that are purified and used as donor DNA to transform an appropriate reference strain (e.g. *A. baylyi* ADP1-ISx). These PCR products, which carry known mutations, are tested for the ability to affect the phenotype of the recipient. This example illustrates the use of different PCR products (a, b, and c) in individual and combined transformations. The transformed cells are first incubated under selective conditions. (B) Cells are then transferred to 96-well plate with various concentrations of the selective compound. Cell growth is monitored by optical density (OD$_{600}$). A negative control should be included that uses comparable conditions with a DNA fragment that should not confer benefit (random DNA or a fragment that has the same sequence in the donor and the recipient). Comparably treated cells to which no DNA is added could serve as an additional negative control. (C) The growth curves indicate which mutations (and which combinations) confer selective growth advantage.

ADP1-derived strains can also be used as recipients. Add purified DNA (0.5–2 µL) onto single colonies and mix well by pipetting up and down. Incubate overnight at 30 °C.

5. Scrape the colonies treated with DNA and suspend in 1–2 mL MM supplemented with a low concentration of the selective substrate (for example, 5 mM of aromatic compounds). Prepare a control culture using a colony without DNA treatment. Another good control is to use a colony and add DNA for transformation that would not be expected to confer a growth benefit

(such as a fragment identical in sequence between the mutant isolate and the recipient strain). Incubate the suspensions at 30 °C with aeration (shaking at 300 rpm) for up to 10 h.

6. Use the suspensions to inoculate 200 μL of MM supplemented with elevated concentrations of the selective substrate (for ferulate, use e.g. 20, 40, 60, and 80 mM) in a 96-well plate. Prepare triplicate cultures for each concentration and mutated allele/combination. Incubate the plate in a microplate reader (e.g. Tecan Spark multimode microplate reader, Tecan, Switzerland) at 30 °C and monitor optical density for 24–48 h.

7. If the cells treated with the mutated allele(s) show improved growth over the controls at the elevated aromatic concentrations, 5 μL of the cells can be taken from the well to further inoculate 5 mL of MM containing the same (or higher) concentration of the corresponding aromatic substrate for further mutant enrichment and storage.

8. Streak cells from the liquid culture on LB agar. Verify the mutated alleles in clones using PCR analysis and/or sequencing.

4.7 Modified gap-repair method

A recombinational capture method, termed gap repair (Gregg-Jolly & Ornston, 1990), allows very large genomic segments to be cloned on a plasmid, as illustrated in Fig. 1C. Compared to PCR-based cloning, this method avoids complications related to polymerase mediated replication fidelity and problems with generating large-sized PCR products. In mutants isolated by adaptive laboratory evolution and EASy, the gap-repair method can help identify whether mutations within a target region are sufficient to confer a selected phenotype (Bedore, 2021). As the first step in such analyses, the *A. baylyi* region of interest is captured on a plasmid that replicates in *E. coli*. In this step, a plasmid-borne drug marker is used (such as ampicillin resistance in Fig. 1C). Next, the purified plasmid can be linearized by restriction enzyme digestion in a region that does not carry *A. baylyi* DNA (i.e. in the backbone of the plasmid). If the linearized plasmid DNA also carries a drug-resistance marker within the region of interest (shown as Km^R in this example), it is easy to move the captured region into different genetic backgrounds by transformation and drug selection. For example, a linearized version of the 'gap-repaired' plasmid could be used as donor DNA to transform a strain that did not undergo laboratory evolution. After selection for allelic replacement, a transformant would have the genomic configuration shown in Fig. 1D, wherein, the only mutations are those introduced in the region of interest from the mutant. Using a comparable series of steps, the chromosomal region of a strain prior to evolution could be introduced into an evolved mutant such that the only mutations are outside the region of interest (Fig. 1E). In this fashion, the significance of mutations localized to specific chromosomal regions can be investigated.

The original gap-repair method was developed to capture chromosomal DNA on a vector that replicates and is stably maintained in *A. baylyi*, pRK415 (Keen, Tamaki, Kobayashi, & Trollinger, 1988). More recently, better results have been obtained using a smaller plasmid that is not stably maintained in *A. baylyi*. With a linearized 'capture plasmid' derived from a pUC18 or pUC19 vector (Yanisch-Perron, Vieira, & Messing, 1985), homologous recombination in *A. baylyi* generates sufficient 'gap-repaired' plasmid to be isolated for use in the subsequent transformation of *E. coli*. Thus, a plasmid resulting from homologous recombination in *A. baylyi* can be characterized and propagated in *E. coli*. The following protocol using pUC18 can be modified as needed to employ different vectors that replicate in *E. coli*.

4.7.1 Design of a 'capture-plasmid' for gap repair

As in the design of any plasmid intended to promote homologous recombination with chromosomal DNA, the choice of *A. baylyi* DNA to clone is important (grey boxes depicted on the plasmid, Fig. 1C). The appropriate *A. baylyi* sequences can be added to a cloning vector (e.g. pUC18) using any method such as splicing by overlap PCR, restriction cloning, in vitro assembly, in vivo assembly, or commercial synthesis (Horton et al., 1990; Kostylev et al., 2015). Several factors to consider in the design of the capture plasmid are indicated below.

1. The plasmid-borne *A. baylyi* sequences, which correspond to those in the chromosome upstream and downstream of the target, should be sufficiently large. Compared to other protocols so far described, longer sequences (1–2 kbp on each side of the target region) may be needed to accommodate the large chromosomal segment separating these sequences in the recipient strain. Long stretches of sequence identity facilitate plasmid-chromosomal alignment for the necessary homologous recombination events.

2. A sequence that can be cleaved by a restriction enzyme needs to be engineered between the two regions of *A. baylyi* DNA on the plasmid. This sequence must be a unique site such that cleavage with the appropriate enzyme will yield a linear fragment, as depicted in Fig. 1C. Both ends of the linear fragment will be generated by digestion at this restriction site (RS in Fig. 1C).

3. Restriction site analysis of the entire region should be considered. If the gap-repaired plasmid is later to be used as linear donor DNA to introduce the captured region into a different recipient chromosome, there must be a site or sites in the backbone for subsequent restriction digestion (indicated by a black triangle in Fig. 1C). These site(s) must not be present within the *A. baylyi* DNA. Depending on the size and sequence of the region of interest, it may be difficult to identify appropriate sites. If no appropriate sites are identified, it is possible to introduce such a site at the junction between the *A. baylyi* DNA and the plasmid DNA to allow cleavage at this boundary. It is important to consider this possibility at the outset such that the capture plasmid will be useful for all intended purposes.

4.7.2 Transformation of an A. baylyi *recipient with a linearized capture plasmid*

1. Digest the capture plasmid with an appropriate restriction enzyme to linearize it (RS in Fig. 1C). Note: if both the capture and gap-repaired plasmids confer the identical drug resistance (e.g. Ap^R), it is important to ensure complete digestion of the capture plasmid to prevent uncut plasmid from being selected at the end of the experiment. If the gap-repaired plasmid will confer additional drug resistance (e.g. Ap^R and Km^R), this pattern can be used to distinguish uncut capture plasmid from the gap-repaired plasmid generated by recombination.

2. It is helpful to use any type of DNA clean up kit and to elute the digested plasmid in a small volume of isotonic solution (the same range of DNA amounts and concentrations as described above for typical allelic replacement experiments will work). Save DNA until needed.

3. Inoculate a 5-mL culture of the *A. baylyi* recipient strain in non-selective medium (MM and 20 mM pyruvate or alternative carbon source) and culture with aeration (shaking at 250 rpm) at 30 °C (or 37 °C) overnight.

4. Subculture 120 μL of overnight culture into 3 mL of fresh medium and incubate cells with aeration (shaking at 250 rpm) at 30 °C (or 37 °C) for 2–3 h.

5. Use the spot transformation method (Fig. 2C) and drop different amounts of recipient cells and donor DNA in multiple spots on an LB plate. For example, use some samples of recipient cells that have not been concentrated and others that have been concentrated by centrifugation.

6. Incubate the LB plate spotted with different mixtures of cells and DNA at 30 °C (or 37 °C) for 6–8 h.

7. Take as many cells as possible from the spots and transfer by heavy patching to plate(s) of LB with appropriate antibiotics associated with the drug-resistance markers on the expected gap-repaired plasmid. However, keep in mind that the ColE1 origin of replication in pUC18 does not lead to stable plasmid maintenance.

8. Incubate plate overnight at 30 °C (or 37 °C).

4.7.3 Isolation of gap-repaired plasmid and use in E. coli *transformation*

1. Scrape as many *A. baylyi* recipient cells as possible from the incubated plate (LB and antibiotics) and suspend in 20 mL LB plus antibiotics. This slightly larger volume of cells than typically used for plasmid minipreps helps to recover sufficient plasmid DNA to transform *E. coli*.

2. Grow culture with aeration (shaking at 250 rpm) at 30 °C (or 37 °C) for 12–24 h.

3. Pellet all cells in the 20-mL culture by centrifugation, and isolate plasmid DNA using any miniprep protocol. Expect relatively low plasmid concentrations (based on the origin of replication) and suspend in a small volume.

4. Use the isolated plasmid to transform *E. coli* (any host strain) using chemically competent (or electrocompetent) cells. Follow typical methods and appropriate antibiotics to select plasmid-carrying *E. coli*.

5. Isolate and characterize plasmids from drug-resistant *E. coli* transformants.

6. Once the gap-repaired plasmid is isolated and checked by DNA sequencing, it may be digested and used as donor DNA in allelic replacement experiments following methods described earlier.

5 Summary and concluding remarks

We describe a variety of methods and applications for natural transformation and allelic replacement in *A. baylyi*. Although the ease and efficiency of such methods for genome editing are unrivalled, this naturally competent bacterium remains relatively obscure. It is often confused with similarly named bacteria or with a problematic pathogenic species that is not highly competent for natural transformation, *Acinetobacter baumannii*. However, the potential benefits for *A. baylyi* to become better known and more commonly chosen as a model organism are highlighted by vast amounts of DNA sequence data that continue to accumulate in databases. Synthetic biology and adaptive laboratory evolution, coupled with affordable and quick whole genome sequencing, are identifying mutations at a rate that exceeds our ability to understand genetic variation. *A. baylyi* offers simple methods to analyse mutations and to engineer strains with multiple and targeted mutations. These rapid techniques, which help link genotypic and phenotypic changes, can be used to extract critical biological information from DNA sequence data.

Acknowledgements

I. Pardo wishes to thank the Spanish National Research Council, Reina Sofía Foundation, and Primafrío Foundation for funding under agreement no. 20210510. Research at the University of Georgia U.S., described in this chapter, was funded by grants from the National Science Foundation (MCB2225858) and the U.S. Department of Energy, Office of Science, Office of Biological and Environmental Research, Genomic Science Program (DE-SC0022220). S. Santala would like to thank the Novo Nordisk Foundation (grant NNF21OC0067758) and the Academy of Finland (grant no. 334822 and 347204). V. Santala is grateful to the Academy of Finland (no. 310188).

References

Bedore, S. R. (2021). *Metabolic expansion in Acinetobacter baylyi ADP1 for enhanced aromatic compound catabolism* (Doctoral Dissertation). U.S: University Of Georgia.

Bedore, S. R., Schmidt, A. L., Slarks, L. E., Duscent-Maitland, C. V., Elliott, K. T., Andresen, S., et al. (2022). Regulation of L- and D-aspartate transport and metabolism in *Acinetobacter baylyi* ADP1. *Applied and Environmental Microbiology*, 88(15), E0088322. https://doi.org/10.1128/Aem.00883-22.

Biggs, B. W., Bedore, S. R., Arvay, E., Huang, S., Subramanian, H., Mcintyre, E. A., et al. (2020). Development of a genetic toolset for the highly Engineerable and metabolically versatile *Acinetobacter baylyi* ADP1. *Nucleic Acids Research*, 48(9), 5169–5182. https://doi.org/10.1093/Nar/Gkaa167.

De Berardinis, V., Durot, M., Weissenbach, J., & Salanoubat, M. (2009). *Acinetobacter baylyi* ADP1 as a model for metabolic system biology. *Current Opinion in Microbiology*, *12*(5), 568–576. https://doi.org/10.1016/J.Mib.2009.07.005.

Earnest, M. J., & Rosenbaum, N. J. (1993). Bacterial transformation. In C. A. Goldman, P. L. Hauta, M. A. O'donnell, S. E. Andrews, & R. Van Der Heiden (Eds.), *Tested studies for laboratory teaching*. Toronto: Assoc. Bio. Lab. Ed.

Elliott, K. T., & Neidle, E. L. (2011). *Acinetobacter baylyi* ADP1: Transforming the choice of model organism. *IUBMB Life*, *63*(12), 1075–1080. https://doi.org/10.1002/Iub.530.

Eraso, J. M., & Kaplan, S. (1994). *prrA*, a putative response regulator involved in oxygen regulation of photosynthesis gene expression in *Rhodobacter sphaeroides*. *Journal of Bacteriololy*, *176*(1), 32–43. https://doi.org/10.1128/jb.176.1.32-43.1994.

Fellay, R., Frey, J., & Krisch, H. (1987). Interposon mutagenesis of soil and water bacteria: A family of DNA fragments designed for in vitro insertional mutagenesis of Gram-negative bacteria. *Gene*, *52*(2–3), 147–154. https://doi.org/10.1016/0378-1119(87)90041-2.

Gerischer, U., & Ornston, L. N. (1995). Spontaneous mutations in *PcaH* and *-G*, structural genes for Protocatechuate 3,4-dioxygenase in *Acinetobacter calcoaceticus*. *Journal of Bacteriology*, *177*(5), 1336–1347. https://doi.org/10.1128/Jb.177.5.1336-1347.1995.

Gerischer, U., & Ornston, L. N. (2001). Dependence of linkage of alleles on their physical distance in natural transformation of *Acinetobacter* sp. strain ADP1. *Archives of Microbiology*, *176*(6), 465–469. https://doi.org/10.1007/S00203-001-0353-7.

Gralton, E. M., Campbell, A. L., & Neidle, E. L. (1997). Directed introduction of DNA cleavage sites to produce a high-resolution genetic and physical map of the *Acinetobacter* sp. strain ADP1 (BD413UE) chromosome. *Microbiology (Reading)*, *143*(Pt. 4), 1345–1357. https://doi.org/10.1099/00221287-143-4-1345.

Gregg-Jolly, L. A., & Ornston, L. N. (1990). Recovery of DNA from the *Acinetobacter calcoaceticus* chromosome by gap repair. *Journal of Bacteriology*, *172*(10), 6169–6172. https://doi.org/10.1128/Jb.172.10.6169-6172.1990.

Hartmans, S., Smits, J. P., Van Der Werf, M. J., Volkering, F., & De Bont, J. A. (1989). Metabolism of styrene oxide and 2-phenylethanol in the styrene-degrading *Xanthobacter* strain 124X. *Applied and Environmental Microbiology*, *55*(11), 2850–2855. https://doi.org/10.1128/Aem.55.11.2850-2855.1989.

Horton, R. M., Cai, Z. L., Ho, S. N., & Pease, L. R. (1990). Gene splicing by overlap extension: Tailor-made genes using the polymerase chain reaction. *BioTechniques*, *8*(5), 528–535.

Jiang, X., Palazzotto, E., Wybraniec, E., Munro, L. J., Zhang, H., Kell, D. B., et al. (2020). Automating cloning by natural transformation. *ACS Synthetic Biology*, *9*(12), 3228–3235. https://doi.org/10.1021/Acssynbio.0c00240.

Jones, R. M., & Williams, P. A. (2003). Mutational analysis of the critical bases involved in activation of the AreR-regulated Sigma54-dependent promoter in *Acinetobacter* Sp. strain ADP1. *Applied and Environmental Microbiology*, *69*(9), 5627–5635. https://doi.org/10.1128/Aem.69.9.5627-5635.2003.

Juni, E. (1972). Interspecies transformation of *Acinetobacter*: Genetic evidence for a ubiquitous genus. *Journal of Bacteriology*, *112*(2), 917–931.

Keen, N. T., Tamaki, S., Kobayashi, D., & Trollinger, D. (1988). Improved broad-host-range plasmids for DNA cloning in gram-negative bacteria. *Gene*, *70*(1), 191–197. https://doi.org/10.1016/0378-1119(88)90117-5.

Kok, R. G., Young, D. M., & Ornston, L. N. (1999). Phenotypic expression of PCR-generated random mutations in a *Pseudomonas putida* gene after its introduction into an *Acinetobacter* chromosome by natural transformation. *Applied and Environmental Microbiology*, *65*(4), 1675–1680. https://doi.org/10.1128/Aem.65.4.1675-1680.1999.

Kostylev, M., Otwell, A. E., Richardson, R. E., & Suzuki, Y. (2015). Cloning should be simple: *Escherichia coli* Dh5alpha-mediated assembly of multiple DNA fragments with short end homologies. *PLoS One*, *10*(9), E0137466. https://doi.org/10.1371/Journal. Pone.0137466.

Luo, J., Efimova, E., Volke, D. C., Santala, V., & Santala, S. (2022). Engineering cell morphology by CRISPR interference in *Acinetobacter baylyi* ADP1. *Microbial Biotechnology*, *15*, 2800–2818. https://doi.org/10.1111/1751-7915.14133.

Luo, J., Mcintyre, E. A., Bedore, S. R., Santala, V., Neidle, E. L., & Santala, S. (2022). Characterization of highly ferulate-tolerant *Acinetobacter baylyi* ADP1 isolates by a rapid reverse engineering method. *Applied and Environmental Microbiology*, *88*(2), E0178021. https://doi.org/10.1128/Aem.01780-21.

Melnikov, A., & Youngman, P. J. (1999). Random mutagenesis by recombinational capture of PCR products in *Bacillus subtilis* and *Acinetobacter calcoaceticus*. *Nucleic Acids Research*, *27*(4), 1056–1062. https://doi.org/10.1093/Nar/27.4.1056.

Metzgar, D., Bacher, J. M., Pezo, V., Reader, J., Doring, V., Schimmel, P., et al. (2004). *Acinetobacter* sp. ADP1: An ideal model organism for genetic analysis and genome engineering. *Nucleic Acids Research*, *32*(19), 5780–5790. https://doi.org/10.1093/Nar/Gkh881.

Neidle, E. L., & Ornston, L. N. (1986). Cloning and expression of *Acinetobacter calcoaceticus* catechol 1,2-dioxygenase structural gene *CatA* in *Escherichia coli*. *Journal of Bacteriology*, *168*(2), 815–820.

Palmen, R., Vosman, B., Buijsman, P., Breek, C. K., & Hellingwerf, K. J. (1993). Physiological characterization of natural transformation in *Acinetobacter calcoaceticus*. *Journal of General Microbiology*, *139*(2), 295–305. https://doi.org/10.1099/00221287-139-2-295.

Pardo, I., Bedore, S. R., Tumen-Velasquez, M. P., Duscent-Maitland, C. V., Baugh, A. C., Santala, S., et al. (2023). Natural transformation as a tool in *Acinetobacter baylyi*: Evolution by amplification of gene copy number. *Methods in Microbiology*, *52*.

Santala, S., & Santala, V. (2021). *Acinetobacter baylyi* ADP1—Naturally competent for synthetic biology. *Essays in Biochemistry*, *65*(2), 309–318. https://doi.org/10.1042/Ebc20200136.

Seaton, S. C., Elliott, K. T., Cuff, L. E., Laniohan, N. S., Patel, P. R., & Neidle, E. L. (2012). Genome-wide selection for increased copy number in *Acinetobacter baylyi* ADP1: Locus and context-dependent variation in gene amplification. *Molecular Microbiology*, *83*(3), 520–535.

Stoudenmire, J. L., Schmidt, A. L., Tumen-Velasquez, M. P., Elliott, K. T., Laniohan, N. S., Whitley, S. W., et al. (2017). Malonate degradation in *Acinetobacter baylyi* ADP1: Operon organization and regulation by MdcR. *Microbiology*, *163*(5), 789–803. https://doi.org/10.1099/Mic.0.000462.

Suárez, G. A., Dugan, K. R., Renda, B. A., Leonard, S. P., Gangavarapu, L. S., & Barrick, J. E. (2020). Rapid and assured genetic engineering methods applied to *Acinetobacter baylyi* ADP1 genome streamlining. *Nucleic Acids Research*, *48*(8), 4585–4600. https://doi.org/10.1093/Nar/Gkaa204.

Suárez, G. A., Renda, B. A., Dasgupta, A., & Barrick, J. E. (2017). Reduced mutation rate and increased transformability of transposon-free *Acinetobacter baylyi* ADP1-ISx. *Applied and Environmental Microbiology*, *83*(17). https://doi.org/10.1128/Aem.01025-17. E01025–01017.

Tumen-Velasquez, M., Johnson, C. W., Ahmed, A., Dominick, G., Fulk, E. M., Khanna, P., et al. (2018). Accelerating pathway evolution by increasing the gene dosage of chromosomal segments. *Proceedings of the National Academy of Sciences of the United States of America*, *115*(27), 7105–7110. https://doi.org/10.1073/Pnas.1803745115.

Yanisch-Perron, C., Vieira, J., & Messing, J. (1985). Improved M13 phage cloning vectors and host strains: Nucleotide sequences of the M13mp18 and PUC19 vectors. *Gene, 33*(1), 103–119.

Young, D. M., Kok, R. G., & Ornston, L. N. (2002). Phenotypic expression of polymerase chain reaction-generated random mutations in a foreign gene after its introduction into an *Acinetobacter* chromosome by natural transformation. *Methods in Molecular Biology, 182*, 103–115. https://doi.org/10.1385/1-59259-194-9:103.

Young, D. M., Parke, D., & Ornston, L. N. (2005). Opportunities for genetic investigation afforded by *Acinetobacter baylyi*, a nutritionally versatile bacterial species that is highly competent for natural transformation. *Annual Review of Microbiology, 59*(1), 519–551.